Cold Hardiness in Plants

Molecular Genetics, Cell Biology and Physiology

Cold Hardiness in Plants

Molecular Genetics, Cell Biology and Physiology

Edited by

Tony H.H. Chen

Oregon State University
Corvallis, Oregon, USA

M. Uemura

Cryobiosystem Research Centre
Iwate University
Morioka, Japan

and

S. Fujikawa

Department of Environmental Resources
Graduate School of Agriculture
Hokkaido University
Sapporo, Japan

CABI Publishing

CABI Publishing is a division of CAB International

CABI Publishing
CAB International
Wallingford
Oxfordshire OX10 8DE
UK

Tel: +44 (0)1491 832111
Fax: +44 (0)1491 833508
E-mail: cabi@cabi.org
Website: www.cabi-publishing.org

CABI Publishing
875 Massachusetts Avenue
7th Floor
Cambridge, MA 02139
USA

Tel: +1 617 395 4056
Fax: +1 617 354 6875
E-mail: cabi-nao@cabi.org

A catalogue record for this book is available from the British Library, London, UK.

Library of Congress Cataloging-in-Publication Data

International Plant Cold Hardiness Seminar (7th : 2004 : Hokkaido University)
Cold hardiness in plants : molecular genetics, cell biology, and physiology / edited by T.H.H. Chen, M. Uemura, S. Fujikawa.
p. cm.
Includes bibliographical references.
ISBN-13: 978-0-85199-059-0 (alk. paper)
ISBN-10: 0-85199-059-2 (alk. paper)
1. Plants--Frost resistance--Congresses. I. Chen, Tony H. H. II. Uemura, M. (Matsuo) III. Fujikawa, S. (Seizo) IV. Title.
QK756.158 2004
581.4'2--dc22

2005011507

ISBN-13: 978 0 85199 059 0
ISBN-10: 0 85199 059 2

Typeset by SPI Publisher Services, Pondicherry, India.
Printed and bound in the UK by Biddles Ltd, King's Lynn.

Contents

Contributors

M.K. Aalto, *Viikki Biocentre, Department of Biological and Environmental Sciences, Division of Genetics, POB 56, FI-00014, Helsinki, Finland.*

K. Amundsen, *Department of Crop and Soil Sciences, Michigan State University, East Lansing, MI 48824, USA.*

K. Arakawa, *Department of Environmental Resources, Graduate School of Agriculture, Hokkaido University, Sapporo 060-8589, Japan.*

C. Benedict, *Umeå Plant Science Centre, Department of Plant Physiology, Umeå University, S-901 87 Umeå, Sweden.*

A. Bertrand, *Soils and Crops Research and Development Centre, Agriculture and Agri-Food Canada, 2560 Hochelaga Blvd, Sainte-Foy, QC, Canada G1V 2J3.*

R. Bhalerao, *Umeå Plant Science Centre, Department of Forest Genetics, Swedish University of Agricultural Sciences, S-901 83 Umeå, Sweden.*

G. Bryant, *Department of Applied Physics, RMIT University, Melbourne 3001, Australia.*

Y. Castonguay, *Soils and Crops Research and Development Centre, Agriculture and Agri-Food Canada, 2560 Hochelaga Blvd, Sainte-Foy, QC, Canada G1V 2J3.*

Y. Chang, *Department of Horticulture, Oregon State University, Corvallis, OR 97331-7304, USA.*

T.H.H. Chen, *Department of Horticulture, College of Agriculture, Oregon State University, Corvallis, OR 97331, USA.*

H. Cheng, *Biotechnology Research Institute, Chinese Academy of Agricultural Sciences, Beijing 100081, China.*

J. Cloutier, *Soils and Crops Research and Development Centre, Agriculture and Agri-Food Canada, 2560 Hochelaga Blvd, Sainte-Foy, QC, Canada G1V 2J3.*

D. Cook, *USDA-ARS, Natural Products Utilization Research Unit, University, MS 38677, USA.*

L.L.D. Cooper, *Department of Crop and Soil Science, Oregon State University, Corvallis, OR 97331, USA.*

J. Danyluk, *Département des Sciences Biologiques, Université du Québec à Montréal, Case Postale 8888, Succursale Centre-ville, Montréal, Québec, Canada H3C 3P8.*

T. Filichkin, *Department of Crop and Soil Science, College of Agriculture, Oregon State University, Corvallis, OR 97331, USA.*
C. Finn, *US Department of Agriculture–Agricultural Research Service, Corvallis, OR 97330, USA.*
S.G. Fowler, *MSU-DOE Plant Research Laboratory, Michigan State University, East Lansing, MI 48824, USA.*
E. Francia, *Experimental Institute for Cereal Research – C.R.A., Section of Fiorenzuola, I-29017 Fiorenzuola d'Arda, Italy.*
S. Fujikawa, *Department of Environmental Resources, Graduate School of Agriculture, Hokkaido University, Sapporo 060–8589, Japan.*
P.M. Hayes, *Department of Crop and Soil Science, Oregon State University, Corvallis, OR 97331, USA.*
A.G. Heyer, *Max-Planck-Institut für Molekulare Pflanzenphysiologie, D-14424 Potsdam, Germany.*
D.K. Hincha, *Max-Planck-Institut für Molekulare Pflanzenphysiologie, D-14424 Potsdam, Germany.*
M. Hundertmark, *Max-Planck-Institut für Molekulare Pflanzenphysiologie, D-14424 Potsdam, Germany.*
V. Hurry, *Umeå Plant Science Centre, Department of Plant Physiology, Umeå University, S-901 87 Umeå, Sweden.*
Z. Jeknić, *Department of Horticulture, College of Agriculture, Oregon State University, Corvallis, OR 97331, USA.*
N.A. Kane, *Département des Sciences Biologiques, Université du Québec à Montréal, Case Postale 8888, Succursale Centre-ville, Montréal, Québec, Canada H3C 3P8.*
I. Karsai, *Agricultural Research Institute of the Hungarian Academy of Sciences, H-2462 Martonvásár, Hungary.*
Y. Kawamura, *Cryobiosystem Research Centre, Iwate University, Morioka 020-8550, Japan.*
K.L. Koster, *Department of Biology, The University of South Dakota, Vermillion, SD 57069, USA.*
S. Laberge, *Soils and Crops Research and Development Centre, Agriculture and Agri-Food Canada, 2560 Hochelaga Blvd, Sainte-Foy, QC, Canada G1V 2J3.*
D.A Laurie, *John Innes Centre, Norwich Research Park, Colney, Norwich NR4 7UH, UK.*
L. Marquez-Cedillo, *Department of Crop and Soil Science, College of Agriculture, Oregon State University, Corvallis, OR 97331, USA.*
R. Meng, *Department of Horticulture, Oregon State University, Corvallis, Oregon 97331-7304, USA.*
R. Michaud, *Soils and Crops Research and Development Centre, Agriculture and Agri-Food Canada, 2560 Hochelaga Blvd, Sainte-Foy, QC, Canada G1V 2J3.*
A. Minami, *Cryobiosystem Research Centre, Faculty of Agriculture, Iwate University, Morioka 020-8550, Japan.*
M. Nagao, *Cryobiosystem Research Centre, Faculty of Agriculture, Iwate University, Morioka 020-8550, Japan.*
C. Nakagawara, *Cryobiosystem Research Centre, Iwate University, Morioka 020-8550, Japan.*
E.T. Palva, *Viikki Biocentre, Department of Biological and Environmental Sciences, Division of Genetics, POB 56, FI-00014, Helsinki, Finland.*
E.J. Park, *Department of Horticulture, College of Agriculture, Oregon State University, Corvallis, OR 97331, USA.*

N. Pecchioni, *Facoltà di Agraria, Università di Modena e Reggio Emilia, I-42100 Reggio Emilia, Italy.*

M.T. Pino, *Department of Horticulture, College of Agriculture, Oregon State University, Corvallis, OR 97331, USA.*

F. Sarhan, *Département des Sciences Biologiques, Université du Québec à Montréal, Case Postale 8888, Succursale Centre-ville, Montréal, Québec, Canada H3C 3P8.*

H. Saruyama, *Hokkaido Green-Bio Institute, Naganuma, Hokkaido 069-1301, Japan.*

Y. Sato, *National Agricultural Research Centre for Hokkaido Region, Hitsujigaoka 1, Toyohira-ku, Sapporo 062-8555, Japan.*

K. Shinozaki, *Laboratory of Plant Molecular Biology, RIKEN Tsukuba Institute, Tsukuba, Ibaraki 305-0074, Japan.*

J.S. Skinner, *Department of Horticulture, College of Agriculture, Oregon State University, Corvallis, OR 97331, USA.*

A.M. Stanca, *Experimental Institute for Cereal Research – C.R.A., Section of Fiorenzuola, I-29017 Fiorenzuola d'Arda, Italy.*

E.J. Stockinger, *Department of Horticulture and Crop Science, 1680 Madison Avenue, The Ohio State University/OARDC, Wooster, OH 44691, USA.*

P. Szűcs, *Agricultural Research Institute of the Hungarian Academy of Sciences, H-2462 Martonvásár, Hungary.*

D. Takezawa, *Department of Regulation Biology, Faculty of Science, Saitama University, Saitama 338-8570, Japan.*

M.F. Thomashow, *MSU-DOE Plant Research Laboratory, Michigan State University, East Lansing, MI 48824, USA.*

Y. Tominaga, *Département des Sciences Biologiques, Université du Québec à Montréal, Montréal, Québec, Canada H3C 3P8.*

M. Uemura, *Cryobiosystem Research Centre, Iwate University, Morioka 020-8550, Japan.*

N. Ukaji, *Department of Environmental Resources, Graduate School of Agriculture, Hokkaido University, Sapporo 060-8589, Japan.*

J.T. Vogel, *MSU-DOE Plant Research Laboratory, Michigan State University, East Lansing, MI 48824, USA.*

J. von Zitzewitz, *Department of Crop and Soil Science, College of Agriculture, Oregon State University, Corvallis, OR 97331, USA.*

K. Yamaguchi-Shinozaki, *Biological Resources Division, Japan International Research Centre for Agricultural Sciences (JIRCAS), Tsukuba, Ibaraki 305-8686, Japan.*

K. Yamane, *Department of Environmental Resources, Graduate School of Agriculture, Hokkaido University, Sapporo 060-8589, Japan.*

E. Zuther, *Max-Planck-Institut für Molekulare Pflanzenphysiologie, D-14424 Potsdam, Germany.*

Preface

We compiled this volume based on presentations at the 7th International Plant Cold Hardiness Seminar (7th IPCHS) held at the Conference Hall of Hokkaido University, Sapporo, Japan, from 10 to 15 July 2004. The overall goal of this seminar was to present the latest research findings on plant freezing and chilling stress from laboratories all around the world, and to provide a forum for exchange of ideas among fellow researchers. The majority of presentations at the 7th IPCHS centred on various aspects of molecular genetics and, in many cases, the utilization of transgenic plants to further our understanding of plant cold hardiness at the molecular level. It is hoped that this volume will provide a glimpse of current advances in plant cold hardiness research.

We would like to thank the individual chapter authors for their cooperation in preparing their manuscripts for publication. We also wish to thank the following for their financial support of the 7th IPCHS: National Agricultural Research Centre for Hokkaido Region; The Japanese Society for the Promotion of Sciences (JSPS); Graduate School of Agriculture, Hokkaido University; Sapporo International Communication Plaza Foundation; The Akiyama Foundation; Japan Plant Science Foundation; Inoue Foundation for Science; Novartis Foundation Japan for the Promotion of Science; The Kao Foundation for Arts and Sciences; The Kajima Foundation; Amino Up Chemical Co., Ltd; Tanaka Co., Ltd; HOKUREN Federation of Agricultural Cooperatives; JEOL Ltd; QIAGEN K.K.; TOMY SEIKO Co., Ltd; and Kyokko Trading Co., Ltd. Our special thanks to Mrs Yasuko Uemura, Dr Daisuke Takezawa and laboratory members of Woody Plant Biology, Graduate School of Agriculture, Hokkaido University for their assistance in organizing this seminar. Finally, we would like to thank Dr Jeffrey Skinner and Ms Lee Ann Julson for their assistance in editing the manuscripts for publication in this volume.

Tony H.H. Chen
Matsuo Uemura
Seizo Fujikawa

1 Global Analysis of Gene Networks to Solve Complex Abiotic Stress Responses

K. SHINOZAKI[1,2,*] AND K. YAMAGUCHI-SHINOZAKI[3,4]

[1]Laboratory of Plant Molecular Biology, RIKEN Tsukuba Institute, Tsukuba, Ibaraki 305-0074, Japan; [2]Plant Functional Genomics Research Group, RIKEN Genomic Sciences Centre, Yokohama, Kanagawa 230-0045, Japan; [3]Laboratory of Plant Molecular Physiology, Graduate School of Agricultural and Life Sciences, The University of Tokyo, Tokyo 113-8657, Japan; [4]Biological Resources Division, Japan International Research Centre for Agricultural Sciences (JIRCAS), Tsukuba, Ibaraki 305-8686 Japan

Introduction

Environmental abiotic stresses, such as drought and cold stresses, have severe effects on plant growth and crop production. Plants respond and adapt to these abiotic stresses in order to survive unfavourable environmental conditions. Abiotic stress induces various biochemical and physiological responses in plants (Bray *et al.*, 2000). Accumulation of various substances, such as sugars, sugar alcohols and proline, is observed during these stresses in various plants. These molecules are thought to function in osmotic adjustment. Plant hormone, abscisic acid (ABA), is produced under stress conditions and plays important roles in response to and tolerance against abiotic stress.

Many genes with various functions have been described that respond to drought stress at transcriptional level (Thomashow, 1999; Shinozaki and Yamaguchi-Shinozaki, 2000; Zhu, 2002; Shinozaki *et al.*, 2003). Their gene products function in stress tolerance and response. Stress-inducible genes have been used to improve stress tolerance of plants by gene transfer. It is important to analyse functions of stress-inducible genes not only for further understanding of molecular mechanisms of stress tolerance and response of higher plants, but also for improvement of stress tolerance of crops by gene manipulation.

Recently, transcriptome analyses using microrarray technology have revealed that many genes are induced by abiotic stress to function in stress response and

*Corresponding author (e-mail: sinozaki@rtc.riken.go.jp; Fax: +81-29-836-9060)

tolerance (Shinozaki *et al.*, 2003). These gene products are involved not only in the protection of cells against stresses, but also in the regulation of gene expression and the signal transduction pathways in abiotic stress response. Most of the stress-inducible genes are also induced by ABA. Dehydration triggers the production of ABA that induces various genes. The existence of ABA-independent as well as ABA-dependent signal transduction cascades upstream of stress-inducible gene expression has been described (Shinozaki and Yamaguchi-Shinozaki, 2000; Zhu, 2002; Shinozaki *et al.*, 2003). *Cis*- and *trans*-acting elements that function in ABA-independent and ABA-responsive gene expression by drought stress have been precisely determined. Various transcription factors function in stress-responsive gene expression, which suggests complex regulatory mechanism of gene expression in response to drought stress. Details of molecular mechanisms regulating genes to stress still remain to be solved concerning signal transduction cascades. In this chapter, molecular processes of drought and cold stress response and tolerance are described. Functions of drought-inducible genes, regulation of their gene expression and signal transduction pathways in drought stress response and tolerance are also discussed.

Identification of Drought- and Cold-inducible Genes

Many plant genes are induced by environmental stresses such as drought, low temperature and high salinity (Thomashow, 1999; Shinozaki and Yamaguchi-Shinozaki, 2000; Zhu, 2002; Shinozaki *et al.*, 2003). Recently, expression profiles of genes under drought, cold and high salinity stress conditions were analysed using microrarray and GeneChip (Seki *et al.*, 2001, 2002a,b; Fowler and Thomashow, 2002). We have identified 299 drought-inducible genes, 54 cold-inducible genes, 213 high salinity stress-inducible genes and 245 ABA-inducible genes using a cDNA microarray containing *c.* 7000 independent *Arabidopsis* full-length cDNA groups (Seki *et al.*, 2001, 2002a,b). More than half of drought-inducible genes overlap with high salinity- and ABA-inducible genes, which indicates the existence of significant crosstalk among drought, high salinity and ABA responses. By contrast, only 10% of drought-inducible genes overlap with cold-inducible genes. Among the stress-inducible genes, many transcription factor genes were found, suggesting that various transcriptional regulatory mechanisms function in the drought-, cold- or high salinity-stress signal transduction pathways.

Functions of Drought- and Cold-inducible Genes

Genes induced during drought and cold stress conditions encode proteins that function not only in the protection of cells from stress, but also in the gene expression and signal transduction in stress response (Thomashow, 1999; Shinozaki and Yamaguchi-Shinozaki, 2000; Zhu, 2002; Shinozaki *et al.*, 2003). These gene products are classified into two groups (Fig. 1.1). The first group includes proteins that probably function in stress tolerance, such as chaperones, late embryogenesis abundant (LEA) proteins, osmotin, antifreeze proteins, mRNA-binding proteins, key enzymes for osmolytes biosynthesis, water channel proteins, sugar and proline transporters, detoxification enzymes and various proteases. The second group contains protein factors involved in further regulation of signal transduction and gene expression that probably function in stress response, such as protein kinases, transcription factors and enzymes in phospholipids metabolism. Existence of a variety of stress-inducible genes suggests complex responses of plants to abiotic stress.

Multiple Regulatory Systems of Stress-inducible Gene Expression

Many drought- and cold-inducible genes are also induced by exogenous ABA treatment, which indicates important roles of ABA in stress-responsive gene expression. However, many stress-inducible genes are regulated by ABA-independent processes. Cold-inducible genes are mainly induced by ABA-independent processes whereas many cold-inducible genes are induced by ABA treatment. Analysis of stress-inducible gene expression has revealed the existence of ABA-independent, as well as ABA-dependent, transcriptional regulatory systems in abiotic stress-inducible genes (Thomashow, 1999; Shinozaki and Yamaguchi-Shinozaki, 2000; Zhu, 2002; Shinozaki *et al.*, 2003). It has been shown that at least four independent regulatory systems function in the activation of stress-inducible genes under dehydration conditions (Fig. 1.2). Two of them are ABA-dependent and the remaining two are ABA-independent.

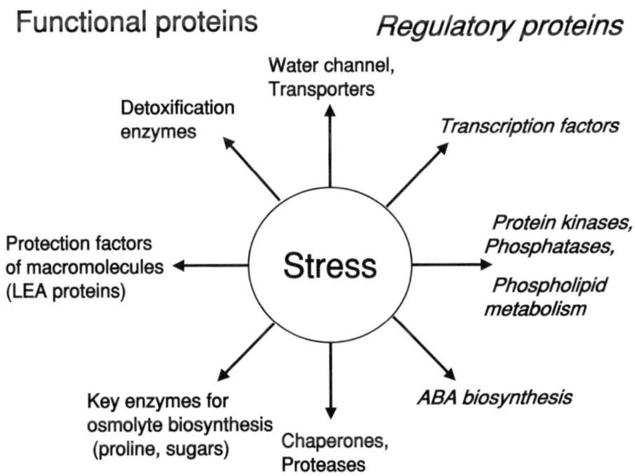

Fig. 1.1. Functions of drought and cold stress-inducible genes in stress tolerance and response. Gene products are classified into two groups. The first group includes proteins that probably function in stress tolerance (functional proteins), and the second group contains protein factors involved in further regulation of signal transduction and gene expression that probably function in stress response (regulatory proteins).

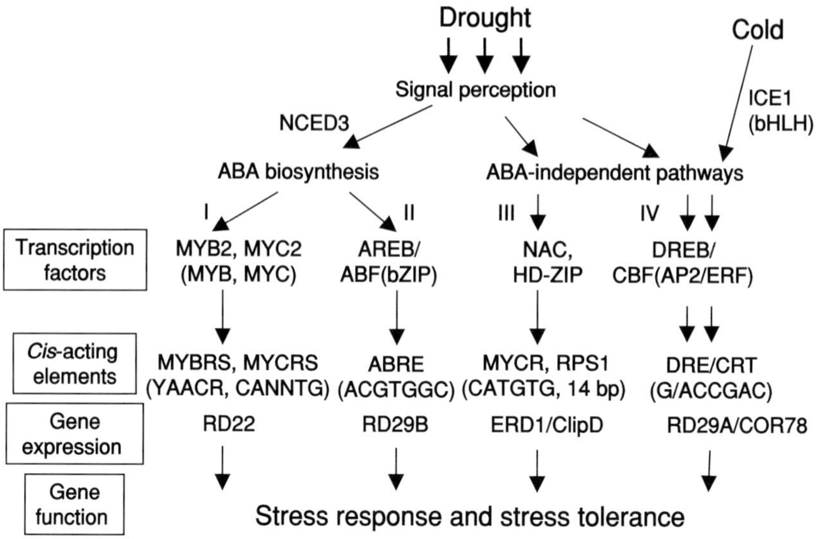

Fig. 1.2. Signal transduction pathways from the perception of drought and cold stress signals to gene expression. At least four signal transduction pathways exist (I–IV): two are abscisic acid (ABA)-dependent (I and II) and two are ABA-independent (III and IV). Protein biosynthesis is required in one of the ABA-dependent pathways in which MYB an MYC transcription factors are involved (I). In another ABA-dependent pathway, ABRE functions as an ABA-responsive element and does not require protein biosynthesis (II). This is a major regulatory system in ABA-dependent gene expression in which bZIP transcription factors function. In one of the ABA-independent pathways, DRE/CRT is involved in the regulation of genes not only by drought and salt, but also by cold stress (IV). In this pathway, AP2 transcription factors, named DREB or CBF, are major transcription factors involved in this process. Another ABA-independent pathway is controlled by drought and salt, but not by cold (III). In this process, NAC and HD-ZIP transcription factors are involved in gene expression.

ABA-independent Gene Expression in Response to Drought and Cold Stress

There are at least two ABA-independent regulatory systems in drought- and cold-inducible gene expression. One of the ABA-independent pathways of drought stress response overlaps with that of cold stress response (Fig. 1.2). A *cis*-acting element including A/GCCGAC, named the dehydration responsive element (DRE) and C-repeat (CRT), is essential for the regulation of many stress-inducible genes under drought, low temperature and high salinity stress in an ABA-independent pathway (Yamaguchi-Shinozaki and Shinozaki, 1994). All the DRE/CRT-binding proteins (DREBs and CBFs) contain a conserved DNA-binding motif (AP2/ERF motif) (Stockinger *et al.*, 1997; Liu *et al.*, 1998). Two independent families of DREB proteins, DREB1/CBF and DREB2, mainly function as *trans*-acting factors in two separate signal transduction pathways under low-temperature and dehydration conditions, respectively (Liu *et al.*, 1998). Overproduction of the DREB1A and CBF1/DREB1B cDNAs driven by the 35S CaMV promoter in transgenic plants significantly improved stress tolerance to drought and freezing (Jaglo-Ottosen *et al.*, 1998; Liu *et al.*,1998; Kasuga *et al.*, 1999). Microarray analysis identified many genes with various functions as DREB/CBF target genes, which suggests overexpression of many genes function in drought stress tolerance (Seki *et al.*, 2001; Fowler and Thomashow, 2002; Maruyama *et al.*, 2004).

Recently, a gene for ICE1 (inducer of CBF expression) that regulates expression of CBF/DREB1 in response to cold stress has been identified by map-based cloning of a gene for *Arabidopsis ice1* mutant that affect the expression of the CBF3/DREB1A promoter::luciferase transgene (Chinnusamy *et al.*, 2003). The *ICE1* gene encodes a MYC-type bHLH transcription factor that regulates the expression of CBF3/DREB1A-regulated but not other CBF/DREB1-regulated genes (Fig. 1.2). Improved freezing tolerance was obtained by overexpression of *ICE1* in transgenics, which also supports an important role of *ICE1* in cold stress response (Chinnusamy *et al.*, 2003, Zarka *et al.*, 2003).

In the ABA-independent pathways, there are several drought-inducible genes that do not respond to either cold or ABA treatment, which suggests that there is a specific pathway that functions in one of the dehydration stress responses (Fig. 1.2). These genes include *RD19* and *RD21* that encode different cysteine proteases, and *ERD1* that encodes a ClipD protease regulatory subunit. Recently, it was shown that NAC transcription factors function in the stress-inducible expression of the *ERD1/ClipD* gene (Simpson *et al.*, 2003; Tran *et al.*, 2004).

ABA-dependent Gene Expression in Response to Drought Stress

ABA is synthesized in response to drought and high salinity stress but not to cold stress. Many stress-inducible genes are regulated by endogenous ABA accumulated during drought and high salinity stress. Recently, genes involved in ABA biosynthesis have been identified based on genetic and genomics analysis. Several genes involved in ABA biosynthesis are induced by drought and high salinity but not by cold stress (Zhu, 2002; Shinozaki *et al.*, 2003). This indicates important roles of ABA in drought stress responses. Among the genes involved in ABA biosynthesis, an *Arabidopsis* gene for nine-*cis*-epoxycarotenoid dioxygenase (NCED3), a key enzyme for ABA biosynthesis, is strongly induced by drought stress.

ABRE is a major *cis*-acting element in ABA-responsive gene expression (Fig. 1.1). Two ABRE motifs are important *cis*-acting elements in ABA-responsive expression of *Arabidopsis RD29B* gene (Uno *et al.*, 2000). Basic leucine zipper (bZIP) transcription factors, AREB/ABF, bind to ABRE and activate ABA-dependent gene expression (Choi *et al.*, 2000; Uno *et al.*, 2000). The AREB/ABF proteins need ABA-mediated signals for their activation, because of their reduced activity in the ABA-deficient *aba2* and ABA-insensitive *abi1* mutants and their enhanced activity in the ABA-hypersensitive *era1* mutant (Uno *et al.*, 2000). This is probably due to ABA-dependent phosphorylation of the AREB/ABF proteins. Overexpression of *ABF3* or *AREB2/ABF4* caused ABA-hypersensitivity, reduced transpiration rate and enhanced drought tolerance of the transgenics (Kang *et al.*, 2002).

The induction of the drought-inducible *RD22* gene is mediated by ABA and requires protein biosynthesis for its ABA-dependent expression (Shinozaki *et al.*, 2003). A MYC transcription factor, *RD22BP1 (AtMYC2)*, and a MYB transcription factor, *AtMYB2*, were shown to bind *cis*-elements in the *RD22* promoter and cooperatively activate *RD22* (Abe *et al.*, 2003). These MYC and MYB proteins are synthesized after the accumulation of endogenous ABA, which indicates their role in a rather later stage of stress responses. Microarray analysis revealed target genes of MYC/MYB overexpressing transgenics, such as alcohol dehydrogenase and ABA- or jasmonic acid (JA)-inducible genes (Abe *et al.*, 2003). Overexpression of both AtMYC2 and AtMYB2 not only revealed ABA-hypersensitive phenotype, but also improved osmotic-stress tolerance of the transgenic plants. Recently, AtMYC2 has been identified as a JASMONATE-INSENSITIVE 1 involved in jasmonate-regulated defense response (Lorenzo *et al.*, 2004).

Other transcription factors with various DNA-binding domains, such as NAC, HD-ZIP, Zn finger and so on, have been identified to function in ABA- and/or stress-responsive gene expression (Shinozaki *et al.*, 2003; Fujita *et al.*, 2004; Sakamoto *et al.*, 2004).

Signal Perception and Signal Transduction in Drought Stress Response

Signal transduction pathways from the sensing of dehydration signals or osmotic change to the expression of various genes, and the signalling molecules that function in stress signalling, have not been extensively studied in plants. Signal transduction pathways in drought stress response have been studied based on the knowledge in yeast and animal systems, as shown in Fig. 1.3 (Shinozaki and Yamaguchi-Shinozaki, 2000; Zhu, 2002; Shinozaki *et al.*, 2003).

Two component systems function in sensing osmotic stress in bacteria and yeasts. Plants, as well as cyanobacteria, contain many genes encoding sensor-histidine kinases and response-regulator homologues, which suggests the involvement of similar osmosensing mechanisms in higher plants. One of the histidine kinases, ATHK1, is shown to function as osmosensor in *Arabidopsis*. Other sensing mechanisms may function during drought stress responses, such as mechanical sensors of cytoskeletons and sensors of superoxides produced by stress. During stomata closure, the level of cytoplasmic Ca^{2+} is increased, which suggests that Ca^{2+} functions as a second messenger in the osmotic stress response. Phospholipids play important roles in calcium signalling.

ABA plays important roles in drought stress responses. ABA is involved not only in stomata closure, but also in induction of many genes. Several mutants in ABA signalling have been identified and their genes encode protein phosphatases, farnesyl transferase and several factors involved in mRNA processing and protein degradation. These suggest that protein modification and mRNA processing are involved in ABA signalling (Shinozaki and Yamaguchi-Shinozaki, 2000; Zhu, 2002; Shinozaki *et al.*, 2003).

Recently, ABA-activated SnRK2 protein kinase (OST1/SRK2E) was shown to function in an ABA-signal transduction pathway in stomata closure (Mustilli *et al.*, 2002; Yoshida *et al.*, 2002). SRK2E/OST1 is activated by ABA treatment and

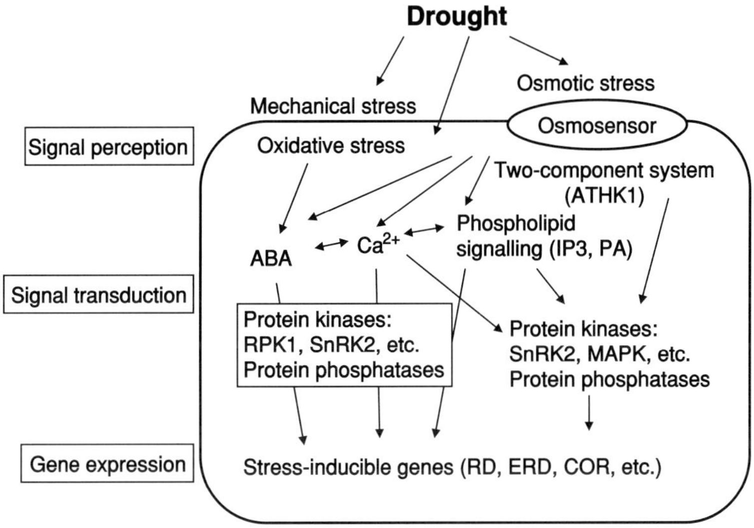

Fig. 1.3. Second messengers and protein factors involved in the signal perception and the signal transduction in drought stress response. Two-component histidine kinase, ATHK1, is thought to function as an osmosensor in plants. Calcium and phospholipids are the most probable cellular second messengers of the drought stress signal. The phosphorylation process functions in water stress and ABA signal transduction pathways. RPK1 functions in the early process of ABA signalling. SnRK2 family protein kinases function in ABA and osmotic stress signalling pathways. ABA plays important roles in the regulation of gene expression as well as in physiological responses during water stress.

osmotic stress. *SRK2E/OST1* is mainly expressed in guard cells. The *ost1/srk2e* mutant showed wilty phenotype, which indicates an important role of SRK2E/OST1 in ABA signalling in stomata closure. Recently, it was shown that SnRK2C, another SnRK2, is also activated by osmotic stress and ABA treatment. Overexpression of SRK2C improved drought stress tolerance (Umezawa *et al.*, 2004). SRK2C is mainly involved in osmotic stress signalling in root tissues.

More recently, it was shown that one of the receptor-like kinases, named RPK1, functions in an early process of ABA signalling. RPK1 is a member of the LRR-RLK family of *Arabidopsis* and its expression is upregulated by ABA (Osakabe *et al.*, 2005). We found that RPK1 is localized in plasma membranes. The repression of *RPK1* expression in *Arabidopsis* resulted in the decrease of sensitivity to ABA in the various physiological events including germination, growth rate and stomata closure. Many ABA-inducible genes are downregulated in these RPK1 antisense transgenics and eliminate mutant plants, suggesting that RPK1 is involved in the main ABA signalling pathway in *Arabidopsis*.

Various signalling molecules seem to be functioning in ABA signalling, such as phospholipids and cyclic ADP ribose. MAP kinase cascades and calcium-dependent protein kinases are suggested to be involved in drought stress response and ABA signalling (Shinozaki and Yamaguchi-Shinozaki, 2000; Zhu, 2002; Shinozaki *et al.*, 2003). Complex signalling cascades are thought to function in molecular responses to drought stress. Molecular analysis of the signalling process is in progress based on forward and reverse genetics.

Perspectives

Various stress-inducible genes have been identified by systematic analysis of gene expression using microarray. However, functions of the gene products have not been fully understood. In functional genomics, the reverse genetic approach, as well as forward genetics, becomes more important in understanding complex molecular processes in environmental stress responses. Efficient gene disruption methods, as well as transgenic approaches using RNAi and overexpressors, will contribute to the precise understanding of molecular networks in drought and cold stress response and tolerance. Global analysis of gene expression and protein–protein interaction during signal transduction is necessary for this purpose. Metabolome analysis is also important in understanding metabolites' changes during stress conditions.

References

Abe, H., Urao, T., Ito, T., Seki, M., Shinozaki, K. and Yamaguchi-Shinozaki, K. (2003) *Arabidopsis* AtMYC2 (bHLH) and AtMYB2 (MYB) function as transcriptional activators in abscisic acid signaling. *Plant Cell* 15, 63–78.

Bray, E., Bailey-Serres, J. and Weretilnyk, E. (2000) Responses to abiotic stresses. In: Buchanan, B.B., Gruissem, W. and Jones, R.L. (ed.) *Biochemistry & Molecular Biology of Plants.* American Society of Plant Physiologists, Rockville, Maryland, pp.1158–1203.

Chinnusamy, V., Ohta, M., Kanrar, S., Lee, B., Hong, X., Agarwal, M. and Zhu, J.-K. (2003) ICE1: a regulator of cold-induced transcriptome and freezing tolerance in *Arabidopsis. Genes & Development* 17, 1043–1054.

Choi, H., Hong, J., Ha, J., Kang, J. and Kim, S.Y. (2000) ABFs, a family of ABA-responsive element binding factors. *The Journal of Biological Chemistry* 275, 1723–1730.

Fowler, S. and Thomashow, M.F. (2002) *Arabidopsis* transcriptome profiling indicates that multiple regulatory pathways are activated during cold acclimation in addition to the CBF cold response pathway. *Plant Cell* 14, 1675–1690.

Fujita, M., Fujita, Y., Maruyama, K., Seki, M., Hiratsu, K., Ohme-Takagi, M., Tran, L.-S.P., Yamaguchi-Shinozaki, K. and Shinozaki, K. (2004) A dehydration-induced NAC protein, RD26, is involved in a novel ABA-dependent stress-signaling pathway. *Plant Journal* 39, 863–876.

Jaglo-Ottosen, K.R., Gilmour, S.J., Zarka, D.G., Schabenberger, O. and Thomashow, M.F. (1998) *Arabidopsis* CBF1 overexpression induces *cor* genes and enhances freezing tolerance. *Science* 280, 104–106.

Kang, J.Y., Choi, H., Im, Y. and Kim, S.Y. (2002) *Arabidopsis* basic leucine zipper proteins that mediate stress-responsive abscisic acid signaling. *Plant Cell* 14, 343–357.

Kasuga, M., Liu, Q., Miura, S., Yamaguchi-Shinozaki, K. and Shinozaki, K. (1999) Improving plant drought, salt, and freezing tolerance by gene transfer of a single stress-inducible transcription factor. *Nature Biotechnology* 17, 287–291.

Liu, Q., Kasuga, M., Sakuma, Y., Abe, H., Miura, S., Yamaguchi-Shinozaki, K. and Shinozaki, K. (1998) Two transcription factors, DREB1 and DREB2, with an EREBP/AP2 DNA binding domain separate two cellular signal transduction pathways in drought- and low-temperature-responsive gene expression, respectively, in *Arabidopsis*. *Plant Cell* 10, 1391–1406.

Lorenzo, O., Chico, J.M., Sánchez-Serrano, J.J. and Solano, R. (2004) *JASMONATE-INSENSITIVE1* encodes a MYC transcription factor essential to discriminate between different Jasmonate-regulated defense responses in *Arabidopsis*. *Plant Cell* 16, 1938–1950.

Maruyama, K., Sakuma, Y., Kasuga, M., Ito, Y., Seki, M., Goda, H., Shimada, Y., Yosihda, S., Shinozaki, K. and Yamaguchi-Shinozaki, K. (2004) Identification of cold-inducible downstream genes of the *Arabidopsis* DREB1A/CBF3 transcriptional factor using two microarray systems. *Plant Journal* 38, 982–993.

Mustilli, A.C., Merlot, S., Vavasseur, A., Fenzi, F. and Girandat, J. (2002) *Arabidopsis* OST1 protein kinase mediates the regulation of stomatal aperture by abscisic acid and acts upstream of reactive oxygen species production. *Plant Cell* 14, 3089–3099.

Osakabe, Y., Seki, M., Maruyama, K., Shinozaki, K. and Yamaguchi-Shinozaki, K. (2005) An LRR receptor kinase, RPK1, is a key membrane-bound regulator of early abscisic acid signaling in *Arabidopsis*. *The Plant Cell* 17, 1105–1119.

Sakamoto, H., Maruyama, K., Meshi, T., Iwabuchi, M., Shinozaki, K. and Yamaguchi-Shinozaki, K. (2004) *Arabidopsis* Cys2/His2-type zinc-finger proteins function as transcription repressors under drought-, cold-, and high-salinity-stress conditions. *Plant Physiology* 136, 2734–2746.

Seki, M., Narusaka, M., Abe, H., Kasuga, M., Yamaguchi-Shinozaki, K., Carninci, P., Hayashizaki, Y. and Shinozaki, K. (2001) Monitoring the expression pattern of 1300 *Arabidopsis* genes under drought and cold stresses using a full-length cDNA microarray. *Plant Cell* 13, 61–72.

Seki, M., Narusaka, M., Ishida, J., Nanjo, T., Fujita, M., Oono, Y., Kamiya, A., Nakajima, M., Enju, A., Sakurai, T., Satou, M., Akiyama, K., Taji, T., Yamaguchi-Shinozaki, K., Carninci, P., Kawai, J., Hayashizaki, Y. and Shinozaki, K. (2002a) Monitoring the expression profiles of 7000 *Arabidopsis* genes under drought, cold and high-salinity stresses using a full-length cDNA microarray. *Plant Journal* 31, 279–292.

Seki, M., Ishida, J., Narusaka, M., Fujita, M., Nanjo, T., Umezawa, T., Kamiya, A., Nakajima, M., Enju, A., Sakurai, T., Satou, M., Akiyama, K., Yamaguchi-Shinozaki, K., Carninci, P., Kawai, J., Hayashizaki, Y. and Shinozaki, K. (2002b) Monitoring the expression pattern of around 7,000 *Arabidopsis* genes under ABA treatments using a full-length cDNA microarray. *Functional & Integrative Genomics* 2, 282–291.

Shinozaki, K. and Yamaguchi-Shinozaki, K. (2000) Molecular responses to dehydration and low temperature: differences and cross-talk between two stress signaling pathways. *Current Opinion in Plant Biology* 3, 217–223.

Shinozaki, K., Yamaguchi-Shinozaki, K. and Seki, M. (2003) Regulatory network of gene expression in the drought and cold stress responses. *Current Opinion in Plant Biology* 6, 410–417.

Simpson, S., Nakashima, K., Narusaka, Y., Seki, M., Shinozaki, K. and Yamaguchi-Shinozaki, K. (2003) Two different novel *cis*-acting elements of *erd1*, a *clpA* homologous *Arabidopsis* gene function in induction by dehydration stress and dark-induced senescence. *Plant Journal* 33, 259–270.

Stockinger, E.J., Gilmour, S.J. and Thomashow, M.F. (1997) *Arabidopsis thaliana CBF1* encodes an AP2 domain-containing transcription activator that binds to the C-repeat/DRE, a *cis*-acting DNA regulatory element that stimulates transcription in response to low temperature and water deficit. *Proceedings of National Academy of Sciences USA* 94, 1035–1040.

Thomashow, M.F. (1999) Plant cold acclimation: freezing tolerance genes and regulatory mechanisms. *Annual Review of Plant Physiology and Plant Molecular Biology* 50, 571–599.

Tran, L.-S.P., Nakashima, K., Sakuma, Y., Simpson, S.D., Fujita, Y., Maruyama, K., Fujita, M., Seki, M., Shinozaki, K. and Yamaguchi-Shinozaki, K. (2004) Functional analysis of *Arabidopsis* NAC transcription factors controlling expression of *erd1* gene under drought stress. *Plant Cell* 16, 2482–2498.

Umezawa, T., Yoshida, R., Maruyama, K., Yamaguchi-Shinozaki, K. and Shinozaki, K. (2004) SRK2C, a SNF1-related protein kinase 2, improves drought stress tolerance by controlling stress-responsive gene expression in *Arabidopsis thaliana*. *Proceedings of National Academy of Sciences USA* 101, 17306–17311.

Uno, Y., Furihata, T., Abe, H., Yoshida, R., Shinozaki, K. and Yamaguchi-Shinozaki, K. (2000) *Arabidopsis* basic leucine zipper transcription factors involved in an abscisic acid-dependent signal transduction pathway under drought and high-salinity conditions. *Proceedings of National Academy of Sciences USA* 97, 11632–11637.

Yamaguchi-Shinozaki, K. and Shinozaki, K. (1994) A novel *cis*-acting element in an *Arabidopsis* gene is involved in responsiveness to drought, low temperature, or high-salt stress. *Plant Cell* 6, 251–264.

Yoshida, R., Hobo, T., Ichimura, K., Mizoguchi, T., Takahashi, F., Aronso, J., Ecker, J.R. and Shinozaki, K. (2002) ABA-activated SnRK2 protein kinase is required for dehydration stress signaling in *Arabidopsis*. *Plant Cell Physiology* 43, 1473–1483.

Zarka, D.G., Vogel, J.T., Cook, D. and Thomashow, M.F. (2003) Cold induction of *Arabidopsis* CBF genes involves multiple ICE (inducer of CBF expression) promoter elements and a cold-regulatory circuit that is desensitized by low temperature. *Plant Physiology* 133, 910–918.

Zhu, J.K. (2002) Salt and drought stress signal transduction in plants. *Annual Review of Plant Biology* 53, 247–273.

2 The CBF Cold Response Pathways of *Arabidopsis* and Tomato

J.T. Vogel,[1] D. Cook,[1,2] S.G. Fowler[1] and M.F. Thomashow[1,3,*]

[1]MSU-DOE Plant Research Laboratory, Michigan State University, East Lansing, MI 48824, USA; [2]USDA-ARS, Natural Products Utilization Research Unit, University, MS 38677, USA; [3]Department of Crop and Soil Sciences, Michigan State University, East Lansing, MI 48824, USA

Introduction

Plants must cope with a variety of abiotic stresses in the environment including extremes in temperature, lack of water and flooding. In many regions of the Earth, the ability to survive freezing temperatures is an essential trait. Many plants from temperate regions, for instance, are able to sense low, non-freezing temperatures and activate mechanisms that lead to an increase in freezing tolerance, a phenomenon known as cold acclimation (Thomashow, 1999; Smallwood and Bowles, 2002). In contrast, plants from warm regions of the Earth, such as banana and rice, are unable to cold acclimate and often suffer injury, and even death, upon exposure to chilling temperatures between 0°C and 10°C (Thomashow and Browse, 1999).

For decades, understanding the mechanisms of chilling and freezing tolerance has been the major goal of investigators studying abiotic stress responses. A recent important advance made with *Arabidopsis* has been the discovery of a stress response pathway, the CBF cold response pathway that has a role in cold acclimation (Thomashow, 2001; Shinozaki *et al.*, 2003). It has been established that *Arabidopsis* encodes a small family of cold-responsive transcriptional activators known either as CBF1, CBF2 and CBF3 (Stockinger *et al.*, 1997; Gilmour *et al.*, 1998; Medina *et al.*, 1999) or DREB1b, DREB1c and DREB1a (Liu *et al.*, 1998; Kasuga *et al.*, 1999), respectively (the CBF designation will be used throughout this chapter). The CBF transcription factors, which are members of the AP2/EREBP family of DNA-binding proteins (Riechmann and Meyerowitz, 1998), recognize

*Corresponding author (e-mail: thomash6@msu.edu; Fax: +1-517-353-9168)

the cold- and dehydration-responsive DNA regulatory element designated the CRT (C-repeat)/DRE (dehydration-responsive element) (Baker *et al.*, 1994; Yamaguchi-Shinozaki and Shinozaki, 1994). CRT/DRE elements are present in the promoters of many cold- and dehydration-responsive genes of *Arabidopsis*, including those designated *COR* (cold-regulated) (Thomashow, 1999). The *CBF1-3* genes are induced within 15 min of a plant's exposure to low, non-freezing temperatures followed at about 2 h by induction of a group of *COR* genes that contain the CRT/DRE-regulatory element, i.e. the 'CBF regulon' (Gilmour *et al.*, 1998; Liu *et al.*, 1998). Expression of the CBF regulon results in an increase in freezing tolerance (Jaglo-Ottosen *et al.*, 1998; Liu *et al.*, 1998; Kasuga *et al.*, 1999; Gilmour *et al.*, 2000; Gilmour *et al.*, 2004) and appears to have a role in chilling tolerance as well (Zhu *et al.*, 2004). We are far from a complete understanding of how the CBF regulon functions to enhance freezing and chilling tolerance. However, it is clear that multiple mechanisms are involved, including the synthesis of low-molecular weight cryoprotective molecules such as proline, sucrose and raffinose (Gilmour *et al.*, 2000; Taji *et al.*, 2002) and the synthesis of hydrophilic proteins that have cryoprotective properties, such as COR15a (Artus *et al.*, 1996; Steponkus *et al.*, 1998).

It is now clear that the CBF cold response pathway is not limited to *Arabidopsis*. For example, it has been shown that *Brassica napus*, a crucifer like *Arabidopsis* that increases in freezing tolerance in response to low temperature, has *COR* genes that encode CBF proteins (Jaglo *et al.*, 2001; Gao *et al.*, 2002). Moreover, constitutive overexpression of *Arabidopsis CBF1*, *2* or *3* in transgenic *B. napus* plants has been shown to activate expression of homologues of *Arabidopsis COR* genes that have CRT/DRE elements in their promoters and result in an increase in freezing tolerance (Jaglo *et al.*, 2001). Other plants that can cold acclimate and are more distantly related to *Arabidopsis* than *B. napus*, such as wheat and barley, have also been shown to have cold-inducible genes that encode CBF proteins and have homologues of *Arabidopsis COR* genes that include CRT/DRE elements in their promoters (Dunn *et al.*, 1998; Vazquez-Tello *et al.*, 1998; Jaglo *et al.*, 2001; Choi *et al.*, 2002; Xue, 2002). Thus, although additional work needs to be done to detail their comparative compositions, CBF cold response pathways appear to be conserved among plants that cold acclimate. Moreover, CBF pathways are not limited to plants that cold acclimate. For instance, cold-inducible genes that encode CBF proteins are found in chilling- and freezing-sensitive plants such as tomato (Jaglo *et al.*, 2001), rice (Dubouzet *et al.*, 2003) and maize (Qin *et al.*, 2004). An important question that now needs to be addressed is whether 'deficiencies' in the CBF cold response pathways of these plants contribute to their chilling and freezing sensitivities.

In this chapter, we review recent experiments from our laboratory designed for the further understanding of the CBF cold response pathways in *Arabidopsis* and tomato. The studies with *Arabidopsis* focus on determining the extent to which the CBF cold response pathway 'configures' the low-temperature transcriptome and metabolome. The studies with tomato were designed to determine whether it has a 'complete' CBF cold response pathway and whether differences between the pathways in tomato and *Arabidopsis* might explain, in part, the cold-sensitive phenotype of tomato.

Research Findings

Role of the CBF cold response pathway in configuring the low-temperature transcriptome of *Arabidopsis*

The CBF cold response pathway is not the only pathway activated at low temperature. Transcriptome-profiling experiments sampling roughly one-third of the *Arabidopsis* genome indicated that extensive changes in gene expression occurred during cold acclimation and that the majority of *COR* genes could not be assigned to the CBF regulon (Seki *et al.*, 2001; Fowler and Thomashow, 2002; Kreps *et al.*, 2002; Seki *et al.*, 2002; Maruyama *et al.*, 2004). To further explore the extent to which the CBF cold response pathway configures the low-temperature transcriptome of *Arabidopsis* (ecotype Col-0), we first identified a core set of cold-responsive transcripts (Vogel *et al.*, 2005) using the *Arabidopsis* ATH1 GeneChip from Affymetrix (Redman *et al.*, 2004). This chip contains probe sets representing 23,423 genes (based on TAIR's 1 July 2004 annotation). Profiling experiments were performed on plants exposed to low temperature for 1 h, 24 h and 7 days grown in duplicate using two different culture conditions (soil and solid media). This resulted in the identification of 514 transcripts (2.5-fold cut-off, $P \leq 0.05$) that were cold-responsive in all four replicates (two per culture condition) (Vogel *et al.*, 2005). Of these, 302 were upregulated and 212 were downregulated in response to low temperature. We designated these genes the 'cold standard' (COS) set of cold-responsive genes. The COS gene set almost certainly does not include all cold-responsive genes, as root tissue was not harvested from the soil-grown plants and the selection criteria used were stringent. Identifying all *COR* genes, however, was not the purpose of these experiments. Rather, it was to identify a set of genes that we could be confident were cold-induced and would serve as a robust resource to begin to decipher the low-temperature regulatory network of *Arabidopsis*.

Once the COS gene set was identified, we moved on to determining which of these genes were members of the CBF regulon. COS genes were assigned to the CBF regulon if the levels of their corresponding transcripts increased or decreased at least 2.5-fold in two independent transgenic *Arabidopsis* lines that constitutively expressed *CBF2* under control of the cauliflower mosaic virus (CaMV) 35S promoter. Other transcript-profiling experiments using the '8K' Affymetrix GeneChip had indicated that there were no obvious differences in genes affected by constitutive overexpression of *CBF1, 2* or *3* (Gilmour *et al.*, 2004). Thus, it was judged that *CBF2* overexpression would probably capture most, if not all, genes that belonged to the CBF regulon.

Of the 514 COS genes, 93 were assigned to the CBF regulon (Vogel *et al.*, 2005). Of these 93 CBF regulon COS genes, the large majority, 85 (91%), were cold-induced. A scan of the promoters of these 85 genes indicated that 68 (80%) had one or more CRT/DRE elements, (A/G)CCGAC, present within 1 kb upstream of the start of the protein-coding sequence. Thus, these genes were likely to be direct targets of CBF2. Those CBF2-regulated COS genes without CRT/DRE elements were presumably regulated by other genes controlled by CBF2. Bioinformatic analysis of the sequences 1 kb upstream of the 85 CBF regulon COS genes also revealed enrichment for the sequence (G/T)(A/G)CCGACNT(A/C), which may represent a larger consensus sequence for the CRT/DRE element.

The 85 cold-induced COS genes that were assigned to the CBF regulon represented a diverse range of functions including roles in metabolism, transcription, intercellular communication and signalling, transport, energy, protein processing, cellular biogenesis and stress (Table 2.1). Further, a distinguishing characteristic of this group of genes was that they comprised the majority of genes that were most highly induced in response to low temperature (Fig. 2.1). Of the 25 COS genes that were upregulated at least 15-fold at 24 h, 21 (84%) were members of the CBF regulon; of those upregulated five- to tenfold, 32/66 (49%) were assigned to the CBF regulon. Conversely, approximately 90% of the COS genes that were induced less than fivefold were not assigned to the CBF regulon. Finally, an additional distinguishing feature of the CBF regulon COS genes was their enrichment among genes that remained upregulated at 7 days. Of these 67 'long-term' upregulated COS genes (≥2.5-fold increase at 7 days), 37 (55%) were members of the CBF regulon.

Taken together, the results of the expression-profiling experiments indicate that the CBF transcription factors have a prominent role in regulating the expression of those cold-responsive COS genes that are both highly induced and long-term upregulated in response to low temperature. However, it is also clear that additional cold-response pathways participate in configuring the low-temperature transcriptome. This is indicated by the finding that the transcript levels for 183 (36%) of the total 514 COS genes showed no detectable change in either of the *CBF2* overexpressing transgenic lines profiled. Thus, it would appear that these *COR* genes fall outside of the CBF regulon. While most of these genes are induced less than fivefold in response to low temperature, 33 were induced more than fivefold. Identifying the transcription factors that regulate the expression of these genes is required to completely define the low-temperature regulatory network of *Arabidopsis* and may reveal additional transcription factors with important roles in cold acclimation.

Role of the CBF cold response pathway in configuring the low-temperature metabolome of *Arabidopsis*

It is well established that biochemical changes are associated with the process of cold acclimation in plants (Levitt, 1980). This includes the accumulation of sugars and amino acids such as sucrose and proline, which have been shown to have cryoprotective properties (Rudolph and Crowe, 1985; Strauss and Hauser, 1986). Until recently, the analysis of the metabolic changes associated with cold acclimation has been limited to targeted studies of individual metabolites. However, the recent development of novel technologies to assess global metabolic changes now makes a more comprehensive analysis possible (Fiehn *et al.*, 2000). Using these technologies in collaboration with Oliver Fiehn (Max Planck Institute for Molecular Plant Physiology), an initial step to determine the global metabolic changes that occur with cold acclimation in *Arabidopsis* was taken (Cook *et al.*,

Table 2.1. Cold-induced COS genes assigned to the CBF regulon.

				Average fold change[b]				A/GCCGAC[c]
Probe set[a]	AGI	Annotation	Subrole	CBF2	1 h	24 h	7 d	1 kb upstream
Metabolism								
265119_at	AT1G62570	Flavin-containing monooxygenase (similar to glutamate synthase)	Amino acid	12.2	–0.6	**30.4**	**9.7**	0
251775_s_at	AT3G55610	Delta-1-pyrroline-5-carboxylate synthetase (P5CS2)	Amino acid	3.3	0.0	**6.1**	2.3	1
264511_at	AT1G09350	Putative galactinol synthase	Carbohydrate	346.5	–0.6	**113.5**	**40.7**	3
257876_at	AT3G17130	Similar to invertase/pectin methylesterase inhibitor	Carbohydrate	14.4	–0.7	**9.4**	2.5	1
245998_at	AT5G20830	Sucrose synthase I (SUS1)	Carbohydrate	2.6	0.0	**10.4**	**4.9**	1
263789_at	AT2G24560	Putative GDSL-motif lipase/hydrolase	Lipid	258.5	1.2	**23.5**	4.1	2
245533_at	AT4G15130	Putative phosphocholine cytidylyltransferase (AtCCT2)	Lipid	4.7	1.2	**6.6**	**3.8**	1
261048_at	AT1G01420	Similar to UTP-glucose glucosyltransferases	Secondary	6.3	0.6	**4.8**	2.8	1
266532_at	AT2G16890	Putative phenylpropenoid glucosyltransferase	Secondary	2.6	2.0	**14.5**	**7.9**	0
253879_s_at	AT4G27570	UDP rhamnose-anthocyanidin-3-glucoside rhamnosyltransferase	Secondary	4.5	0.6	**7.3**	2.5	0
Transcription								
260776_at	AT1G14580	Zinc finger (C2H2 type) protein family	mRNA synthesis	3.4	0.0	**5.3**	2.5	0
245807_at	AT1G46768	AP2 domain transcription factor (RAP2.1)	mRNA synthesis	5.3	–5.8	**10.4**	**5.6**	2

(Continued)

Table 2.1. *(continued)*

Probe set[a]	AGI	Annotation	Subrole	Average fold change[b]				A/GCCGAC[c]
				CBF2	1 h	24 h	7 d	1 kb upstream
259971_at	AT1G76580	Squamosa promoter binding protein-related	mRNA synthesis	3.3	0.6	**4.5**	1.9	1
266514_at	AT2G47890	CONSTANS B-box zinc finger family protein	mRNA synthesis	3.6	0.0	**5.5**	2.7	1
251793_at	AT3G55580	Regulator of chromosome condensation RCC1 family	mRNA synthesis	5.5	2.6	**38.1**	**41.9**	0
253722_at	AT4G29190	Zinc finger (CCCH type) protein family	mRNA synthesis	2.9	2.2	**4.1**	2.6	1
253219_at	AT4G34990	MYB family transcription factor (MYB32)	mRNA synthesis	3.1	–1.3	**2.7**	1.5	0
245711_at	AT5G04340	Putative C2H2 zinc finger transcription factor	mRNA synthesis	3.5	**5.7**	**3.7**	0.9	2
Intercellular communication and signal transduction								
246922_at	AT5G25110	Serine/threonine protein kinase-like	Intracellular signalling	11.5	0.5	**91.1**	**47.9**	0
246756_at	AT5G27930	Protein phosphatase 2C (PP2C), putative	Signal transduction	3.7	0.0	**2.9**	1.7	0
Energy								
264953_at	AT1G77120	Alcohol dehydrogenase	Fermentation	6.1	–0.7	**11.8**	2.9	1
253416_at	AT4G33070	Pyruvate decarboxylase-1 (Pdc1)	Fermentation	3.7	–0.9	**22.6**	7.1	1
Transport								
247937_at	AT5G57110	Ca^{2+}-transporting ATPase (ACA8)	Calcium transport	2.6	–0.2	**7.6**	**3.6**	1

245427_at	AT4G17550	Glycerol-3-phosphate permease-like protein	Sugar transport	6.7	0.7	**15.2**	**4.8**	2
260410_at	AT1G69870	Putative proton-dependent oligopeptide transport (POT) protein	Transport	3.7	0.8	**7.3**	2.8	1
263574_at	AT2G16990	Putative tetracycline transporter protein	Transport	14.9	–1.2	**3.3**	1.9	2
266225_at	AT2G28900	Putative membrane channel protein	Transport	7.8	–0.1	**6.1**	3.5	2
250151_at	AT5G14570	Similar to trans-membrane nitrate transporter protein	Transport	3.1	0.6	**5.2**	2.8	1
Protein processing and fate								
262440_at	AT1G47710	Serpin, putative	Protein modification	3.8	–1.1	**5.3**	2.0	2
251899_at	AT3G54400	Aspartyl protease family	Proteolysis	3.0	0.0	**4.8**	2.8	1
262644_at	AT1G62710	Vacuolar cystein proteinase (beta-VPE)	Targeting and sorting	2.7	–0.1	**8.9**	**5.2**	0
Stress related								
264787_at	AT2G17840	ERD7	Ageing	3.6	0.8	**9.4**	**3.2**	1
263951_at	AT2G35960	Putative Harpin-induced protein	Defense	2.7	0.5	**4.4**	2.7	0
260556_at	AT2G43620	Putative endochitinase	Defense	40.4	1.5	**12.0**	**9.3**	1
253104_at	AT4G36010	Thaumatin-like protein	Defense	5.0	**3.2**	**13.2**	2.5	3
258893_at	AT3G05660	Disease resistance protein	Intracelluar signalling	20.5	0.5	**7.3**	**5.7**	2
259426_at	AT1G01470	LEA protein (LEA14)	LEA/dehydrin	10.3	0.1	**8.6**	**3.4**	3
259570_at	AT1G20440	Dehydrin (COR47)	LEA/dehydrin	5.7	0.1	**13.5**	**4.8**	3
259516_at	AT1G20450	Dehydrin (ERD10)	LEA/dehydrin	8.6	1.6	**15.3**	**6.9**	2

(Continued)

Table 2.1. *(continued)*

Probe set[a]	AGI	Annotation	Subrole	Average fold change[b]				A/GCCGAC[c]
				CBF2	1 h	24 h	7 d	1 kb upstream
256310_at	AT1G30360	Dehydrin (ERD4)	LEA/dehydrin	2.6	1.1	**2.8**	0.7	1
263495_at	AT2G42530	COR15b	LEA/dehydrin	53.8	0.2	**36.3**	**25.8**	2
263497_at	AT2G42540	COR15a	LEA/dehydrin	484.8	0.0	**115.9**	**87.8**	2
252102_at	AT3G50970	Dehydrin (Xero2)	LEA/dehydrin	79.0	0.7	**52.3**	**25.3**	3
246481_s_at	AT5G15960	KIN1	LEA/dehydrin	21.1	0.0	**8.3**	**7.4**	1
248337_at	AT5G52310	COR78	LEA/dehydrin	57.7	0.8	**51.3**	**24.1**	4
262452_at	AT1G11210	Expressed protein (COR35)	Unknown	3.5	4.3	**33.4**	**14.0**	0
256114_at	AT1G16850	Expressed protein (COR17)	Unknown	380.4	0.4	**218.0**	**60.8**	3
259789_at	AT1G29395	Similar to cold acclimation protein WCOR414 (WCOR414-TM1)	Unknown	20.0	–0.1	**9.4**	**7.3**	1
262050_at	AT1G80130	Expressed protein (COR33.5)	Unknown	6.1	0.6	**24.1**	**26.3**	1
265480_at	AT2G15970	Similar to cold acclimation protein WCOR413 (WCOR413-PM1)	Unknown	4.0	–1.1	**5.0**	**3.3**	2
267261_at	AT2G23120	Expressed protein (COR8.5)	Unknown	8.6	1.4	**6.2**	**3.4**	1
254818_at	AT4G12470	pEARLI 1-like protein	Unknown	11.7	1.1	**38.1**	**34.1**	0
254805_at	AT4G12480	pEARLI 1	Unknown	9.7	0.6	**21.4**	**47.2**	0
254832_at	AT4G12490	pEARLI 1-like protein	Unknown	4.1	–0.1	**3.0**	**10.2**	0
253627_at	AT4G30650	Hydrophobic protein (LTI6A/RCI2A)	Unknown	11.7	0.0	**8.9**	**7.2**	1
253595_at	AT4G30830	Expressed protein (COR42)	Unknown	66.3	0.8	**91.6**	40.9	2

Cellular biogenesis								
259173_at	AT3G03640	Beta-glucosidase (GLUC)	Cell wall	3.2	0.4	**4.2**	1.8	2
258719_at	AT3G09540	Putative pectate lyase	Cell wall	4.5	–0.6	**6.9**	3.0	2
251229_at	AT3G62740	Beta-glucosidase-like protein	Cell wall	7.8	–0.2	**12.1**	12.2	1
254662_at	AT4G18270	Glycosyltransferase family 4 (ATTRANS11)	Cell wall	2.6	0.0	**2.9**	1.6	1
252997_at	AT4G38400	Expansin protein family (EXPL2)	Cell wall	3.3	1.9	**9.6**	2.4	3
247478_at	AT5G62360	Similar to pectinesterase	Cell wall	3.9	1.4	**8.3**	2.0	1
Cellular organization								
254085_at	AT4G24960	Similar to abscisic acid-induced protein (HVA22D)	Vesicular trafficking	13.1	0.0	**8.6**	2.1	3
DNA replication and cell division								
257237_at	AT3G14890	DNA nick sensor, putative	DNA synthesis	3.1	–1.1	**7.6**	2.8	2
Unknown role								
260727_at	AT1G48100	Glycoside hydrolase family 28 protein	Unknown	16.2	0.6	**15.8**	**3.3**	0
262881_at	AT1G64890	Expressed protein (integral membrane/transporter family)	Unknown	2.8	0.0	**9.6**	**3.4**	0
252956_at	AT4G38580	Farnesylated protein (ATFP6)	Unknown	3.2	1.1	**10.0**	**4.5**	2
250279_at	AT5G13200	Like ABA-responsive protein	Unknown	7.0	1.5	**4.4**	2.7	1
248467_at	AT5G50800	Nodulin MtN3-like protein	Unknown	3.1	–0.1	**4.4**	**7.0**	0
245749_at	AT1G51090	Similar to proline-rich protein	Unknown	56.5	1.3	**29.3**	4.4	2
251927_at	AT3G53990	Universal stress protein (USP) family protein	Unknown	3.4	1.1	**4.4**	2.1	2

(Continued)

Table 2.1. *(continued)*

Probe set[a]	AGI	Annotation	Subrole	Average fold change[b]				A/GCCGAC[c]
				CBF2	1 h	24 h	7 d	1 kb upstream
262164_at	AT1G78070	WD-40 repeat family protein	Unknown	4.8	0.6	**5.1**	2.7	2
Unknown protein								
264516_at	AT1G10090	Expressed protein	Unknown	3.1	–0.6	**3.4**	1.4	2
264458_at	AT1G10410	Expressed protein	Unknown	2.9	1.2	**6.4**	3.0	1
262496_at	AT1G21790	Expressed protein	Unknown	4.0	0.1	**5.8**	2.1	1
264989_at	AT1G27200	Expressed protein	Unknown	5.2	1.6	**7.5**	**3.9**	2
261566_at	AT1G33230	Expressed protein	Unknown	3.0	–0.1	**4.0**	2.2	1
245200_at	AT1G67850	Expressed protein	Unknown	3.7	1.1	**3.7**	2.8	0
260264_at	AT1G68500	Expressed protein	Unknown	14.8	1.2	**17.8**	3.2	1
263352_at	AT2G22080	Expressed protein	Unknown	2.9	0.6	**4.3**	1.8	1
251753_at	AT3G55760	Expressed protein	Unknown	9.9	–0.1	**10.4**	2.1	1
254850_at	AT4G12000	Expressed protein	Unknown	4.3	0.5	**3.8**	1.7	1
253425_at	AT4G32190	Expressed protein	Unknown	2.7	0.5	**4.2**	1.5	2
246125_at	AT5G19875	Expressed protein	Unknown	4.1	0.6	**3.6**	2.3	1
249750_at	AT5G24570	Expressed protein	Unknown	5.0	–0.1	**3.8**	2.5	1

For details on the experiments that generated these data, see Vogel *et al.* (2005).
[a]AGI identifier numbers, annotations and subrole assignments for each probe set were derived from data provided by Affymetrix (http://www.affymetrix.com/), TAIR (http://www.arabidopsis.org/), TIGR (http://www.tigr.org/tdb/e2k1/ath1/) and MIPS (http://mips.gsf.de/proj/thal/db/index.html).
[b]The average fold change values are given for the two CBF2 overexpressing lines (E2 and E24) profiled and the four replicates of cold-treated wild-type *Arabidopsis*. Values in bold indicate times at which the transcript was labelled as cold-responsive; see Vogel *et al.* (2005) for the selection criteria.
[c]The number of CRT/DRE elements present 1 kb upstream of the transcript listed was based on datasets available from TAIR using the TIGR 5.0 annotation.

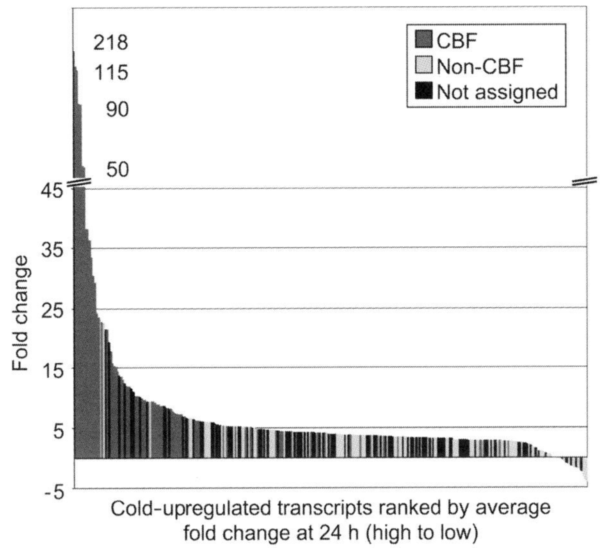

Fig. 2.1. A majority of the transcripts that exhibit the greatest fold increase in response to low temperature are members of the CBF regulon. Cold-upregulated transcripts were ranked by average fold change after 24 h at 4°C. Those that were assigned to the CBF regulon (CBF) are coloured grey; those that were assigned as not being members of the CBF regulon (non-CBF) are coloured light grey; and those that could not be assigned (not assigned) to either of these classes are coloured black.

2004). A total of 434 low-molecular weight carbohydrates, amines, organic acids and other polar molecules were monitored in the analysis. In addition, the role of the CBF cold response pathway in configuring these changes was assessed by analysing the changes that occurred in the metabolome of non-acclimated transgenic *Arabidopsis* plants that constitutively overexpressed *CBF3*.

The analysis indicated that cold acclimation in *Arabidopsis* (ecotype Ws-2) is associated with extensive changes in the metabolome profile (Cook *et al.*, 2004). Using a rigorous statistical cut-off ($P < 0.001$), the results indicated that of the 434 metabolites monitored, 325 (75%) increased in cold-acclimated plants. The increases observed varied from less than twofold to greater than 25-fold, with 114 (35%) increasing at least fivefold (Table 2.2). Of the metabolites that increased with cold acclimation, 75 were identified and 250 were categorized according to class (carbohydrate, amine, organic acid or other polar molecules). However, a significant number, 116 (36%), were unidentified.

Several of the identified metabolites had previously been shown to increase in *Arabidopsis* plants upon exposure to low temperature including the amino acid, proline, and sugars such as glucose, fructose, inositol, galactinol, raffinose and sucrose. However, the majority of the identified metabolites had not previously been shown to increase in *Arabidopsis*. This includes trehalose, putrescine and ascorbate, which have potential roles in cold tolerance. In addition, increases observed in ornithine and citrulline, precursors to polyamine biosynthesis, suggest

Table 2.2. Number of metabolites that increase in *Arabidopsis* (ecotype Ws-2) with cold acclimation.

Fold change in peak area in cold ($P < 0.001$)[a]	Number of metabolites
≥ 25	26
≥ 5 to < 25	88
> 1 to < 5	211
Total	325

[a]A total of 434 metabolites were monitored using GC-time-of-flight mass spectrometry, 325 of which increased the indicated amounts in response to low temperature (14 days).

upregulation of the urea cycle; and increases in α-ketoglutarate, fumarate, malate and citrate, precursors to amino acid biosynthesis, suggest upregulation of the TCA cycle.

To explore the extent to which the low-temperature metabolome of *Arabidopsis* is conditioned by the CBF cold response pathway, we compared the metabolomes of non-acclimated and cold-acclimated wild-type *Arabidopsis* Ws-2 plants with the metabolome of non-acclimated transgenic *Arabidopsis* Ws-2 plants overexpressing *CBF3* (Cook *et al.*, 2004). The results indicated that expression of the CBF regulon brought about extensive changes in the metabolome that were similar to those that occurred in response to low temperature. Of the 325 metabolites that were identified as increasing in response to low temperature, 256 (79%) were also found to increase in response to overexpression of *CBF3* ($P < 0.001$). Moreover, of the 114 metabolites that increased more than fivefold in response to low temperature, 102 (90%) also increased in response to *CBF3* overexpression. These results suggested that the metabolome of non-acclimated *CBF3*-overexpressing plants was similar to that of cold-acclimated plants. To test this further, the levels of the 434 metabolites in the non-acclimated *CBF3*-overexpressing plants were compared to those in non-acclimated and cold-acclimated Ws-2 plants and the results were subjected to hierarchical clustering. The analysis indicated that the metabolome of the non-acclimated *CBF3*-overexpressing plants more closely resembled the metabolome of cold-acclimated wild-type plants than the metabolome of non-acclimated wild-type plants.

Taken together, the results of the metabolite-profiling experiments indicated that extensive changes occurred in the metabolome of *Arabidopsis* Ws-2 plants in response to low temperature and that the CBF cold response pathway has a prominent role in bringing about these changes.

The CBF cold response pathway of tomato

As noted earlier, in contrast to *Arabidopsis*, tomato (*Lycopersicon esculentum/Solanum lycopersicum*) is a chilling-sensitive plant that cannot cold acclimate. Given the role of the CBF cold response pathway in chilling and freezing tolerance in *Arabidopsis*, it was of interest to determine whether tomato has a CBF cold response pathway, and if so, whether 'deficiencies' in it might contribute to the cold-sensitive phenotype of tomato. Towards this end, it should be asked first whether tomato has functional *CBF* genes. It was found that tomato has three genes encoding CBF-like proteins, *LeCBF1–3*, that, as in *Arabidopsis*, are located in tandem array in the genome (Jaglo *et al.*, 2001; Zhang *et al.*, 2004). All three tomato CBF proteins contain the CBF family 'signature sequences' flanking the AP2/ERF domain (Jaglo *et al.*, 2001), and alignments of the tomato and *Arabidopsis* CBF1–3 proteins reveal additional highly conserved regions outside of the AP2/ERF domain (Jaglo *et al.*, 2001; Zhang *et al.*, 2004). Overall, the amino acid sequences of LeCBF1–3 are 70–84% identical to each other and 51–59% identical to CBF1–3 proteins from *Arabidopsis*.

Expression analysis studies showed that in a similar manner to *CBF1–3* in *Arabidopsis*, transcripts for the tomato *LeCBF1* gene accumulate rapidly (within 15 min) upon transferring tomato plants to low temperature and also after mechanical agitation (Jaglo *et al.*, 2001; Zhang *et al.*, 2004). In contrast, neither *LeCBF2* nor *LeCBF3* was induced in response to low temperature. They were, however, responsive to mechanical agitation, indicating that these genes could still be activated under certain conditions. None of the tomato *CBF* genes was upregulated in response to high salinity, drought or treatment with ABA.

To test whether the cold-responsive tomato gene, *LeCBF1*, encodes a functional homologue of the *Arabidopsis* CBF proteins, transgenic *Arabidopsis* plants that constitutively expressed *LeCBF1* under the control of the CaMV 35S promoter were created (Zhang *et al.*, 2004). The transgenic plants exhibited stunted growth and delayed flowering, a phenotype similar to that described when *CBF* genes are overexpressed in *Arabidopsis* (Liu *et al.*, 1998; Kasuga *et al.*, 1999; Gilmour *et al.*, 2004). Furthermore, constitutive expression of *LeCBF1* in transgenic *Arabidopsis* plants activated expression of *COR* genes and increased plant freezing tolerance in the absence of a low-temperature stimulus (Zhang *et al.*, 2004). These results indicated that *LeCBF1* encodes a functional homologue of the *Arabidopsis CBF1–3* genes.

The finding that tomato is able to activate expression of at least one functional *CBF* gene in response to low temperature indicated that upstream components of the CBF cold response pathway were functional in tomato. That is, tomato could sense low temperature and activate expression of a *CBF* gene that encoded a functional protein. The question then became whether downstream components of the CBF cold response pathway were present and functional in tomato. To determine whether the CBF regulon of tomato was similar to that of *Arabidopsis*, transgenic tomato plants constitutively overexpressing either *AtCBF3* or *LeCBF1* under control of the CaMV 35S promoter were created and tested for changes in gene expression and freezing tolerance.

The results indicated that tomato did have a CBF regulon, but that it was much smaller than that of *Arabidopsis*. In particular, transcriptome analysis using a cDNA microarray surveying approximately 25% of the tomato genome identified only three genes that were members of the tomato CBF regulon (i.e. were induced by both low temperature and overexpression of *AtCBF3* and *LeCBF1*).

Two of these genes encoded dehydrins (genes encoding dehydrins are also found in the *Arabidopsis* CBF regulon) and the third encoded a putative proteinase inhibitor. Limited sequence analysis of the promoter regions of the two dehydrin genes indicated the presence of putative CRT/DRE elements. Further analysis indicated that eight tomato genes that were likely homologues of *Arabidopsis* CBF regulon genes were not responsive to *CBF* overexpression in tomato. Finally, constitutive expression of *AtCBF3* or *LeCBF3* did not result in increased freezing tolerance in the transgenic tomato plants. A similar result was described by Hsieh *et al.* (2002) for constitutive expression of *AtCBF1* in tomato.

Taken together, the results obtained with tomato indicate that it has a complete CBF cold response pathway; i.e. tomato can sense low temperature, activate the expression of a *CBF* gene encoding a functional protein and has a CBF regulon that includes genes with putative functional CRT/DRE elements and encoding dehydrin proteins. However, it also appears that the CBF regulon of tomato is much smaller, and, consequently, is less functionally diverse than that of *Arabidopsis*.

Concluding Remarks

A central goal of the studies described above was to determine the extent to which the CBF cold response pathway ‘configures’ the low-temperature transcriptome and metabolome of *Arabidopsis*. The results provide significant insights into this issue. The expression-profiling experiments indicate that the CBF transcription factors have a prominent role in regulating the expression of those cold-responsive genes that are the most highly induced in response to low temperature and remain upregulated for extended periods of time (7 days in these experiments). In addition, the metabolite-profiling experiments indicate that extensive changes occur in the metabolome of *Arabidopsis* Ws-2 plants in response to low temperature, and that the CBF cold response pathway has a prominent role in bringing about these changes. In sum, the results provide additional evidence that the CBF cold response pathway has a central role in the cold acclimation response in *Arabidopsis*.

While it is evident that the CBF cold response pathway is a major player in cold acclimation, it is also apparent that many cold-induced genes fall outside of the CBF regulon. Key challenges now are to determine what transcription factors control the expression of the ‘non-CBF regulon’ genes; to define the regulons controlled by these transcription factors; and to determine whether these regulons contribute significantly to cold tolerance. Towards these ends, Vogel *et al.* (2005) identified a number of transcription factors that are expressed in parallel with the *CBF1–3* genes and demonstrated that one of these transcription factors, ZAT12, a C2H2 zinc finger protein (Meissner and Michael, 1997), contributes to conditioning the low-temperature transcriptome and freezing tolerance. Overexpression of *ZAT12* in transgenic *Arabidopsis* plants led to the assigning of 24 genes to the ZAT12 regulon and demonstrates that expression of the ZAT12 regulon results in a small, but detectible, increase in plant freezing tolerance (Vogel *et al.*, 2005). Similarly, Zhu *et al.* (2004) have identified a transcription factor, HOS9, a homeodomain protein, that appears to control expression of certain *COR* genes and to contribute to freezing tolerance. Continued efforts along these lines should, over the next few years, lead to the construction of a low-temperature 'wiring diagram' that includes the identification of regulatory networks that function in *Arabidopsis* cold acclimation.

In regard to the low-temperature metabolome studies, we would like to integrate the metabolite data with the transcriptome results. At present, however, such efforts are severely limited by the fact that, of the estimated 5000 primary and secondary metabolites present in an *Arabidopsis* leaf, only about 10% have been identified (Bino *et al.*, 2004). Indeed, of the 325 metabolites that we found increased upon cold acclimation, only 75 (23%) were conclusively identified (Cook *et al.*, 2004). Similarly, efforts to integrate metabolome and transcriptome data are limited by the fact that a large percentage (approximately one-third) of the identified transcripts are not yet assigned functions. It is also the case that the linking of individual metabolites to particular transcripts or pathways is complicated by the fact that individual metabolites can be members of multiple pathways and individual enzymes can catalyse the synthesis of multiple metabolites. Despite these limitations, we have learned that cold acclimation is associated with extensive changes in the metabolome and that the CBF cold response pathway has a prominent role in bringing about these changes. It is also the case that improved technologies and methods for metabolome analysis are intensively being developed as well as bioinformatic approaches to integrate datasets (e.g. Scholz *et al.*, 2004; Thimm *et al.*, 2004). This is an exciting area of research that should continue to blossom in the coming years.

The second major goal of the research described here was to develop a better understanding of the CBF cold response pathway of tomato. It was found that tomato can sense low temperature and activate expression of a *CBF* gene encoding a functional CBF protein. In addition, we found that expression of the tomato *CBF* gene leads to induction of target genes that have putative functional CRT/DRE elements, i.e. tomato has a CBF regulon. Thus, it can be concluded that tomato has a 'complete' CBF cold response pathway. However, the results also indicate that the CBF regulon of tomato is much smaller, and consequently is less functionally diverse, than the *Arabidopsis* CBF regulon. Whereas the *Arabidopsis* CBF regulon appears to include more than 100 genes (Fowler and Thomashow, 2002; Seki *et al.*, 2002; Maruyama *et al.*, 2004; Vogel *et al.*, 2005), the CBF regulon of tomato may comprise only about 10–15 genes (three genes were assigned to the tomato CBF regulon after surveying approximately 25% of the genome). What could account for this large difference in regulon composition? One possibility is that tomato might have only a few genes with functional CRT/DRE elements within their promoters. Alternatively, the differences in CBF regulon composition might reflect differences in the protein factors required for CBF to function. For instance, in *Arabidopsis*, the *SFR6* gene is required for CBF to activate expression of its target genes (Knight *et al.*, 1999; Boyce *et al.*, 2003). Perhaps tomato encodes a weak allele of *SFR6* or some other gene required for full CBF activity, resulting in weak induction of most CBF target genes. Distinguishing between these and other possibilities is a goal that is being pursued.

Acknowledgements

Research conducted in the laboratory of M.F. Thomashow was made possible by grants from the NSF Plant Genome Program (DBI 0110124), the DOE (DEFG0291ER20021) and the Michigan Agricultural Experiment Station. D. Cook was a recipient of an Alfred Toepfer award from the von Humboldt Foundation. J.T. Vogel was a recipient of a US Department of Education Graduate Assistantship in Areas of National Need (GAANN) fellowship.

References

Artus, N.N., Uemura, M., Steponkus, P.L., Gilmour, S.J., Lin, C. and Thomashow, M.F. (1996) Constitutive expression of the cold-regulated *Arabidopsis thaliana COR15a* gene affects both chloroplast and protoplast freezing tolerance. *Proceedings of the National Academy of Sciences USA* 93, 13404–13409.

Baker, S.S., Wilhelm, K.S. and Thomashow, M.F. (1994) The 5′-region of *Arabidopsis thaliana cor15a* has *cis*-acting elements that confer cold-, drought- and ABA-regulated gene expression. *Plant Moecular Biology* 24, 701–713.

Bino, R.J., Hall, R.D., Fiehn, O., Kopka, J., Saito, K., Draper, J., Nikolau, B.J., Mendes, P., Roessner-Tunali, U., Beale, M.H., Trethewey, R.N., Lange, B.M., Wurtele, E.S. and Sumner, L.W. (2004) Potential of metabolomics as a functional genomics tool. *Trends in Plant Science* 9, 418–425.

Boyce, J.M., Knight, H., Deyholos, M., Openshaw, M.R., Galbraith, D.W., Warren, G. and Knight, M.R. (2003) The *sfr6* mutant of *Arabidopsis* is defective in transcriptional activation via CBF/DREB1 and DREB2 and shows sensitivity to osmotic stress. *Plant Journal* 34, 395–406.

Choi, D.W., Rodriguez, E.M. and Close, T.J. (2002) Barley *Cbf3* gene identification, expression pattern, and map location. *Plant Physiology* 129, 1781–1787.

Cook, D., Fowler, S., Fiehn, O. and Thomashow, M.F. (2004) A prominent role for the CBF cold response pathway in configuring the low-temperature metabolome of *Arabidopsis*. *Proceedings of the National Academy of Sciences USA* 101, 15243–15248.

Dubouzet, J.G., Sakuma, Y., Ito, Y., Kasuga, M., Dubouzet, E.G., Miura, S., Seki, M., Shinozaki, K. and Yamaguchi-Shinozaki, K. (2003) *OsDREB* genes in rice, *Oryza sativa* L., encode transcription activators that function in drought-, high-salt- and cold-responsive gene expression. *Plant Journal* 33, 751–763.

Dunn, M.A., White, A.J., Vural, S. and Hughes, M.A. (1998) Identification of promoter elements in a low-temperature-responsive gene (*blt*4.9) from barley (*Hordeum vulgare* L.). *Plant Molecular Biology* 38, 551–564.

Fiehn, O., Kopka, J., Dormann, P., Altmann, T., Trethewey, R.N. and Willmitzer, L. (2000) Metabolite profiling for plant functional genomics. *Nature Biotechnology* 18, 1157–1161.

Fowler, S. and Thomashow, M.F. (2002) *Arabidopsis* transcriptome profiling indicates that multiple regulatory pathways are activated during cold acclimation in addition to the CBF cold response pathway. *Plant Cell* 14, 1675–1690.

Gao, M.J., Allard, G., Byass, L., Flanagan, A.M. and Singh, J. (2002) Regulation and characterization of four *CBF* transcription factors from *Brassica napus*. *Plant Molecular Biology* 49, 459–471.

Gilmour, S.J., Zarka, D.G., Stockinger, E.J., Salazar, M.P., Houghton, J.M. and Thomashow, M.F. (1998) Low temperature regulation of the *Arabidopsis* CBF family of AP2 transcriptional activators as an early step in cold-induced *COR* gene expression. *Plant Journal* 16, 433–442.

Gilmour, S.J., Sebolt, A.M., Salazar, M.P., Everard, J.D. and Thomashow, M.F. (2000) Overexpression of the *Arabidopsis CBF3* transcriptional activator mimics multiple biochemical changes associated with cold acclimation. *Plant Physiology* 124, 1854–1865.

Gilmour, S.J., Fowler, S.G. and Thomashow, M.F. (2004) *Arabidopsis* transcriptional activators CBF1, CBF2, and CBF3 have matching functional activities. *Plant Molecular Biology* 54, 767–781.

Hsieh, T.H., Lee, J.T., Chang, Y.Y. and Chan, M.T. (2002) Tomato plants ectopically expressing *Arabidopsis CBF1* show enhanced resistance to water deficit stress. *Plant Physiology* 130, 618–626.

Jaglo, K.R., Kleff, S., Amundsen, K.L., Zhang, X., Haake, V., Zhang, J.Z., Deits, T. and Thomashow, M.F. (2001) Components of the *Arabidopsis* C-repeat/dehydration-responsive element binding factor cold-response pathway are conserved in *Brassica napus* and other plant species. *Plant Physiology* 127, 910–917.

Jaglo-Ottosen, K.R., Gilmour, S.J., Zarka, D.G., Schabenberger, O. and Thomashow, M.F. (1998) *Arabidopsis CBF1* overexpression induces *COR* genes and enhances freezing tolerance. *Science* 280, 104–106.

Kasuga, M., Liu, Q., Miura, S., Yamaguchi-Shinozaki, K. and Shinozaki, K. (1999) Improving plant drought, salt, and freezing tolerance by gene transfer of a single stress-inducible transcription factor. *Nature Biotechnology* 17, 287–291.

Knight, H., Veale, E.L., Warren, G.J. and Knight, M.R. (1999) The *sfr6* mutation in *Arabidopsis* suppresses low-temperature induction of genes dependent on the CRT/DRE sequence motif. *Plant Cell* 11, 875–886.

Kreps, J.A., Wu, Y., Chang, H.S., Zhu, T., Wang, X. and Harper, J.F. (2002) Transcriptome changes for *Arabidopsis* in response to salt, osmotic, and cold stress. *Plant Physiology* 130, 2129–2141.

Levitt, J. (1980) *Responses of Plants to Environmental Stress. Chilling, Freezing and High Temperature Stresses*, 2nd edn. Academic Press, New York.

Liu, Q., Kasuga, M., Sakuma, Y., Abe, H., Miura, S., Yamaguchi-Shinozaki, K. and Shinozaki, K. (1998) Two transcription factors, DREB1 and DREB2, with an EREBP/AP2 DNA binding domain separate two cellular signal transduction pathways in drought- and low-temperature-responsive gene expression, respectively, in *Arabidopsis*. *Plant Cell* 10, 1391–1406.

Maruyama, K., Sakuma, Y., Kasuga, M., Ito, Y., Seki, M., Goda, H., Shimada, Y., Yoshida, S., Shinozaki, K. and Yamaguchi-Shinozaki, K. (2004) Identification of cold-inducible downstream genes of the *Arabidopsis* DREB1A/CBF3 transcriptional factor using two microarray systems. *Plant Journal* 38, 982–993.

Medina, J., Bargues, M., Terol, J., Perez-Alonso, M. and Salinas, J. (1999) The *Arabidopsis CBF* gene family is composed of three genes encoding AP2 domain-containing proteins whose expression is regulated by low temperature but not by abscisic acid or dehydration. *Plant Physiology* 119, 463–470.

Meissner, R. and Michael, A.J. (1997) Isolation and characterisation of a diverse family of *Arabidopsis* two and three-fingered C2H2 zinc finger protein genes and cDNAs. *Plant Molecular Biology* 33, 615–624.

Qin, F., Sakuma, Y., Li, J., Liu, Q., Li, Y.Q., Shinozaki, K. and Yamaguchi-Shinozaki, K. (2004) Cloning and functional analysis of a novel DREB1/CBF transcription factor involved in cold-responsive gene expression in *Zea mays* L. *Plant Cell Physiology* 45, 1042–1052.

Redman, J.C., Haas, B.J., Tanimoto, G. and Town, C.D. (2004) Development and evaluation of an *Arabidopsis* whole genome Affymetrix probe array. *Plant Journal* 38, 545–561.

Riechmann, J.L. and Meyerowitz, E.M. (1998) The AP2/EREBP family of plant transcription factors. *Biological Chemistry* 379, 633–646.

Rudolph, A.S. and Crowe, J.H. (1985) Membrane stabilization during freezing: the role of two natural cryoprotectants, trehalose and proline. *Cryobiology* 22, 367–377.

Scholz, M., Gatzek, S., Sterling, A., Fiehn, O. and Selbig, J. (2004) Metabolite fingerprinting: detecting biological features by independent component analysis. *Bioinformatics* 20, 2447–2454.

Seki, M., Narusaka, M., Abe, H., Kasuga, M., Yamaguchi-Shinozaki, K., Carninci, P., Hayashizaki, Y. and Shinozaki, K. (2001) Monitoring the expression pattern of 1300 *Arabidopsis* genes under drought and cold stresses by using a full-length cDNA microarray. *Plant Cell* 13, 61–72.

Seki, M., Narusaka, M., Ishida, J., Nanjo, T., Fujita, M., Oono, Y., Kamiya, A., Nakajima, M., Enju, A., Sakurai, T., Satou, M., Akiyama, K., Taji, T., Yamaguchi-Shinozaki, K., Carninci, P., Kawai, J., Hayashizaki, Y. and Shinozaki, K. (2002) Monitoring the expression profiles of 7000 *Arabidopsis* genes under drought, cold and high-salinity stresses using a full-length cDNA microarray. *Plant Journal* 31, 279–292.

Shinozaki, K., Yamaguchi-Shinozaki, K. and Seki, M. (2003) Regulatory network of gene expression in the drought and cold stress responses. *Current Opinion in Plant Biology* 6, 410–417.

Smallwood, M. and Bowles, D.J. (2002) Plants in a cold climate. *Philosophical Transactions of the Royal Society of London. Series B, Biological Science* 357, 831–847.

Steponkus, P.L., Uemura, M., Joseph, R.A., Gilmour, S.J. and Thomashow, M.F. (1998) Mode of action of the *COR15a* gene on the freezing tolerance of *Arabidopsis thaliana. Proceedings of the National Academy of Sciences USA* 95, 14570–14575.

Stockinger, E.J., Gilmour, S.J. and Thomashow, M.F. (1997) *Arabidopsis thaliana CBF1* encodes an AP2 domain-containing transcriptional activator that binds to the C-repeat/DRE, a *cis*-acting DNA regulatory element that stimulates transcription in response to low temperature and water deficit. *Proceedings of the National Academy of Sciences USA* 94, 1035–1040.

Strauss, G. and Hauser, H. (1986) Stabilization of lipid bilayer vesicles by sucrose during freezing. *Proceedings of the National Academy of Sciences USA* 83, 2422–2426.

Taji, T., Ohsumi, C., Iuchi, S., Seki, M., Kasuga, M., Kobayashi, M., Yamaguchi-Shinozaki, K. and Shinozaki, K. (2002) Important roles of drought- and cold-inducible genes for galactinol synthase in stress tolerance in *Arabidopsis thaliana. Plant Journal* 29, 417–426.

Thimm, O., Blasing, O., Gibon, Y., Nagel, A., Meyer, S., Kruger, P., Selbig, J., Muller, L.A., Rhee, S.Y. and Stitt, M. (2004) MAPMAN: a user-driven tool to display genomics data sets onto diagrams of metabolic pathways and other biological processes. *Plant Journal* 37, 914–939.

Thomashow, M.F. (1999) Plant cold acclimation: freezing tolerance genes and regulatory mechanisms. *Annual Review of Plant Physiology and Plant Molecular Biology* 50, 571–599.

Thomashow, M.F. (2001) So what's new in the field of plant cold acclimation? Lots! *Plant Physiology* 125, 89–93.

Thomashow, M.F. and Browse, J. (1999) Plant cold tolerance. In: Shinozaki, K. and Yamaguchi-Shinozaki, K. (eds) *Drought, Salt, Cold and Heat Stress: Molecular Responses in Higher Plants.* R.G. Landes Company, Austin, Texas, pp. 61–80.

Vazquez-Tello, A., Ouellet, F. and Sarhan, F. (1998) Low temperature-stimulated phosphorylation regulates the binding of nuclear factors to the promoter of *Wcs120*, a cold-specific gene in wheat. *Molecular and General Genetics* 257, 157–166.

Vogel, J.T., Zarka, D.G., Van Buskirk, H.A., Fowler, S.G. and Thomashow, M.F. (2005) Roles of the CBF2 and ZAT12 transcription factors in configuring the low temperature transcriptome of *Arabidopsis. Plant Journal* 41, 195–211.

Xue, G.P. (2002) An AP2 domain transcription factor *HvCBF1* activates expression of cold-responsive genes in barley through interaction with a (G/a)(C/t)CGAC motif. *Biochimica et Biophysica Acta* 1577, 63–72.

Yamaguchi-Shinozaki, K. and Shinozaki, K. (1994) A novel *cis*-acting element in an *Arabidopsis* gene is involved in responsiveness to drought, low-temperature, or high-salt stress. *Plant Cell* 6, 251–264.

Zhang, X., Fowler, S.G., Cheng, H., Lou, Y., Rhee, S.Y., Stockinger, E.J. and Thomashow, M.F. (2004) Freezing-sensitive tomato has a functional CBF cold response pathway, but a CBF regulon that differs from that of freezing-tolerant *Arabidopsis*. *Plant Journal* 39, 905–919.

Zhu, J.H., Shi, H.Z., Lee, B.H., Damsz, B., Cheng, S., Stirm, V., Zhu, J.K., Hasegawa, P.M. and Bressan, R.A. (2004) An *Arabidopsis* homeodomain transcription factor gene, *HOS9*, mediates cold tolerance through a CBF-independent pathway. *Proceedings of the National Academy of Sciences USA* 101, 9873–9878.

3 Barley Contains a Large *CBF* Gene Family Associated with Quantitative Cold-tolerance Traits

J.S. SKINNER,[1,2,*] J. VON ZITZEWITZ,[2] L. MARQUEZ-CEDILLO,[2] T. FILICHKIN,[2] P. SZŰCS,[3] K. AMUNDSEN,[4] E.J. STOCKINGER,[5] M.F. THOMASHOW,[4] T.H.H. CHEN[1] AND P.M. HAYES[2]

[1]Department of Horticulture, College of Agriculture, Oregon State University, Corvallis, OR 97331, USA; [2]Department of Crop and Soil Science, College of Agriculture, Oregon State University, Corvallis, OR 97331, USA; [3]Agricultural Research Institute of the Hungarian Academy of Sciences, H-2462 Martonvásár, Hungary; [4]Department of Crop and Soil Sciences, Michigan State University, East Lansing, MI 48824, USA; [5]Department of Horticulture and Crop Science, The Ohio State University/OARDC, Wooster, OH 44691, USA

Introduction

Low-temperature tolerance within the *Triticeae*

Plants display a broad capacity range to survive cold and freezing conditions (Thomashow, 1999). Most current work into the molecular basis of plant cold tolerance has been performed in the model non-crop plant *Arabidopsis*. Increasingly, these observations are being tested in crop plants. The grasses, or *Poaceae*, contain the economically most important crop plant family members, including all the cereal crops utilized by humans. These include rice (*Oryza sativa*), maize (*Zea mays*) and members of the *Triticeae*, which includes the important members wheat (*Triticum aestivum*), barley (*Hordeum vulgare*) and rye (*Secale cereale*).

Within the cereals, a broad range of cold tolerance, from completely sensitive to extreme cold hardy, is observed. At one extreme are the cereals of subtropical origin such as rice and maize, which are sensitive to cold and will not survive freezing temperatures. These characteristics make these systems of limited use for

*Corresponding author (e-mail: skinnerj@science.oregonstate.edu; Fax: +1-541-737-3479)

dissecting the molecular basis of freezing tolerance within the cereals. In contrast, within the *Triticeae*, a wide range of phenotypic variation for cold tolerance occurs, and is exemplified at the other extreme by rye, which can withstand temperatures of −30°C when fully cold-acclimated (Thomashow, 1999). Our group has been using barley as a model to study the mechanisms by which the *Triticeae*, and cold-tolerant cereals in general, adapt to and survive cold and freezing conditions.

Domesticated barley, *Hordeum vulgare* subsp. *vulgare*, is an economically important crop model for the study and dissection of the molecular, genetic and physiological components of *Triticeae* winter hardiness. Winter hardiness consists of three major traits: low-temperature (LT) tolerance, vernalization (VRN) response and photoperiod (PPD) sensitivity. The specific combination of these three traits determines the growth habit of a genotype (below). Barley is a self-pollinated diploid and abundant genetic variation for cold tolerance is found within its primary gene pool. Relative to the cold-tolerant cereals, barley is unique in the broad array of ever-expanding tools that are currently available for genetic and molecular analysis (reviewed in Hayes *et al.*, 2003), which include multiple doubled haploid mapping populations, near isogenic lines, arrayed large insert BAC clones (Yu *et al.*, 2000), a large expressed sequence tag (EST) database, and a 22K microarray gene chip (Close *et al.*, 2004), among others.

The *Triticeae* form a homogeneous genetic system with a high degree of synteny. Comparative genetics studies confirm that the genetic determinants of winter hardiness appear to be conserved among members that possess this trait, and, therefore, results from one species are frequently applicable to other members of the cereal tribe (Dubcovsky *et al.*, 1998; Mahfoozi *et al.*, 2000). Furthermore, all barley linkage maps are collinear and allow linkage map position alignments of loci and quantitative trait locus (QTL) mapped in different populations, utilizing the BIN map concept of Kleinhofs and Graner (2001). In barley, as in other members of the *Triticeae*, genetic variation for growth habit occurs. QTL analysis tools have revealed that a limited number of conserved genome regions are responsible for the components of winter hardiness. A region is present on the long arm of chromosome 5H where QTL for VRN response, LT tolerance and PPD sensitivity are co-localized in the *Triticeae* (Pan *et al.*, 1994; Cattivelli *et al.*, 2002); this cluster was first observed in the Dicktoo × Morex (DM) barley population as coincident LT tolerance and PPD sensitivity QTLs (Hayes *et al.*, 1993). While VRN, LT and PPD are interrelated (Limin and Fowler, 2002), these three trait phenotypes occur in all possible combinations within the barley germplasm (Karsai *et al.*, 2001), suggesting the interrelationships of these traits may be attributable to linkage rather than to pleiotropy.

Relative to the LT, VRN and PPD trait combinations, three winter hardiness growth habit classes – winter, facultative and spring – are defined for the barley germplasm (Karsai *et al.*, 2001). Winter habits are VRN responsive, PPD sensitive and the most LT tolerant. Higher LT-tolerance capacity is necessary in winter habit genotypes so the plants can survive the prolonged LT exposure needed to fulfil the VRN requirement. In contrast, spring habits lack a VRN response, are insensitive to short-day PPD, and are the most sensitive to LT, and thus have a growth habit that is essentially the inverse of the winter habit. Facultative habits are PPD sensitive and can be as LT tolerant as winter varieties, but lack a VRN requirement. The facultative growth habit may actually be a winter genotype subclass in which the *Vrn-H2* candidate gene has been deleted (von Zitzewitz, 2004; Karsai *et al.*, 2005; Cooper *et al.*, Chapter 5, this volume). In cereals, maximum LT tolerance is achieved following cold acclimation, the process by which a plant develops freezing tolerance when exposed to low, but non-freezing, temperatures while still within the vegetative growth state (Hayes *et al.*, 1997).

*CBF*s are key components of the plant low-temperature tolerance pathway

During the cold acclimation process, many changes occur at both the biochemical and physiological levels in a plant (Thomashow, 1999). One major change is the induction of a suite of LT-responsive genes termed *cor* (cold-regulated) genes. The products of these *cor* genes are thought to collectively impart the necessary physiological and biochemical alterations that increase a plant's freezing tolerance (Thomashow, 1999). Many of these *cor* genes contain a CRT (C-repeat) regulatory element having the core sequence motif, CCGAC, in their promoter. This element has been shown to respond to the abiotic stresses of cold, drought and salt exposure (Thomashow, 1999); this motif is also referred to as a DRE (dehydration response element). One major advance in plant cold hardiness research was the identification of the *CBF* (C-repeat binding factor) family of genes from *Arabidopsis* (Stockinger *et al.*, 1997; Gilmour *et al.*, 1998; Liu *et al.*, 1998). *CBF* genes encode transcription factors that are members of the AP2/EREBP family of DNA-binding proteins and specifically bind to the CRT element of *cor* gene promoters in response to cold, inducing their expression (Stockinger *et al.*, 1997; Gilmour *et al.*, 1998; Jaglo *et al.*, 2001). *CBF*s are distinguished from other members of the AP2/EREBP superfamily by the presence of *CBF* signature sequence motifs flanking the AP2 domain (Jaglo *et al.*, 2001). The *Arabidopsis* genome encodes six *CBF* genes and ectopic expression of *AtCBF1–4* in transgenic plants results in activation of components of the *CBF* regulon, including *cor* genes that harbour CRT elements within their promoters (Jaglo-Ottosen *et al.*, 1998; Liu *et al.*, 1998; Gilmour *et al.*, 2000; Jaglo *et al.*, 2001; Haake *et al.*, 2002). EST database searches reveal that *CBF*-like genes are present in most dicot and monocot EST collections (J.S. Skinner, unpublished results), implying the CBF response pathway is widely distributed within higher plants.

CRT motifs are present in the regulatory regions of most barley *cor* genes for which a promoter has been isolated; three *CBF* genes have been reported from barley (Choi *et al.*, 2002; Xue, 2002, 2003) to date. In *Arabidopsis*, *AtCBF1–3* are present as a tandem gene array (Gilmour *et al.*, 1998) that is coincident with a LT-tolerance QTL (Salinas, 2002), implying that genetic variation at *CBF* gene loci could be a basis for genotypical differences in LT tolerance. Together, these results suggest that *CBF* genes are involved in LT tolerance in barley, and could be a basis for the differential genotypic LT capacities. Two important barley populations for mapping winter growth habit traits, including LT tolerance, were available at the initiation of this study and accordingly employed to maximize the barley germplasm variation analysed. The DM population is a barley research community standard for mapping QTL for LT tolerance and PPD sensitivity (Pan *et al.*, 1994; Hayes *et al.*, 1997; Mahfoozi *et al.*, 2000; Fowler *et al.*, 2001; Karsai *et al.*, 2001) as well as candidate genes for stress-related traits (e.g. Dehydrins) (van Zee *et al.*, 1995; Choi *et al.*, 2000). This is a 'facultative' (Dicktoo) × 'spring' (Morex) population with a major LT-tolerance QTL on chromosome 5H (Hayes *et al.*, 1997) and the primary genetic resource for addressing the question 'are *CBF* genes related to LT tolerance in barley?' We also used the 'winter' (Strider) × 'facultative' (88Ab536) (STAB) mapping population in which the parents have similar LT-tolerance values (Karsai *et al.*, 2004). We have found that this population shows significant phenotypic variation for LT tolerance, but there is no QTL on 5H; additional efforts to localize the LT-tolerance QTL are currently in progress. The principal objective is to determine whether allelic variation at *CBF* loci is responsible for genetic variation in the cold tolerance of barley. Towards this end, we assessed the *CBF* gene family size and complexity, genome distribution and relationship with known *Triticeae* LT-tolerance QTLs in barley.

The *CBF* Gene Family in Barley

Barley contains a large *CBF* gene family

Analysis of the AP2 domain-containing clones in the large barley EST collection (350K barley EST sequences as of the 01/09/04 release) revealed nine distinct candidate barley *CBF* genes were represented, which was confirmed by sequencing the cDNA insert of a representative EST clone to each of the nine genes. These EST cDNAs originated from several of the barley varieties that have been utilized in EST projects and subsequently designated as *HvCBF1*-Mx, *HvCBF4B*-Mx, *HvCBF5*-Op, *HvCBF6*-Mx, *HvCBF7*-Mx, *HvCBF8C*-Bk, *HvCBF9*-Mx, *HvCBF10B*-Op and *HvCBF11*-Op (Table 3.1). Barley *CBF* allelic nomenclature is defined below. The alleles to each *HvCBF* were cloned, where possible, from the parental Dicktoo, Morex, Strider and 88Ab536 mapping genotypes for allelic comparisons and linkage mapping studies (Table 3.1). In the course of the allelic isolations, six additional *CBF* genes (*HvCBF2A*, *HvCBF4A*, *HvCBF4D*, *HvCBF8A*, *HvCBF8B* and *HvCBF10A*) were isolated. Thus, a total of 17 distinct barley *CBF* genes were isolated (Table 3.1), demonstrating that the barley *CBF* gene family is substantially larger than in dicots. At least three additional *HvCBF* genes, as well as pseudogenes, are present in barley, based on gDNA library cloning of *HvCBF* loci currently underway (Stockinger *et al.*, Chapter 4, this volume).

Three barley *CBFs*, designated *HvCBF1*, *HvCBF2* and *HvCBF3*, were independently reported on during the course of this work (Choi *et al.*, 2002; Xue, 2002, 2003). To keep the barley nomenclature consistent, we assigned subsequent numbers to each novel barley *CBF* gene in the order of isolation, starting with *HvCBF4*. *HvCBF* gene alleles are designated by a dash following the *HvCBF#* designation and a two-letter abbreviation for the barley variety from which the allelic form was obtained (see Table 3.1). As an example, *HvCBF9*-Dt represents the *HvCBF9* allele from the barley variety Dicktoo (Dt). The 17 barley *CBF* genes

Table 3.1. Cloned barley *CBF* genes and characteristics.

CBF gene	Cloned alleles[b]	HvCBF-sub-group	Length (aa)	pI[c] (full, ACD)	CRT binding[d] 2°C	CRT binding[d] 20°C	*cor* gene activation[e]
HvCBF1	Dt, Mx, St, Ab, Hn	1	217	5.3, 3.4	Yes[f]	Yes[f]	
HvCBF2A	Dt, Mx, St, Ab	4	221–230	5.0, 3.8	Yes	No	
HvCBF2B	St, Hn	4	221	5.1, 3.9	Yes[f]	No[f]	
HvCBF3	Dt, Mx, St, Ab	3	249	5.3, 4.0	Yes	Yes	Yes
HvCBF4A[a]	Dt, Mx	4	225	8.4, 4.1			
HvCBF4B[a]	Dt, Mx, CI	4	225	8.4, 4.1	Yes	No	No
HvCBF4[a]	St	4	225	8.4, 4.1			
HvCBF4D	Ab	4	225	7.0, 4.0			
HvCBF5-Dt	Dt, Mx, St, Ab, Op	1	214	6.5, 3.9			
HvCBF6	Dt, Mx, St, Ab	3	244	5.2, 3.6	Yes	Yes	Yes
HvCBF7	Dt, Mx, St, Ab	1	219	5.7, 3.7	Yes	Yes	
HvCBF8A[a]	Dt, Mx, Ab	3	—[a]	—[a]			
HvCBF8B[a]	Dt, Mx, St, Ab, Bk	3	—[a]	—[a]			
HvCBF8C[a]	Dt, Mx	3	—[a]	—[a]			
HvCBF9	Dt, Mx, St, Ab	4	291	8.9, 4.6			No
HvCBF10A	Dt, Mx, St, Ab	3	241	4.9, 3.8			
HvCBF10B	Dt, St, Ab, Op	3	227–240	5.2, 4.0			
HvCBF11	Dt, Mx, St, Ab, Op	1	218	5.6, 4.0			

[a]All *HvCBF4A/B* and St allele forms encode an identical polypeptide, *HvCBF8* alleles encode frameshift-based pseudogenes.
[b]Allele code: 88Ab536 (Ab), Barke (Bk), CI16151(CI), Dicktoo (Dt), Halcyon (Hn), Morex (Mx), Optic (Op), Strider (St).
[c]Predicted isoelectric point (pI) value: total protein (full), acidic C-terminal domain (ACD); values are allele average.
[d]Ability of Dicktoo allele to bind barley *cor14b* and *Arabidopsis cor15a* CRT motif.
[e]Ability of Dicktoo allele to activate *cor15a* expression in transgenic *Arabidopsis* at 20°C when constitutively overexpressed.
[f]Halcyon allele assayed in Xue (2003).

represent 11 distinct families (designated *HvCBF1–HvCBF11*), with *HvCBF2* (2A, 2B), *HvCBF4* (4A, 4B, 4D), *HvCBF8* (8A, 8B, 8C) and *HvCBF10* (10A, 10B) having more than one subfamily member each.

Is the large barley *CBF* family size characteristic of cereals?

The size and complexity of the barley *CBF* family, relative to dicots, was larger than anticipated. The genome of *Arabidopsis* (Haake *et al.*, 2002; Sakuma *et al.*, 2002), an annual weed, and poplar (Benedict *et al.*, Chapter 12, this volume; J.S. Skinner, unpublished results), a boreal tree, have been sequenced, and each contains six *CBF* genes. To determine whether other cereals contained an increased *CBF* family size, we took advantage of the two reported rice genome draft sequences (Goff *et al.*, 2002; Yu *et al.*, 2002) and the extensive wheat EST collection, and searched for *CBF* genes in these two additional cereals. Fourteen rice *CBF* genes were identified through analysis of the genomic and EST sequence datasets, consisting of ten full-length genes (Fig. 3.1), three pseudogenes and a partial novel *CBF* terminating at the end of a scaffold clone. Four of the rice *CBFs* are identical to those reported by Dubouzet *et al.* (2003), who used the alternate *CBF* designations, *OsDREB1A* through *OsDREB1D*. For consistency, the six additional full-length rice *CBFs* are designated *OsDREB1E–OsDREB1J*. We noted sequence discrepancies between our rice *OsDREB1B* (designated *OsDREB1B.1*; AY785894) and that of Dubouzet *et al.* (2003), which probably represent miscalled bases in the original *OsDREB1B* (AF300972) report. From the wheat species *T. aestivum* and *T. monococcum*, we identified and sequenced full-length EST clones that encode products falling into nearly all the *HvCBF* families; these are designated *TaCBF2*, *TaCBF4-L1* (Like1), *TaCBF5*, *TaCBF6*, *TmCBF7*, *TaCBF9* and *TaCBF11* (Fig. 3.1) to reflect their relationship to their putative barley orthologue/paralogue. A wheat *CBF* (*TaCBF1*) reported by Jaglo *et al.* (2001) is closely related to the *TaCBF4-L1* and *HvCBF4* genes. Two additional wheat EST clones (BE493405, CA593798), most similar to *HvCBF3/HvCBF10* clade, are present in the wheat collection, but were not available and therefore not included in the analysis presented in Fig. 3.1. Based on this EST survey, wheat has a minimum of ten non-homoeologous *CBFs* and, as is the case for barley, is predicted to encode additional members not represented in the EST collection. In sum, these data indicate that the increased *CBF* family size observed for barley is characteristic of cereals.

The phylogentic complexity of the barley *CBF* family is representative of cereals

Phylogenetic analysis of the barley *CBFs* revealed that three major phylogenetic subgroups are present. All of the non-barley monocot *CBFs* fall within one of these three subgroups (Fig. 3.1), demonstrating that the barley *CBF* members are relatively evenly distributed and the barley gene complexity is representative of cereals. We designated each of the three subgroups – HvCBF1-, HvCBF3- and HvCBF4-subgroups – for a representative barley member. The HvCBF3- and

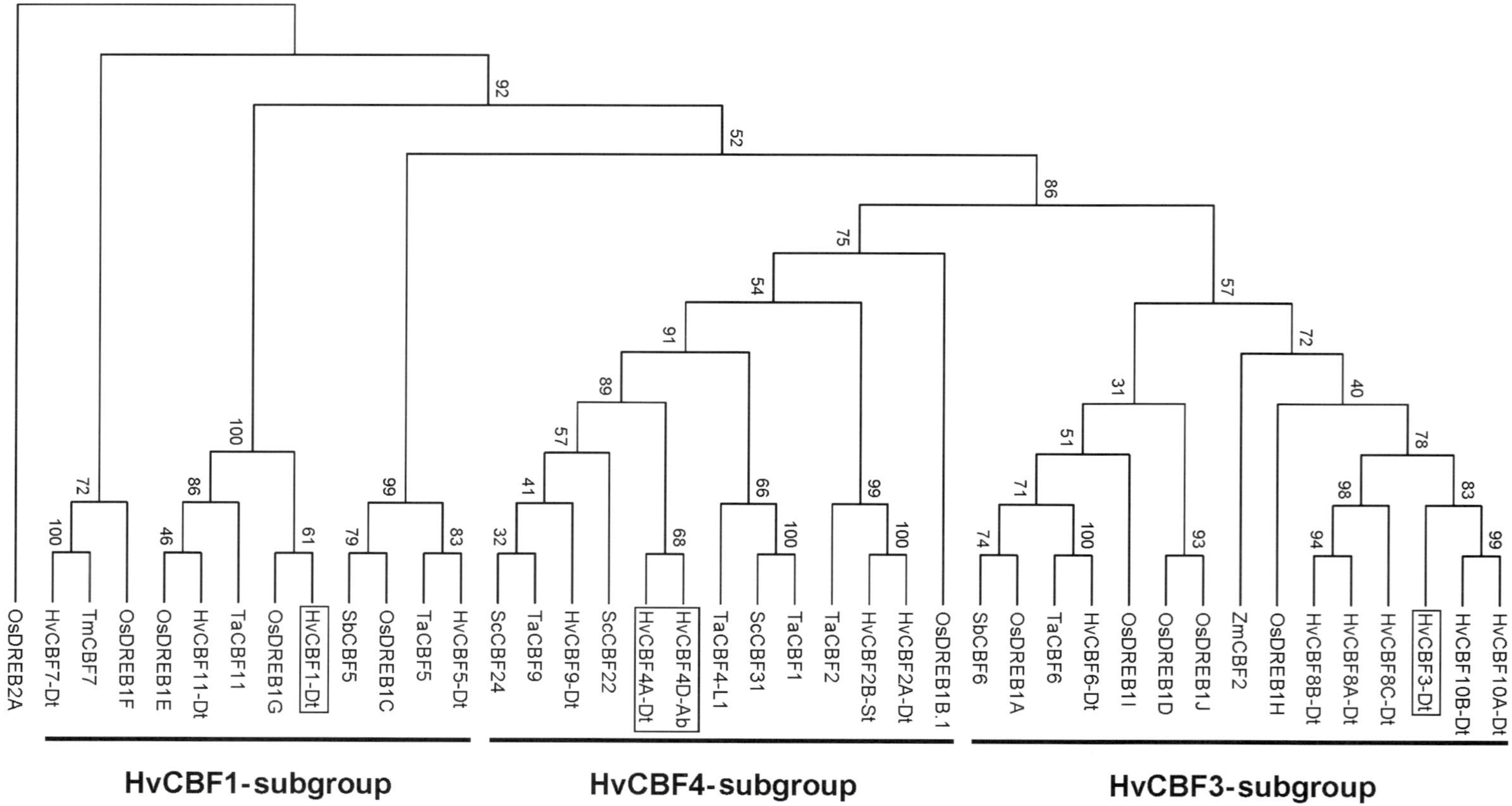

Fig. 3.1. Monocot CBF phylogenetic relationships. A minimum evolution phylogenetic tree was derived from an alignment of the monocot CBF polypeptides; an identical tree topology was obtained utilizing the neighbour joining function on the same alignment. For HvCBF8 members, theoretical polypeptide sequences were generated that account for the pseudogene-based frameshifts (see text). HvCBF4A and HvCBF4B encode identical polypeptides; only HvCBF4A is shown. OsDREB2A is a closely related monocot AP2 domain-containing protein that lacks the flanking CBF signature sequences. Vertical bars denote the respective HvCBF-subgroups and members, while each subgroup-defining member(s) is boxed.

HvCBF4-subgroups are clearly defined phylogenetically and have a single defined root branch each. The HvCBF1-subgroup is ancestral to the HvCBF3- and HvCBF4-subgroups, and while collectively assigned as a single subgroup for convenience, is actually composed of three distinct gene family clusters – the *HvCBF5* family, the *HvCBF7* family and the *HvCBF1/HvCBF11* families (Fig. 3.1). Each of the *HvCBF8* family members of the HvCBF3-subgroup encode frameshift-based pseudogenes and the polypeptides used for generation of the phylogenetic tree (Fig. 3.1), and the AP2 domain alignment (Fig. 3.2C), are based on hypothetical reading frames that adjust for the single nucleotide frameshift deletion(s) found in each pseudogene. The wheat *CBF* complexity is well distributed among these three subgroups. While *TaCBF6* is currently the only HvCBF3-subgroup member for which a full-length coding sequence has been determined, at least two additional HvCBF3-subgroup members, more similar to *HvCBF3* and *HvCBF10A/B*, are also encoded by wheat, based on the EST collection data (above; data not shown). Additional HvCBF3-subgroup members have recently been isolated via BAC screens from the diploid wheat *T. monococcum* (Miller and Dubcovsky, 2005), confirming that this subgroup is also well represented in wheat. Since the genome sequence of two rice species has been determined, the total rice *CBF* number detected is probably close to the entire *CBF* family size for this organism. While *CBFs* corresponding to each of the subgroups are present, the distribution across subgroups is uneven. Whereas the size and complexity of the rice and barley HvCBF1- and HvCBF3-subgroups are similar, rice contains only a single HvCBF4-subgroup member – *OsDREB1B.1* – which is ancestral to the other HvCBF4-subgroup members. These results suggest that while an increase in *CBF* family size, relative to dicots, is consistent among these three cereals – two cold-tolerant and one cold-sensitive – the duplication and divergence events leading to increased complexity of the HvCBF4-subgroup may have occurred after separation of cold-tolerant cereals from the cold-sensitive member. It is possible that this radiation of the HvCBF4-subgroup complexity in cold-tolerant cereals is one basis of the increased LT tolerance found in temperate cereals. None of the AP2 domain-containing monocot proteins harbouring the *CBF* signature motifs fell outside of these three subgroups, implicating that these three clades represent the sum of the major CBF phylogenetic clusters occurring in monocots.

Monocot CBF characteristics

Monocot *CBFs* share a number of common contextual characteristics both *inter se* and with dicot *CBFs* (Fig. 3.2A). Like dicots, all reported monocot *CBF* genes lack introns. The primary protein structure is likewise conserved, which has the following general amino to carboxyl terminal major features: a leader sequence of approximately 15–40 amino acids, an AP2 DNA-binding domain flanked by the conserved signature motif residues characteristic to the CBF subfamily and an acidic C-terminal domain postulated to act as a transcriptional activation region (Fig. 3.2A). The leader region preceding the 5′ CBF motif is the most variable, both in length and composition. The monocot CBFs have an overall acidic character in general (Table 3.1). An exception is found among a subset of

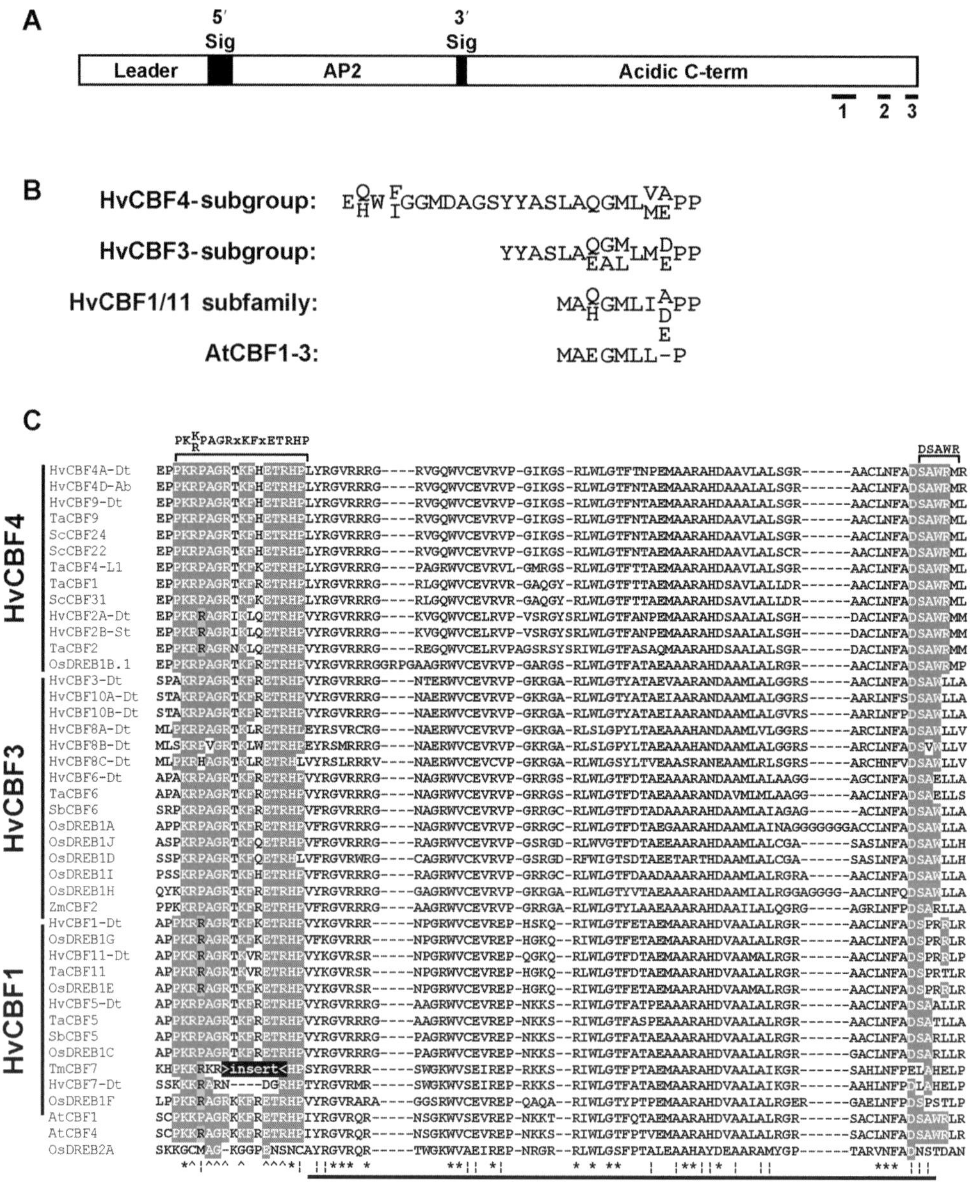

Fig. 3.2. Major monocot CBF features. (A) Basic monocot CBF domain structure. The variable leader, AP2, and acidic C-terminal domains are noted. The AP2 domain-flanking CBF signature motifs (5′ Sig and 3′ Sig) are indicated as black blocks. Three additional regions of note are: (1) a nine-amino-acid block shared between dicot CBFs and all monocot CBFs except those of the HvCBF5 and HvCBF7 clades; (2) an acidic cluster common to monocot CBFs; and (3) the LWS(Y) motif typically found near the end of monocot CBFs. (B) General consensus sequence of a conserved C-terminal domain feature (#1 of panel 3.2A) for each indicated group; the maximum conserved block length is shown per group. (C) Alignment of monocot CBF AP2 domain and flanking signature motifs. All currently reported monocot CBFs, grouped by HvCBF-subgroup membership, were utilized; Dicktoo HvCBF alleles were preferentially

the members of the HvCBF4-subgroup where HvCBF4A/B, HvCBF9 and their close rye/wheat homologues have an overall basic pI (pI range of 8.4–9.4) (Table 3.1, Fig. 3.1, data not shown). Despite the overall pI difference, the C-terminal domain, which has been proposed to function as an activation domain, has maintained a strongly acidic character like the remaining CBFs (Table 3.1). The shared acidic character of this domain appears to be contextual however, as a large block of common homology to account for the negative domain charge is not observed.

CBFs are distinguished from other members of the AP2/EREBP family of transcription factors by the presence of direct flanking signature motifs 5′ (PKK/RPAGRxKFxETRHP) and 3′ (DSAWR) to the AP2 domain (Jaglo *et al.*, 2001). The AP2 DNA-binding domain and flanking CBF signature motifs are the most highly conserved regions among the monocot CBFs (Fig. 3.2C); a 45 amino acid insertion disrupting the 5′ CBF motif of TmCBF7 is not shown. With the exception of HvCBF7 and TmCBF7, which respectively contain an internal deletion and insertion event relative to all other CBFs, the 5′ CBF signature motif is highly conserved. One common 5′ CBF signature motif variation observed among a subset of monocot CBFs is the occurrence at the fourth amino acid position of an arginine (R) in place of the more common proline (PKK/R*P*AGRxKFxETRHP) (Fig. 3.2C). This variation is also commonly observed in dicot CBFs (Jaglo *et al.*, 2001; J.S. Skinner, unpublished results), suggesting that the repeated occurrence of this alternate amino acid is of an unknown functional importance. In contrast to the 5′ motif, the 3′ CBF signature motif (DSAWR) is more variable, with the degree of divergence from the consensus correlating with respective HvCBF-subgroup membership of an HvCBF (Fig. 3.2C).

An approximately nine-amino-acid block in the acidic C-terminal domain is conserved between the HvCBF1/HvCBF11 clade of the HvCBF1-subgroup and the HvCBF3- and HvCBF4-subgroups (Fig. 3.2A, B); additional flanking sequence

Fig. 3.2. *(Continued).* utilized when applicable. The 45-amino-acid insertion event of TmCBF7 that disrupts the 5′ CBF motif is denoted as '>insert<' and black highlighted for alignment purposes. OsDREB2A is a closely related monocot AP2 domain-containing protein that lacks the flanking CBF signature sequences. The 5′ and 3′ CBF signature motifs are bracketed with the respective consensus indicated over the top, while the AP2 domain is underlined. Amino acids matching the 5′ and 3′ CBF signature sequences are backshaded dark grey, while arginines of the alternate motif (see text) are light grey backshaded; residues at the two variable motif positions (x) are not shaded. An asterisk denotes strictly conserved residues among functional monocot CBF polypeptides (i.e. *HvCBF8* pseudogene reconstructions ignored) and a broken bar (¦) highly conserved positions at which only two amino acids occur. An ^ denotes additional 5′ CBF motif positions that would be strictly conserved discounting the mutated motifs of HvCBF7 and TmCBF7.

is conserved between and among the HvCBF3- and HvCBF4-subgroups for this motif (Fig. 3.2B). Intriguingly, this nine-amino-acid block is also shared with dicot CBFs, indicating possible functional significance (Fig. 3.2B). Two additional common contextual features present in the acidic C-terminal domain are: (i) the majority of the monocot CBFs tend to either terminate with the residues LWS(Y), or have this sequence very close to their C-terminal end, and (ii) within 15 residues upstream of this sequence, a small clustered block of 3–5 acidic amino acids is typically observed. These two contextual features are also observed in many dicot CBFs. These three attributes of the acidic C-terminal domain are the only common features/domains shared between the dicot and monocot CBFs outside the AP2 DNA-binding domain and flanking signature motifs. HvCBF1 and its close homologues, but not other monocot CBFs, display additional conserved blocks with dicot CBFs in the acidic C-terminal domain (data not shown).

HvCBF Function

Barley *HvCBFs* display CBF functional characteristics

CBFs activate *cor* gene expression via binding to the CRT motif. To determine whether the genes classified via sequence characteristics as barley *CBFs* actually display CBF functional characteristics too, we tested five – *HvCBF2A*, *HvCBF3*, *HvCBF4*, *HvCBF6*, *HvCBF7* – for their ability to bind three CRT motifs (Table 3.1); *AtCBF1* was also tested as a control. The three CRT motifs all have the CCGAC core, but differ in the surrounding flanking sequence. Two of the CRT motifs were from the barley *cor* genes *cor14b* (Dal Bosco *et al.*, 2003) and *Dhn5* (Choi *et al.*, 1999), and the third was from the *Arabidopsis cor15a* CRT, which has been shown to interact with *AtCBF1* (Stockinger *et al.*, 1997). The tested *HvCBFs* were each alleles from the cold-tolerant Dicktoo variety, representing each of the three phylogenetic subgroups, and *HvCBF7* specifically represents the most ancestral form and contains the most diverged 5′ and 3′ CBF motifs (Figs 3.1, 3.2C). Xue (2003) had previously demonstrated that while binding to the CRT motif of *HVA1* was temperature-independent for *HvCBF1-Hn*, binding of *HvCBF2B-Hn*, an *HvCBF4*-subgroup member, was LT-dependent. Therefore, binding capacity was tested at both warm (20°C) and cold (2°C) temperatures for each CBF to each CRT motif. As shown in Table 3.1, stable complexes were detected at both temperatures for *HvCBF3*-Dt, *HvCBF6*-Dt, *HvCBF7*-Dt and *AtCBF1* (data not shown). These *HvCBFs* represent *HvCBF1*- and *HvCBF3*-subgroup members. In contrast, for *HvCBF2A*-Dt and *HvCBF4*-Dt, both *HvCBF4*-subgroup members, stable binding was detected only at 2°C. The binding was dependent on the CCGAC core in all cases, as the complexes were specifically compared with cold wild-type competitors, but not mutant versions in which the CCGAC core was mutated to AAATA. Differences in binding efficiency relative to each CRT motif were noted for a subset of the CBFs tested, indicating that the contextual differences in the sequence flanking the CRT motif CCGAC core influences binding. These results demonstrate that representative barley CBFs, including the most phylogenetically ancestral form (*HvCBF7*), meet the

CBF functional criterion of specific binding to the CRT motif. It is therefore likely that each of the *HvCBF*s identified in this study (Fig. 3.1) should be capable of specifically binding to the CRT motif core.

We also tested the ability of three barley *HvCBF*s – *HvCBF3*-Dt, *HvCBF4*-Dt and *HvCBF6*-Dt – when constitutively expressed in transgenic *Arabidopsis*, to induce CBF-responsive *cor15a* expression; the *cor15a* CRT motif was specifically bound by each of these *HvCBF*s. Expression of *CBF* genes in *Arabidopsis* under the control of the constitutive CaMV 35S promoter results in the induction of CRT-containing *cor* gene expression at normally non-inducing warm temperatures (Gilmour *et al.*, 1998; Haake *et al.*, 2002). In agreement with the binding observations, while *HvCBF3* and *HvCBF6*, both *HvCBF3*-subgroup members, could induce *cor15a* expression at warm temperatures, *HvCBF4* could not (Table 3.1); similar results were obtained relative to induction of *Arabidopsis cor6.6* (data not shown). Whole-plant freezing tests at −5°C on *Arabidopsis* lines constitutively expressing HvCBF3 show that the non-acclimated *HvCBF3* lines impart similar survival rates to acclimated wild-type *Arabidopsis* plants, in contrast to non-acclimated wild-type *Arabidopsis* plants which are killed (data not shown; Amundsen, 2004). This indicates *HvCBF3* is competent to induce a large enough portion of the *Arabidopsis* CBF regulon to enhance freezing tolerance. *OsDREB1A* of rice, also an HvCBF3-subgroup member (Fig. 3.1), can induce the expression of a series of CRT-containing *cor* genes in transgenic *Arabidopsis* under equivalent conditions (Dubouzet *et al.*, 2003). Preliminary tests with barley *HvCBF9*, wheat *TaCBF1*, and rye *ScCBF22*, *ScCBF24* and *ScCBF31*, all HvCBF4-subgroup members (Fig. 3.1), indicate that they fail to induce *cor* gene expression at warm temperatures also (data not shown). These results indicate that the cold-binding dependency of HvCBF4-subgroup members is probably a cross species-shared trait among the members of this phylogenetic group. As *OsDREB1B.1* is the most ancestral HvCBF4-subgroup member and the sole rice member of this subgroup, it will be interesting to determine if LT binding dependence is specific to just the cold-tolerant cereals.

HvCBF Variation

Sequence variation between *HvCBF* gene alleles is minimal

HvCBF gene alleles were cloned, where possible, from the Dicktoo, Morex, Strider and 88Ab536 parental genotypes (Table 3.1) to investigate the degree of allelic variation in the predicted polypeptides, and whether polypeptide variation could be a basis of LT tolerance differences. These genotypes differ in their LT capacities and are the parents of two linkage mapping populations: DM and STAB. The overall degree of allelic variation was low however, and variations within the coding region were typically composed of randomly dispersed single nucleotide polymorphisms which resulted either in no amino acid change or a conservative amino acid substitution in the case of an altered codon. As an example, the greatest degree of variation was between the Morex and Dicktoo *HvCBF6* alleles where only five, mostly conservative, amino acid substitutions had occurred. These were localized to variable segments of the acidic C-terminal domain and did not alter the domain's acidic character. The allelic variation between the genotypes, while minimal, correlated with HvCBF-subgroup membership, where HvCBF3-subgroup members were the most variable; HvCBF1-subgroup members the most conserved; and HvCBF4-subgroup members intermediate. For most of the *HvCBFs*, variation was typically greatest between the Dicktoo (facultative) and Morex (spring) alleles. In agreement with the hypothesis that facultative genotypes are a winter genotype subclass with a deleted *Vrn-H2* locus (von Zitzewitz., 2004; Cooper *et al.*, Chapter 5, this volume), Dicktoo (facultative) alleles were usually the most similar to those of Strider (winter), frequently being completely identical at the nucleotide level. This overall minor degree of allelic variation suggests that differences in the encoded HvCBF polypeptides are not likely to be a primary source of LT tolerance differences. Gene presence/absence was not a common occurrence and was only noted with confidence for *HvCBF10B*, where *HvCBF10B* is probably deleted from Morex relative to the other genotypes examined. Morex contains the highly similar family member *HvCBF10A*, which is probably substitutable and able to compensate for *HvCBF10B* function.

Genotypic variations are observed in *HvCBF* expression

As the HvCBF polypeptides of the various genotypes were highly similar, we determined if cold-responsive regulation of representative *HvCBFs* from each subgroup differed between the cold-tolerant and cold-sensitive parents of the DM population (Fig. 3.3), which harbours the prominent LT tolerance QTL. *HvCBF2*, *HvCBF4* and *HvCBF8* are multigene family members and the detected signal is probably a composite of the expressed family members for each gene. The genotypic expression patterns for each individual *HvCBF* to cold was analogous between the two genotypes. However, the level and duration of the response differed between the cold-tolerant and cold-sensitive genotype (Fig. 3.3). With the exception of *HvCBF8*, cold exposure resulted in the induction of each *HvCBF* gene tested, where expression was basically absent under warm conditions (Fig. 3.3). In contrast, *HvCBF8* expression was more abundant under warm conditions and substantially downregulated during cold exposure. Cold invoked a strong and rapid, but semi-transient, induction for the HvCBF4-subgroup member genes, where maximal expression was obtained within 4–8 h following exposure to low temperatures. The level of response differed by genotype, with *HvCBF2* and *HvCBF9* more strongly induced in cold-tolerant Dicktoo, while *HvCBF4* was higher in the spring genotype Morex. More importantly, the duration of expression was greater for all three genes in the cold-tolerant Dicktoo genotype, although *HvCBF9*, while maintained longer in Dicktoo (for 24 h), had tapered off by 96 h. *HvCBF2* has the most interesting response of the three, responding rapidly and being maintained throughout the time course in Dicktoo compared to only a negligible induction through the first 8 h of treatment in Morex. In contrast to the HvCBF4-subgroup, differential responses to cold were observed for the HvCBF1- and HvCBF3-subgroup members assayed. Each of the three HvCBF3-subgroup members examined (*HvCBF3*, *HvCBF6* and *HvCBF8*) displayed a distinct expression profile. *HvCBF3* was very weakly expressed and transcript was barely detectable as a slight induction following 1 h of cold exposure in the Dicktoo genotype only (Fig. 3.3). This rapid temporal induction of *HvCBF3* expression is in agreement with the results of Choi *et al.* (2002). *HvCBF6* expression, while stronger than *HvCBF3*, was weak relative to the HvCBF4-subgroup genes and displayed a delayed response to cold treatment, first being just detectable by 8 h and peaking at a sustained level at 24 h. *HvCBF8* gene family transcript was negatively regulated by cold and transcript level was greater in the Dicktoo genotype. If no additional *HvCBF8* genes exist, this result would indicate that at least one of the three *HvCBF8* pseudogenes is expressed; *HvCBF8B* was isolated as a cv. Barke EST (Table 3.1) and is therefore expressed in at least one barley genotype. The two HvCBF1-subgroup genes assayed also show distinct responses to cold. *HvCBF1* was transiently induced by cold, peaking at 4 h. In contrast, transcript levels for *HvCBF7*, the most ancestral barley CBF, steadily increased through the last time point examined (96 h) in response to cold, and were more abundant in Morex.

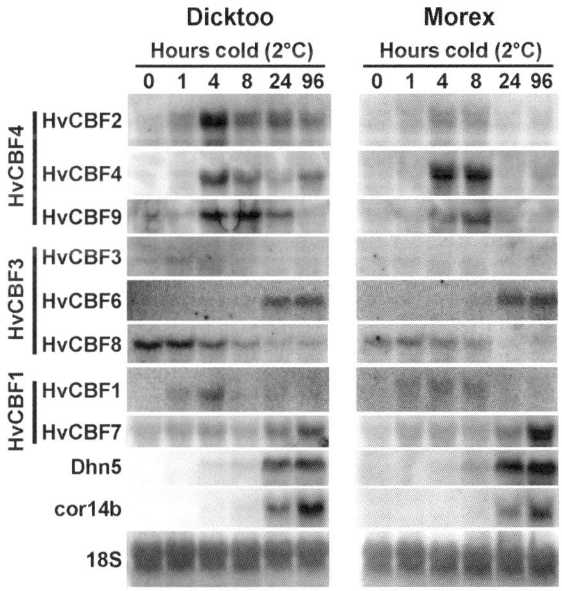

Fig. 3.3. HvCBF responses to cold stress in barley genotypes Dicktoo (facultative winter) and Morex (spring). Dicktoo and Morex are the parental genotypes of the LT-tolerant DM mapping population. 20-μg blotted total RNA from plants exposed to cold (2°C) for the indicated number of hours was hybridized to each indicated *HvCBF* and barley *Dhn5* and *cor14b cor* genes; *HvCBF2A*, *HvCBF4A*, and *HvCBF8B* specifically were utilized for detection of their respective composite gene family profiles. The 0-hour cold sample represents unstressed, warm control conditions. An 18S rRNA gene probe was used as a loading control. HvCBF-subgroup members (vertical bar) are clustered for comparison.

In *Arabidopsis*, *CBF* genes, after being induced by cold, subsequently induce expression of *cor* genes that contain CRT regulatory elements in their promoters (Thomashow, 1999). We demonstrated that multiple HvCBFs can bind the CRT motifs located in the promoters of the two monocot *cor* genes, *cor14b* (Table 3.1) and *Dhn5* (not shown). The cold-induced expression of these two barley *cor* genes follows the induction of the cold-responsive *HvCBF* genes (Fig. 3.3). The observed induction profile of these *cor* genes is consistent with one or more of the cold-induced HvCBFs binding to the *cor14b* and *Dhn5* CRT motifs and activating their expression in an analogous fashion to the dicot model. Interestingly, *cor14b* only contains a single CRT motif, which is located in a 27 bp minimal fragment that controls the LT responsiveness of the gene (Dal Bosco *et al.*, 2003), suggesting it is an HvCBF that is responsible for this LT trait.

HvCBF Mapping

HvCBF gene map positions are coincident with *Triticeae* LT-tolerance QTL

The DM mapping population harbours a major freezing-tolerance QTL on the long arm of chromosome 5H and has been a research community standard for mapping genes relative to winter hardiness QTL traits (Fig. 3.4; Hayes *et al.*, 1997). We mapped or aligned as many of the *HvCBF* genes as possible relative to this population to determine if any were candidates for this prominent LT-tolerance QTL. Of the 17 *HvCBFs* identified in this study, eight were directly mapped in the DM population, three in the STAB population, and three assigned to chromosome arms via wheat:barley addition lines (Table 3.2; Islam *et al.*, 1981). The *HvCBF* genes are located on four of barley's seven chromosomes, with a majority of the genes forming two tandem clusters on the long arm of chromosome 5H (Fig. 3.4). For a subset of the *HvCBF* genes that lacked allelic variation, map position and loci number were inferred via physical linkage to mapped genes based on gDNA phage clones harbouring more than one *CBF* gene (Stockinger *et al.*, Chapter 4, this volume).

Five *HvCBF* genes mapped to chromosomes other than 5H (Table 3.2). In the DM population, the *HvCBF8C* pseudogene mapped to chromosome 2H-S and *HvCBF5* to chromosome 7H-S. The alleles of *HvCBF7*, *HvCBF8A*, *HvCBF8B* and *HvCBF11* were identical in both the DM and STAB populations, necessitating the use of the wheat:barley disomic addition lines (Islam *et al.*, 1981) to assign chromosome position. Using this strategy, *HvCBF7* was assigned to chromosome 6H-L and *HvCBF11* to 2H-L. Cross amplification with the wheat genomic background currently precludes *HvCBF8A* and *HvCBF8B* assignment however. While not polymorphic in the DM population, we were able to map *HvCBF1* to 6H-L in the STAB population. In addition to cold, some of the *HvCBFs* respond to drought and salt stress (data not shown). QTLs for drought and/or salt tolerance have been determined to lie on these non-5H chromosomes (summarized in Cattivelli *et al.*, 2002), though their relationship to *HvCBF* genes is currently unknown. Additionally, small-effect LT-tolerance QTLs have been reported on barley chromosomes 2H and 6H (Tuberosa *et al.*, 1997), but were

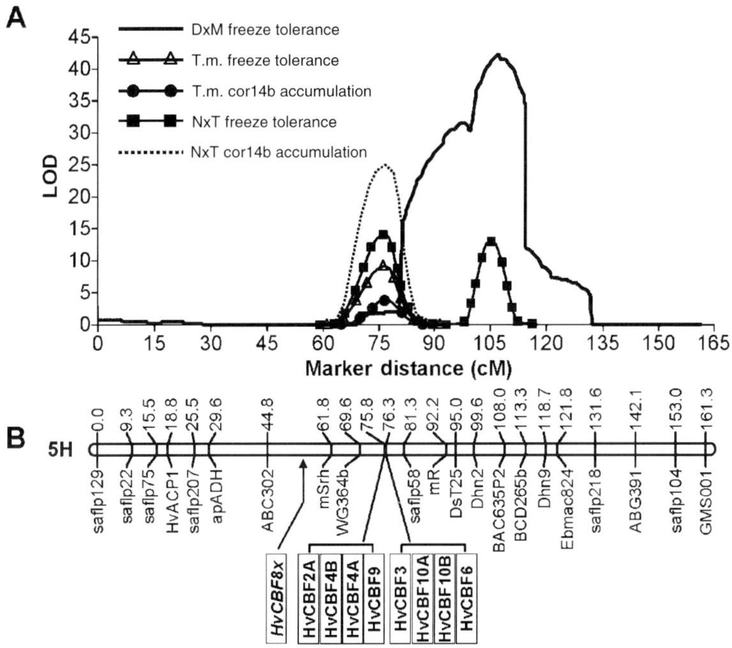

Fig. 3.4. *HvCBF*s mapping to chromosome arm 5H-L. (A) LT-tolerance QTL LOD plot of Dicktoo × Morex mapping population controlled freeze test survival. Marker distance in centimorgans (cM) is indicated. Peaks representing the estimated aligned location on the DM 5H-L map for freezing tolerance and *cor14b* accumulation QTLs from the Nure × Tremois population (N × T; Francia *et al.*, 2004) and a *T. monococcum* population (T.m.; Vágújfalvi *et al.*, 2003) are indicated as rough plots that reflect their reported LOD score peak values. (B) A proportional diagrammatic representation of the chromosome 5H long arm is shown directly under the LOD plot and respective positions of mapped genes/markers are indicated; *HvCBF* genes are boxed. The approximate position of an undetermined *HvCBF8* form (*HvCBF8x*) mapped in the NT population (Francia *et al.*, 2004) is italicized and boxed. Chromosome 5H-L arm is oriented with the centromere to the left.

only marginally significant and accounted for a low percentage (< 10%) of the phenotypic variance. The positional relationship of the non-5H CBFs to these abiotic stress QTL effects is currently under investigation.

Eight of the *HvCBF*s (*2A*, *3*, *4A*, *4B*, *6*, *9*, *10A* and *10B*) map to a localized region of the long arm of chromosome 5H (5H-L), the region where major effect LT-tolerance QTL are reported in the *Triticeae* (Fig. 3.4, Table 3.2; Hayes *et al.*, 1997; Vágújfalvi *et al.*, 2003; Francia *et al.*, 2004). These genes are all members of the HvCBF3- and HvCBF4-subgroups (Fig. 3.1). *HvCBF*s *2A*, *4A*, *4B* and *9* form an inseparable cluster, while *HvCBF*s *3*, *6*, *10A* and *10B* form a second unbroken cluster less than 1 cM distant. The map-based clustering of some of these genes is in agreement with their occurrence as physically tandem sets (below).

Table 3.2. Mapping summary of barley *CBF* genes.

Gene	HvCBF-subgroup	Mapping genotype(s)[a]	Chromosome[b]
HvCBF1	1	STAB	6H-L[1,3]
HvCBF2A	4	Dicktoo	5H-L[4]
HvCBF2B	4	Betzes	5H-L[3]
HvCBF3	3	DM, STAB	5H-L[1]
HvCBF4A/B[c]	4	DM	5H-L[1]
HvCBF5	1	DM	7H-S[1]
HvCBF6	3	DM, STAB	5H-L[1,2]
HvCBF7	1	Betzes	6H-L[3]
HvCBF8A	3	N/D	N/D
HvCBF8B	3	N/D	N/D
HvCBF8C	3	DM	2H-S[1]
HvCBF9	4	DM	5H-L[1]
HvCBF10A	3	DM	5H-L[1]
HvCBF10B	3	Dicktoo	5H-L[4]
HvCBF11	1	Betzes	2H-L[3]

[a]Barley genotype used to determine/infer map or chromosome location: Dicktoo × Morex population (DM), Strider × 88Ab536 population (STAB), wheat:barley (cv. Betzes) addition lines, Dicktoo gDNA phage clone (Dicktoo).
[b]Chromosome location or map position determined by: [1]CAPs assay, [2]InDel, [3]wheat:barley addition lines, [4]inferred via phage clone gene linkage.
[c]*HvCBF4A* and *HvCBF4B* are co-amplified yet unseparable by recombination in our full mapping population (n = 236) and therefore map as a single locus.
N/D: not determined.

The two *HvCBF* clusters map ~20 cM proximal to the peak of the LT-tolerance QTL in the DM population, suggesting they are not a strong candidate for the QTL effect. However, they are still within a LOD3 significance threshold for the LT-tolerance QTL, suggesting one or more may be making a contribution towards the overall effect (Fig. 3.4). A ninth gene, *HvCBF2B*, also localizes to chromosome 5H-L as determined via the wheat:barley addition lines (Table 3.2), but its position relative to the LT-tolerance QTL is currently unknown. It is possible *HvCBF2B* or one of the uncharacterized *HvCBF*s (this chapter; Stockinger *et al.*, Chapter 4, this volume) could map under the DM population QTL peak. Alternatively, a gene other than a *CBF* may be responsible for this QTL, such as a barley counterpart to the CBF-regulating factor *AtICE1* (Chinnusamy *et al.*,

2003), or the LT-tolerance QTL could be a pleiotropic effect of the *Vrn-H1* candidate gene *HvBM5A* (von Zitzewitz, 2004; Cooper *et al.*, Chapter 5, this volume).

In contrast to their weak candidacy for the DM LT-tolerance QTL, the two 5H-L *HvCBF* gene clusters are strong prospective candidates for more recently determined LT-tolerance QTL in the *Triticeae*. During the latter stages of this project, two additional *Triticeae* populations mapping for a new syntenous LT-tolerance QTL were introduced where the parents contrast for LT tolerance and VRN: Nure × Tremois (NT), a 'winter' × 'spring', respectively, barley population (Francia *et al.*, 2004) and a 'winter' × 'spring' *T. monococcum* population (Vágújfalvi *et al.*, 2003). In the NT and *T. monococcum* populations, the syntenous LT-tolerance QTL peak directly corresponds to this *HvCBF*-rich region (Fig. 3.4), making these two clusters strong gene candidates for the effect. The NT population has two distinct LT-tolerance QTLs on 5H-L. One is coincident with the LT-tolerance QTL of the DM population, while the second sits 20 cM proximal (towards the centromere) from the first. In Francia *et al.* (2004), we mapped *HvCBF4A* coincident with the novel LT-tolerance QTL of NT, confirming this QTL peaks sits directly over the location of the two tandem *HvCBF* gene clusters (Fig. 3.4). Approximately 20 cM proximal from the *HvCBF4* cluster, an *HvCBF8* gene was also mapped in the NT population (Francia *et al.*, 2004), suggesting that at least one of the unmapped *HvCBF8* pseudogenes may be located in this region.

Of extra interest, in addition to the LT-tolerance QTLs, QTLs for accumulation of two *cor* gene products are also coincident in the NT population with these *HvCBF* clusters (Fig. 3.4). One of these genes is *cor14b*. This gene: (i) contains a single CRT motif that is responsible for *cor14b* LT responsiveness (Dal Bosco *et al.*, 2003), (ii) the CRT is specifically bound by a number of the HvCBFs (Table 3.1), (iii) *cor14b* expression is induced after expression of *HvCBF*s (Fig. 3.3), and (iv) *HvCBF* genes are coincident with a QTL controlling the cold-responsive accumulation of *cor14b* product (Fig. 3.4; Francia *et al.*, 2004). Together, these results indicate that one or more *HvCBF* genes is probably a key regulator of *cor14b*. We are currently extending the allele cloning (Stockinger *et al.*, Chapter 4, this volume) and expression data reported in this work to the Nure and Tremois genotypes to begin elucidation of the means by which the *HvCBF*s may be contributing to and/or determining the differential freezing-tolerance capacities of these two parents.

AtCBF1-3 of *Arabidopsis* (Gilmour *et al.*, 1998) and *LeCBF1–3* of tomato (Zhang *et al.*, 2004), each occur as a physically tandem array. The occurrence of the two genetically inseparable *HvCBF* gene clusters on 5H-L, as well as the co-localization of *HvCBF1* and *HvCBF7* to 6H-L, suggested *CBF* genes in barley may also occur as physically tandem arrays (Table 3.2). We analysed the *H. vulgare* cv. Morex BAC clones identified by Choi *et al.* (2002) that contain *HvCBF3* cross-hybridizing genes (BAC clones 424C17, 424E16, 572K24, 745C23, 790P15 and 804E19) to determine if more than one *CBF* gene was present and confirm whether multiple *HvCBF* genes were present on a defined segment of DNA. Gene-specific PCR and blot hybridizations established that these BACs form two contigs and carry at least four barley *CBF*s between them: *HvCBF3*, *HvCBF6*,

HvCBF8A and *HvCBF10A* (data not shown), all HvCBF3-subgroup members. Contig 1 was represented by a single BAC clone (745C23) which harboured *HvCBF8A*-Mx alone. The remaining five BAC clones form contig 2 and collectively harbour *HvCBF3*-Mx, *HvCBF10A*-Mx and *HvCBF6*-Mx. This contig is exemplified by BAC clone 804E19, which harbours an ~100–150 kb insert and contains all three *HvCBF*s. This data supports the clustered mapping results of this gene set and verifies the presence of at least one physical tandem array of *CBF* genes in barley. It is currently unknown whether *HvCBF1* and *HvCBF7*, both located on 6H-L (Table 3.2), are tandem also.

Conclusions

Barley harbours a large family of at least 17 *CBF* genes (Table 3.1). The barley genome contains additional *HvCBF*s, including pseudogenes, based on genomic DNA library cloning and analysis currently underway (Stockinger *et al.*, Chapter 4, this volume). The large size and complexity of the barley *CBF* family, in comparison to dicots, was representative of cereals, as all known monocot *CBF*s fell under one of the three phylogenetic HvCBF-subgroups (Fig. 3.1). This implies these three subgroups were established prior to cereal, and possibly monocot, speciation. The CBF signature motifs flanking the AP2 domain identified by Jaglo *et al.* (2001) were useful for distinguishing which monocot AP2 domain genes probably encoded *CBF* family members. These signature motifs are conserved between monocots and dicots, although the function of these conserved motifs is unknown; the 5′ motif does display characteristics of a nuclear localization signal (Stockinger *et al.*, 1997). We validated that the barley AP2 domain-containing genes assigned as CBF family member candidates based on the CBF signature sequence presence was valid via demonstration that a representative subset could: (i) specifically bind the CRT motif of various *cor* genes, and (ii) activate *cor* gene expression *in vivo*. *HvCBF7* belongs to the most ancestral CBF group and its product specifically bound the CRT element. All other HvCBFs tested specifically bound the CRT motif also, although the binding was LT dependent for the HvCBF4-subgroup members assayed. Independent work with HvCBF1-Hn, HvCBF2B-Hn and OsDREB1A is in agreement with these results (Xue, 2003; Dubouzet *et al.*, 2003). Together, this suggests that the genes assigned as CBF family members in this chapter (Fig. 3.1) are all likely to interact specifically with the CRT motif.

Only minor allelic differences were present in the coding region between genotypes differing in LT tolerance, suggesting that if *HvCBF* loci are contributing towards this trait, the encoded polypeptide is probably not the principal variation source. Differences in the amount and duration of *HvCBF* gene expression in response to cold stress appeared to be a more likely source for influencing the LT-tolerance phenotype. For each *HvCBF*, where expression level differed, the duration of expression was consistently greater in the cold-hardy Dicktoo genotype compared to that of cold-sensitive Morex. *HvCBF2* response to cold was the most intriguing, where only a very minor and temporary induction was observed in Morex, as compared to a substantial and sustained response in Dicktoo, suggesting one or both *HvCBF2* genes could be a key contributor towards the LT-tolerance phenotype. The hypothesis that the difference in the magnitude and duration of *HvCBF* gene expression is the important contributing factor to LT-tolerance differences supports the observation that the expression duration of LT-responsive structural genes (i.e. *cor* genes) is likely to be a key component of final LT-tolerance capacity (Mahfoozi *et al.*, 2000; Fowler *et al.*, 2001). Therefore, duration of *CBF* expression could be a likely control point determining the expression characteristics of downstream monocot *cor* genes.

A major question motivating this research was 'are *CBF*s candidate genes for LT-tolerance QTL in the *Triticeae*?' Our initial model was the DM LT-tolerance QTL coincident with the *Vrn-H1* locus on 5H-L (Hayes *et al.*, 1997). While multiple *HvCBF* genes are present on 5H-L, all mapped ~20 cM distal to the peak, but were still within the LOD 3 confidence interval. This suggests that while one or more of these *HvCBF*s could be a component of a multigenic QTL effect, they are probably not the primary component of the DM LT-tolerance QTL. We recently reported that two LT-tolerance QTLs are present on 5H-L in the NT barley population – one coincident with the position of the DM QTL and the second coincident with the 5H-L *HvCBF* gene clusters (Francia *et al.*, 2004). This strongly implicates that one or more of the *HvCBF* genes is a candidate for the second LT-tolerance QTL (Fig. 3.4).

We are systematically identifying the genes that influence the LT-tolerance capacity of barley, as a basis for the *Triticeae*. In this study, our goal was to determine the relationship between *CBF* genes and LT tolerance. We determined that: (i) barley has a large and complex *CBF* family; (ii) the barley *CBF* family is representative in both size and complexity of *CBF* families of other cereals; (iii) barley *HvCBF*s display functional characteristics of CBF factors; (iv) differences in expression, rather than the encoded polypeptide, are a more likely source of *HvCBF* allelic variation affecting LT tolerance; and (v) two *CBF* gene clusters are candidates for *Triticeae* LT-tolerance QTLs. The results collectively suggest that, as in dicots, CBFs are also an important component of the cold response pathway of cereals.

Acknowledgements

This work was supported by NSF Functional Genomics grant 0110124 and the US Barley Genome Project.

References

Amundsen, K.L. (2004) Conservation of the CBF low temperature response pathway in cereals. MS thesis, Michigan State University, East Lansing, Michigan.

Cattivelli, L., Baldi, P., Crosatti, C., Di Fonzo, N., Faccioli, P., Grossi, M., Mastrangelo, A.M., Pecchioni, N. and Stanca, A.M. (2002) Chromosome regions and stress-related sequences involved in resistance to abiotic stress in *Triticeae*. *Plant Molecular Biology* 48, 649–665.

Chinnusamy, V., Ohta, M., Kanrar, S., Lee, B.H., Hong, X., Agarwal, M. and Zhu, J.K. (2003) ICE1: a regulator of cold-induced transcriptome and freezing tolerance in *Arabidopsis. Genes and Development* 17, 1043–1054.

Choi, D.-W., Zhu, B. and Close, T.J. (1999) The barley (*Hordeum vulgare* L.) dehydrin multigene family: sequences, allele types, chromosome assignments, and expression characteristics of 11 *Dhn* genes of cv. Dicktoo. *Plant Physiology* 98, 1324–1247.

Choi, D.-W., Koag, M.C. and Close, T.J. (2000) Map locations of barley *Dhn* genes determined by gene-specific PCR. *Theoretical and Applied Genetics* 101, 350–354.

Choi, D.-W., Rodriguez, E.M. and Close, T.J. (2002) Barley *Cbf3* gene identification, expression pattern, and map location. *Plant Physiology* 129, 1781–1787.

Close, T.J., Wanamaker, S.I., Caldo, R.A., Turner, S.M., Ashlock, D.A., Dickerson, J.A., Wing, R.A., Muehlbauer, G.J., Kleinhofs, A. and Wise, R.P. (2004) A new resource for cereal genomics: 22K barley GeneChip comes of age. *Plant Physiology* 134, 960–968.

Dal Bosco, C., Busconi, M., Govoni, C., Baldi, P., Stanca, A.M., Crosatti, C., Bassi, R. and Cattivelli, L. (2003) *cor* gene expression in barley mutants affected in chloroplast development and photosynthetic electron transport. *Plant Physiology* 131, 793–802.

Dubcovsky, J., Lijavetzky, D., Appendino, L. and Tranquilli, G. (1998) Comparative RFLP mapping of *Triticum monococcum* genes controlling vernalization requirement. *Theoretical and Applied Genetics* 97, 968–975.

Dubouzet, J.G., Sakuma, Y., Ito, Y., Kasuga, M., Dubouzet, E.G., Miura, S., Seki, M., Shinozaki, K. and Yamaguchi-Shinozaki, K. (2003) *OsDREB* genes in rice, *Oryza sativa* L., encode transcription activators that function in drought-, high-salt- and cold-responsive gene expression. *Plant Journal* 33, 751–763.

Fowler, D.B., Breton, G., Limin, A.E., Mahfoozi, S. and Sarhan, F. (2001) Photoperiod and temperature interactions regulate low-temperature-induced gene expression in barley. *Plant Physiology* 127, 1676–1681.

Francia, E., Rizza, F., Cattivelli, L., Stanca, A.M., Galiba, G., Tóth, B., Hayes, P.M., Skinner, J.S. and Pecchioni, N. (2004) Two loci on chromosome 5H determine low-temperature tolerance in a 'Nure' (winter) × 'Tremois' (spring) barley map. *Theoretical and Applied Genetics* 108, 670–680.

Gilmour, S.J., Zarka, D.G., Stockinger, E.J., Salazar, M.P., Houghton, J.M. and Thomashow, M.F. (1998) Low temperature regulation of the *Arabidopsis* CBF family of AP2 transcriptional activators as an early step in cold-induced *COR* gene expression. *Plant Journal* 16, 433–442.

Gilmour, S.J., Sebolt, A.M., Salazar, M.P., Everard, J.D. and Thomashow, M.F. (2000) Overexpression of the *Arabidopsis CBF3* transcriptional activator mimics multiple biochemical changes associated with cold acclimation. *Plant Physiology* 124, 1854–1865.

Goff, S.A., Ricke, D., Lan, T.-H., Presting, G. *et al.* (2002) A draft sequence of the rice genome (*Oryza sativa* L. ssp. *japonica*). *Science* 236, 92–100.

Haake, V., Cook, D., Riechmann, J.L., Pineda, O., Thomashow, M.F. and Zhang, J.Z. (2002) Transcription factor CBF4 is a regulator of drought adaptation in *Arabidopsis. Plant Physiology* 130, 639–648.

Hayes, P.M., Blake, T.K., Chen, T.H.H., Tragoonrung, S., Chen, F., Pan, A. and Liu, B. (1993) Quantitative trait loci on barley (*Hordeum vulgare* L.) chromosome 7 associated with components of winter hardiness. *Genome* 36, 66–71.

Hayes, P.M., Chen, F.Q., Corey, A., Pan, A., Chen, T.H.H., Baird, E., Powell, W., Thomas, W., Waugh, R., Bedo, Z., Karsai, I., Blake, T. and Oberthur, L. (1997) The Dicktoo × Morex population: a model for dissecting components of winter hardiness in barley. In: Li, P.H. and Chen, T.H.H. (eds) *Plant Cold Hardiness.* Plenum Press, New York, pp. 77–87.

Hayes, P.M., Castro, A., Marquez-Cedillo, L., Corey, A., Henson, C., Jones, B.L., Kling, J., Mather, D., Matus, I., Rossi, C. and Sato, K. (2003) Genetic diversity for quantitatively inherited agronomic and malting quality traits. In: Von Bothmer, R., Knupffer, H., van Hintum, T. and Sato, K. (eds) *Diversity in Barley.* Elsevier Science Publishers, Amsterdam, The Netherlands, pp. 201–226.

Islam, A.K.M.R., Shepherd, K.W. and Sparrow, D.H.B. (1981) Isolation and characterization of euplasmic wheat-barley chromosome addition lines. *Heredity* 46, 161–174.

Jaglo, K.R., Kleff, S., Amundsen, K.L., Zhang, X., Haake, V., Zhang, J.Z., Deits, T. and Thomashow, M.F. (2001) Components of the *Arabidopsis* C-repeat/dehydration-responsive element binding factor cold-response pathway are conserved in *Brassica napus* and other plant species. *Plant Physiology* 127, 910–917.

Jaglo-Ottosen, K.R., Gilmour, S.J., Zarka, D.G., Schabenberger, O. and Thomashow, M.F. (1998) *Arabidopsis* CBF1 overexpression induces COR genes and enhances freezing tolerance. *Science* 280, 104–106.

Karsai, I., Meszaros, K., Lang, L., Hayes, P.M. and Bedo, Z. (2001) Multivariate analysis of traits determining adaptation in cultivated barley. *Plant Breeding* 120, 217–222.

Karsai, I., Hayes, P.M., Kling, J., Matus, A., Mészáros, K., Láng, L., Bedo, Z. and Sato, K. (2004) Genetic variation in component traits of heading date in *Hordeum vulgare* subsp. *spontaneum* accessions characterized in controlled environments. *Crop Science* 44, 1622–1632.

Karsai, I., Szücs, P., Mészáros, K., Filichkin, T., Hayes, P.M., Skinner, J.S., Lang, L. and Bedo, Z. (2005) The *Vrn-H2* locus is a major determinant of flowering time in a facultative × winter growth habit barley (*Hordeum vulgare* L.) mapping population. *Theoretical and Applied Genetics* 110, 1458–1466.

Kleinhofs, A. and Graner, A. (2001) An integrated map of the barley genome. In: Phillips, R.L. and Vasil, I.K. (eds) *DNA-Based Markers in Plants*. Kluwer Academic Publishers, Dordrecht, The Netherlands, pp. 187–199.

Limin, A.E. and Fowler, D.B. (2002) Developmental traits affecting low-temperature tolerance response in near-isogenic lines for the vernalization locus *Vrn-A1* in wheat (*Triticum aestivum* L. em Thell). *Annals of Botany* 89, 579–585.

Liu, Q., Kasuga, M., Sakuma, Y., Abe, H., Miura, S., Yamaguchi-Shinozaki, K. and Shinozaki, K. (1998) Two transcription factors, DREB1 and DREB2, with an EREBP/AP2 DNA binding domain separate two cellular signal transduction pathways in drought- and low- temperature-responsive gene expression, respectively, in *Arabidopsis*. *Plant Cell* 10, 1391–1406.

Mahfoozi, S., Limin, A.E., Hayes, P.M., Hucl, P. and Fowler, D.B. (2000) Influence of photoperiod response on the expression of cold hardiness in wheat and barley. *Canadian Journal of Plant Science* 80, 721–724.

Miller, A.K. and Dubcovsky, J. (2005) Characterization of the *Triticum monococcum* frost tolerance locus *Fr-A2*. In: *Plant and Animal Genomes XII Conference*. San Diego, California.

Pan, A., Hayes, P.M., Chen, F., Chen, T.H.H., Blake, T., Wright, S., Karsai, I. and Bedo, Z. (1994) Genetic analysis of the components of winter hardiness in barley (*Hordeum vulgare* L.). *Theoretical and Applied Genetics* 89, 900–910.

Sakuma, Y., Liu, Q., Dubouzet, J.G., Abe, H., Shinozaki, K. and Yamaguchi-Shinozaki, K. (2002) DNA-binding specificity of the ERF/AP2 domain of *Arabidopsis* DREBs, transcription factors involved in dehydration- and cold-inducible gene expression. *Biochemical and Biophysical Research Communications* 290, 998–1009.

Salinas, J. (2002) Genetic approaches confirm an important role for CBFs in cold acclimation and reveal a complex network for regulating CBF expression. In: *XIII International Conference on Arabidopsis Research*. Seville, Spain. Available at http://www.arabidopsis2002.com

Stockinger, E.J., Gilmour, S.J. and Thomashow, M.F. (1997) *Arabidopsis thaliana CBF1* encodes an AP2 domain-containing transcriptional activator that binds to the C-repeat/DRE, a *cis*-acting DNA regulatory element that stimulates transcription in response to low temperature and water deficit. *Proceedings of the National Academy of Sciences USA* 94, 1035–1040.

Thomashow, M.F. (1999) Plant cold acclimation: freezing tolerance genes and regulatory mechanisms. *Annual Reviews of Plant Physiology and Plant Molecular Biology* 50, 571–599.

Tuberosa, R., Galiba, G., Sanguineti, M.C., Noli, E. and Sutka, J. (1997) Identification of QTL influencing freezing tolerance in barley. *Acta Agronomica Hungarica* 45, 413–417.

Vágújfalvi, A., Galiba, G., Cattivelli, L. and Dubcovsky, J. (2003) The cold regulated transcriptional activator *Cbf3* is linked to the frost tolerance locus *Fr-A2* on wheat chromosome 5A. *Molecular Genetics and Genomics* 269, 60–67.

van Zee, K., Chen, F.Q., Hayes, P.M., Close, T.J. and Chen, T.H.H. (1995) Cold-specific induction of a dehydrin gene family member in barley. *Plant Physiology* 108, 1233–1239.

von Zitzewitz, J. (2004) Structural and functional characterization of barley vernalization genes. MS thesis, Oregon State University, Corvallis, Oregon.

Xue, G.P. (2002) Characterization of the DNA-binding profile of barley HvCBF1 using an enzymatic method for rapid, quantitative and high-throughput analysis of the DNA-binding activity. *Nucleic Acids Research* 30, 77.

Xue, G.P. (2003) The DNA-binding activity of an AP2 transcriptional activator HvCBF2 involved in regulation of low-temperature responsive genes in barley is modulated by temperature. *Plant Journal* 33, 373–383.

Yu, Y., Tomkins, J.P., Waugh, R., Frisch, D.A., Kudrna, D., Kleinhofs, A., Brueggeman, R.S., Muehlbauer, G.J., Wise, R.P. and Wing, R.A. (2000) A bacterial artificial chromosome library for barley (*Hordeum vulgare* L.) and the identification of clones containing putative resistance genes. *Theoretical and Applied Genetics* 101, 1093–1099.

Yu, J., Hu, S., Wang, J., Wong, G.K.-S. *et al.* (2002) A draft sequence of the rice genome (*Oryza sativa* L. ssp. *indica*). *Science* 236, 79–92.

Zhang, X., Fowler, S.G., Cheng, H., Lou, Y., Rhee, S.Y., Stockinger, E.J. and Thomashow, M.F. (2004) Freezing-sensitive tomato has a functional CBF cold response pathway, but a CBF regulon that differs from that of freezing-tolerant *Arabidopsis*. *Plant Journal* 39, 905–919.

4

Structural Organization of Barley *CBF* Genes Coincident with a QTL for Cold Hardiness

E.J. STOCKINGER,[1,*] H. CHENG[1,2] AND J.S. SKINNER[3]

[1]*Department of Horticulture and Crop Science, 1680 Madison Avenue, The Ohio State University/OARDC, Wooster, OH 44691, USA;* [2]*Biotechnology Research Institute, Chinese Academy of Agricultural Sciences, Beijing 100081, China;* [3]*Department of Horticulture, 4017 Agricultural & Life Science Building, Oregon State University, Corvallis, OR 97331, USA*

Introduction

Genetic identification of low-temperature tolerance loci in the *Triticeae*

The major temperate-climate cereal crops of the *Triticeae* – wheat (*Triticum* spp.), barley (*Hordeum vulgare*) and rye (*Secale cereale*) – are cool-season annual plants generally classified as being of either winter or spring habit. Increasing the capacity to perform at and survive low temperatures for either of these growth habit forms has tremendous potential to extend both the geographic region and the length of the growing season in which these plants could be grown.

As an agronomic group, winter types are generally more tolerant of colder temperatures than their spring counterparts. Consequently, strategies to identify the genetic components conferring low-temperature tolerance frequently entail making crosses between winter and spring genotypes. The progeny are then evaluated for numerous traits associated with low-temperature tolerance including measurements of crown fructan content (Livingston *et al.*, 1989), LT_{50} values (Sukumaran and Weiser, 1972), chlorophyll fluorescence (Fv/Fm) (Krause and Weis, 1991), scoring plant regrowth after controlled freezes and field survival after overwintering.

The first locus identified in the *Triticeae* conferring cold tolerance, *Fr1* (frost resistance 1), was identified using a population of 22 single chromosome recombinant lines generated between the 5A chromosomes of the English winter wheat

*Corresponding author (e-mail: stockinger.4@osu.edu; Fax: +1-330-263-3887)

 Cold Hardiness in Plants: Molecular Genetics, Cell Biology and Physiology (eds T.H.H. Chen *et al.*)

cultivar Hobbit and *Triticum spelta* (Sutka and Snape, 1989). The *Fr1* allele donated by the Hobbit genome conferred frost tolerance and co-segregated with a recessive *vrn1* allele that conferred the requirement for vernalization, and by definition the winter growth habit. The *T. spelta Fr1* allele, which predisposed the plant to frost susceptibility, co-segregated with a *Vrn1* spring allele such that plants carrying the *T. spelta Vrn1* allele no longer required vernalization (Sutka and Snape, 1989). This co-segregation between *Fr1* and *vrn1* left open the possibility that increased frost resistance and the concomitant requirement for vernalization might be due to the pleiotropic effects of a single gene. However, a single recombinant between *Fr1* and *Vrn1* was subsequently identified using a larger population in the Chinese Spring (CS) background (Fig. 4.1). In this instance, recombination

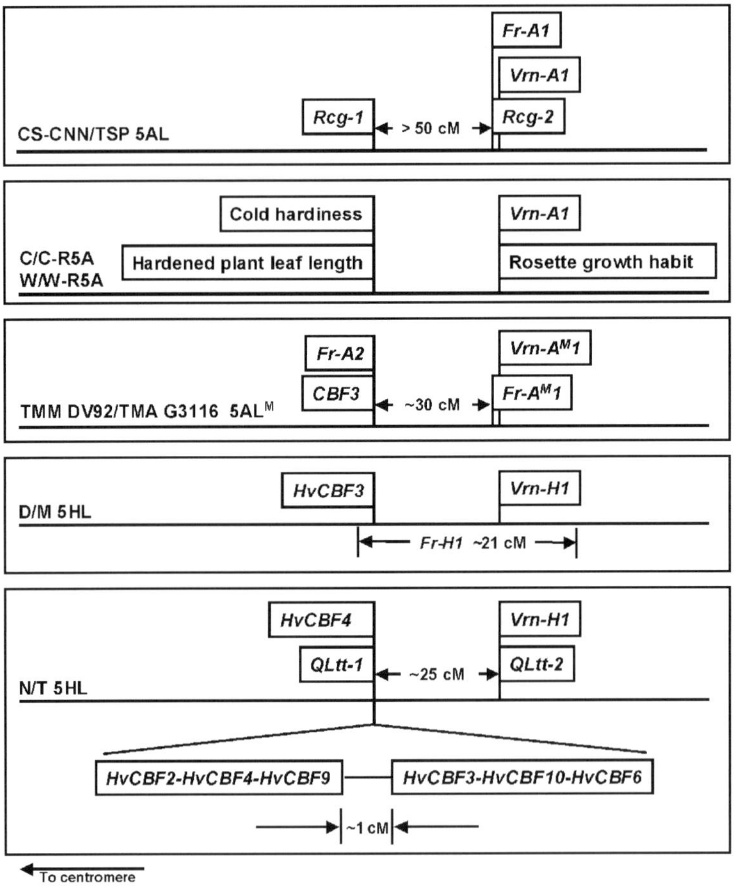

Fig. 4.1. Simple line drawing of the long (L) arm of homoeologous group 5 chromosomes to illustrate identified *Triticeae* genes affecting low-temperature tolerance in relation to the *Vrn-1* locus and relative positions of *CBF*s. *Fr-A1* (Sutka and Snape, 1989), *Rcg-1* and *Rcg-2* (Vágújfalvi *et al.*, 2000), cold hardiness and rosette growth habit (Roberts, 1990), *Fr-A2* (Vágújfalvi *et al.*, 2003), *Fr-H1* (Hayes *et al.*, 1993; Pan *et al.*, 1994), *HvCBF3* (Choi *et al.*, 2002), *HvCBF4* and the *QLtt-1* and *QLtt-2* (Francia *et al.*, 2004) and *HvCBF* clusters (Skinner *et al.*, Chapter 3, this volume).

took place between the 5A chromosomes of the North American winter wheat cultivar 'Cheyenne' (CNN) and *T. spelta* (TSP) (Galiba *et al.*, 1995). Other similar chromosomal substitution strategies have revealed loci controlling frost tolerance on chromosomes 5B and 5D, and these have been assigned *Fr-B1* and *Fr-D1*, respectively, whereas the 5A locus has been reassigned *Fr-A1* (Sutka and Snape, 1989; Snape *et al.*, 1997; Tóth *et al.*, 2003).

In addition to many of the aforementioned traits measured to evaluate low-temperature tolerance, the relative abundance of the *COR14b* (cold-regulated) gene product was also used to score the CS-CNN/TSP 5A chromosome recombinant lines (Vágújfalvi *et al.*, 2000). Briefly, *COR14b* and many other cold-regulated genes show a coordinate upregulation in response to low temperatures, and the gene product abundances exhibit a strong positive correlation with the individual genotype's capacity to cold acclimate and develop freezing tolerance (Houde *et al.*, 1992; Danyluk *et al.*, 1994; Crosatti *et al.*, 1996; Fowler *et al.*, 1996; Limin *et al.*, 1997; Sarhan *et al.*, 1997; Danyluk *et al.*, 1998; Grossi *et al.*, 1998; NDong *et al.*, 2002), and transgenic overexpression of similar genes in *Arabidopsis* results in increased freezing tolerance (Artus *et al.*, 1996; Jaglo-Ottosen *et al.*, 1998; Liu *et al.*, 1998). In the CS-CNN/TSP 5A population, it was revealed that two loci had regulatory control over *COR14b* accumulation and were assigned Regulator for *COR* genes, *Rcg-1* and *Rcg-2*, respectively (Fig. 4.1; Vágújfalvi *et al.*, 2000). Mapping of the two genes indicated that *Rcg-2* co-segregated with *Fr-1*, whereas *Rcg-1* resided proximal to the centromere.

The data indicating that there are two loci on chromosome 5 affecting cold hardiness are not unique to the CS-CNN/TSP 5A recombinants, however. In fact, much earlier, using Rescue (spring type) 5A chromosome substitutions in either the Cadet (spring) or Winalta (winter) backgrounds, it was observed that two linked 5A loci had a major effect upon cold hardiness; one thought to control directly the capacity to cold harden while the second locus, *Vrn1*, or a tightly linked gene, sensed temperature, triggered vernalization and induced cold hardening (Fig. 4.1; Roberts, 1990). More recent genetic studies using diploid wheat lend additional support to the existence of a second major locus on chromosome 5. In the latter, a mapping population consisting of the recombinant inbred lines derived from the cross of *Triticum monococcum* ssp. *aegilopoides* G3116 (winter) and *T. monococcum* ssp. *monococcum* DV92 (spring) quantitative trait locus (QTL) analyses indicated that the only chromosomal region contributing a significant effect was the long arm of 5A approximately 30 cM away from the *Fr-A1*/*Vrn-A*^{M}*1* region, proximal to the centromere, and was subsequently assigned *Fr-A2* (Fig. 4.1; Vágújfalvi *et al.*, 2003). The *Fr-A2* locus also exerts regulatory control over *COR14b* accumulation, and, based on inference with common reference markers used in the different *Triticeae* mapping populations, appears to reside at the same location as *Triticum aestivum Rcg-1*.

Since hexaploid wheat chromosome 5A is the homoeologue of diploid wheat 5A, barley 5H and rye 5R, one might anticipate finding equivalent loci effecting low-temperature tolerance on other *Triticeae* chromosome 5 homoeologues. Indeed this is the case for barley. QTL mapping of winter hardiness and low-temperature survival using two different winter × spring barley mapping populations indicated that only 5H harboured the major genetic determinants. These are the Dicktoo × Morex (DM)

(Hayes *et al.*, 1993) and the Nure × Tremois (NT) (Francia *et al.*, 2004) mapping populations. In the DM population, there is one broad QTL on the long arm of 5H that spans a 21 cM interval (Fig. 4.1). Interestingly, the peak of this QTL coincides precisely with the *Vrn-H1* locus, although neither Dicktoo nor Morex requires vernalization. Based on a subset of common molecular markers used to generate the respective *Triticeae* linkage maps and the inferred map distances between those markers, the DM QTL has been assigned *Fr-H1* (Hayes *et al.*, 1993; Pan *et al.*, 1994). However, 21 cM is a fairly large interval and, as alluded to, it is likely that *Fr-H1* encompasses a region harbouring a number of genes contributing favourable alleles to cold hardiness and winter survival (Hayes *et al.*, 1993; Pan *et al.*, 1994).

In the NT mapping population two low-temperature tolerance QTL (*QLtt*) were identified, both of which are also on the long arm of 5H. One locus, *QLtt-2*, is coincident with *Vrn-H1*. The other locus, *QLtt-1*, maps about 25 cM away and is more proximal to the centromere. *QLtt-1*, similar to *Rcg-1* and *Fr-A2*, also influences *COR14b* gene product accumulation. Thus *Fr-A1*, *Rcg-2*, *QLtt-2* and the *Fr-H1* peak all appear to localize to the same homoeologous group 5 chromosome position, being tightly linked to *Vrn-1*. *QLtt-1*, *Fr-A2* and *Rcg-1* all reside more proximal to the centromere and, perhaps, are also in a homoeologous or genetically collinear position in each of the respective genomes.

One of the more exciting findings associated with *QLtt-1* and *Fr-A2* is that barley ESTs with very high identity to the *Arabidopsis thaliana* C-repeat binding factors (CBFs) co-localize to these map positions (Choi *et al.*, 2002; Vágújfalvi *et al.*, 2003; Francia *et al.*, 2004; Skinner *et al.*, Chapter 3, this volume). The CBFs are transcriptional activator proteins that were first identified in *Arabidopsis* by virtue of their ability to bind to the C-repeat/dehydration responsive element (CRT/DRE), and *trans*-activate gene expression (Stockinger *et al.*, 1997). The CRT/DRE (Baker *et al.*, 1994; Yamaguchi-Shinozaki and Shinozaki, 1994) is a 5 bp *cis*-acting DNA regulatory motif that confers low temperature-regulated expression and is usually present in multiple copies in each of the different *Arabidopsis COR* gene promoters (Fig. 4.2A). When the *CBF*s are overexpressed as transgenes in *Arabidopsis*, the *COR* genes are activated in the absence of any low-temperature stimulus, as are a battery of other genes collectively referred to as the CBF regulon. Additionally, biochemical changes associated with cold acclimation and increased freezing tolerance are also induced and, notably, these plants show increased freezing tolerance (Jaglo-Ottosen *et al.*, 1998; Liu *et al.*, 1998; Gilmour *et al.*, 2000). Since *COR14b* and many of the other *Triticeae* cold-regulated genes harbour multiple CRT/DRE motifs in their upstream regulatory regions (Fig. 4.2B), the co-localization of barley CBFs with the low-temperature tolerance QTL provides very compelling support for the hypothesis that the *Triticeae* CBFs are critical players in winter hardiness and low-temperature survival of this group of plants.

The *Triticeae CBF* gene family

Before delving further into the co-localization of CBFs with the low-temperature tolerance QTL, it is helpful to briefly review some of what is already known about

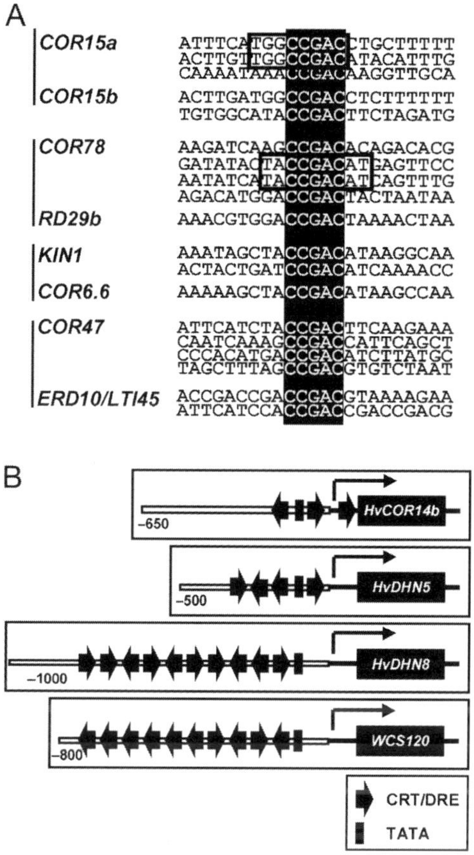

Fig. 4.2. Cold-regulated genes and the preponderance of CRT/DRE motifs. (A) CRT/DRE core motif (white lettering on a black background) and flanking nucleotide sequences in each of the different *Arabidopsis COR* genes. *COR15a* (Baker *et al.*, 1994), *COR15b* (Wilhelm and Thomashow, 1993), *COR78* (*RD29a*) and *RD29b* (Yamaguchi-Shinozaki and Shinozaki, 1994), *KIN1* and *COR6.6* (Wang *et al.*, 1995), *COR47* (Iwasaki *et al.*, 1997) and *ERD10/LTI45* (Welin *et al.*, 1995). The 8 bp C-repeat sequence identified in the *COR15a* promoter (Baker *et al.*, 1994) and the 9 bp dehydration responsive element identified in *COR78/RD29a* (Yamaguchi-Shinozaki and Shinozaki, 1994) are boxed. (B) *Triticeae* cold-regulated genes and CRT/DRE copy number and orientation relative to the TATA box and transcriptional start site in the sequenced genomic regions. Barley *COR14b* (Dal Bosco *et al.*, 2003); Barley *DHN5* and *DHN8* (Choi *et al.*, 1999); Wheat *WCS120* (Vazquez-Tello *et al.*, 1998).

the complexity of the barley CBFs at the level of the cDNA clones in hand and the map position of these clones (Skinner *et al.*, Chapter 3, this volume). Presently, there are estimated to be nearly 20 CBF protein-encoding genes in the barley genome originally identified as unique cDNA clones. About half of these reside in two separate but tightly linked genetic clusters on barley chromosome 5H. Of those residing on 5H, the CBFs can be broken down into seven distinct *HvCBF* gene families, most of which have a second, and frequently third, nearly identical sister gene that have been classified as an A, B or C type, rather than as an additional CBF. The two clusters are estimated to be about 1 cM apart and linkage between the individual genes within each cluster is as yet unbroken (Fig. 4.1). Presently, all CBF genes comprising one cluster fall into a single clade and all CBFs within the second cluster fall into a separate clade, which have been deemed the HvCBF3- and the HvCBF4-subgroups. To avoid confusion, each *HvCBF* is sequentially identified such that numbering starts with *HvCBF4* maintaining the *HvCBF1*, *HvCBF2* and *HvCBF3* designations of published reports describing barley CBF genes (Choi *et al.*, 2002; Xue, 2002, 2003). It is also important to point out that these numbered designations are not meant to imply an orthologous relationship to that of the previously described *Arabidopsis CBF1*, *CBF2*, *CBF3* or *CBF4* genes (Gilmour *et al.*, 1998; Haake *et al.*, 2002).

Strategy for testing whether the *CBFs* form the molecular basis of NT *QLtt-1*

One idea is that the proximal NT *QLtt-1* coincident with the *CBF*s might be the result of a structural difference between these two cultivars at the genetically collinear region encompassing the *HvCBF* gene clusters. Further, since this region did not surface as a major QTL in the DM population, it may be an indication that Dicktoo and Morex do not significantly differ at the *CBF* locus and perhaps are even the same as either Nure or Tremois. The difference between Nure and Tremois could be due to any number of possible scenarios including structural rearrangement of the entire *CBF* locus, a deletion of a portion or of a whole of one or more *CBF* genes, or even more subtle differences such as a premature termination codon in one or more *CBF* genes. In order to investigate these possibilities and gain a better understanding of the *Triticeae CBF* locus structure, we have begun the molecular isolation and DNA sequence determination of the genomic regions encompassing the 5H *CBF*s from the four barley cultivars Dicktoo, Morex, Nure and Tremois. The strategy employed basically consists of the following steps:

- Construction of bacteriophage λ genomic libraries for all four genotypes.
- Screening recombinants using a mixture of 5H *CBF* cDNA probes.
- Sorting clones into classes using restriction enzyme fingerprinting and hybridization.
- Selection of clones that could be classified as unique and/or harbouring one or more 5H *HvCBF* gene family members, and/or an additional *HvCBF* gene not present in the probe used for library screening.
- Subcloning λ clone inserts into pGEM11Z.
- Generation of shotgun subclone libraries for high throughput DNA sequencing.
- Assembly and annotation.

Structure of the Genomic Clones Encompassing the Barley 5H *CBF* Genes

Although the comparative genomic sequence analyses are still a long way off, at the present, a number of observations have been made that provide a glimpse into the structural organization of the *CBF* loci in barley. Since we have made the most progress in our analyses of the genomic regions encompassing the *CBF*s from cultivar Dicktoo, we will limit our discussion to the Dicktoo *CBF*s.

The first observation is that the overall complexity observed at the cDNA level is reiterated at the genomic level. There really are a lot of 5H CBF genes. Similar to both the *Arabidopsis* and tomato (Gilmour *et al.*, 1998; Zhang *et al.*, 2004), none of the barley *CBF* coding sequences is interrupted by introns. However, this is where the similarity of the barley *CBF*s to the two dicot *CBF* genomic sequences ends. In both *Arabidopsis* and tomato, the CBF loci consist of three head-to-tail, tandem-linked genes spanning about 10 kb of genomic DNA (Gilmour *et al.*, 1998; Zhang *et al.*, 2004). For the most part this is not the case with the barley CBFs. In fact, the arrangement of the *HvCBF* genes does not seem to follow a consistent pattern at all, which is briefly summarized below (Fig. 4.3).

HvCBF10A and *HvCBF10B*, two nearly identical *HvCBF*s, are the only two *CBF*s that do show tandem arrangement, having about 5 kb between them. In contrast, *HvCBF2B*, which has 98% sequence identity to *HvCBF2A* (Skinner *et al.*, Chapter 3, this volume), was not physically linked to *HvCBF2A* on any of our genomic clones. Rather, *HvCBF2B* existed as an isolated *CBF* gene. *HvCBF2A* was, however, physically linked to *HvCBF4B* on the same genomic clone but these two genes are inverted relative to one another and are separated by about 15 kb. And although *HvCBF3* resides on the same ~100 kb BAC clone as *HvCBF10A* and *HvCBF6* (Skinner *et al.*, Chapter 3, this volume), none of these is linked to *HvCBF3*

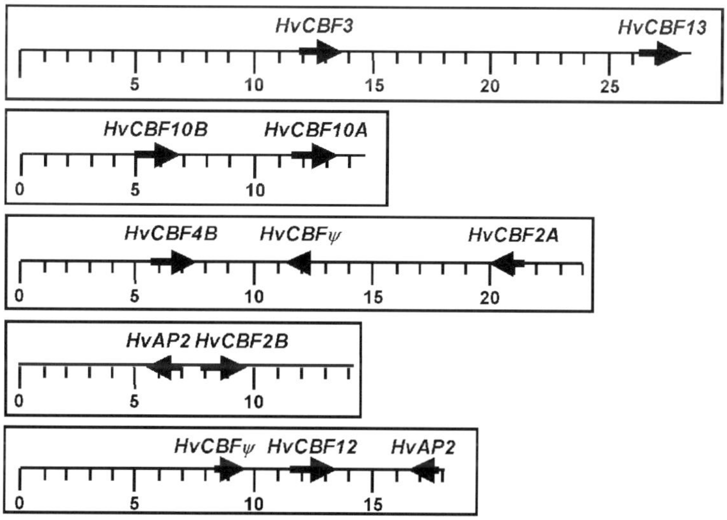

Fig. 4.3. Structure of the genomic regions encompassing the 5H *CBF* genes from barley cultivar Dicktoo. The continuous horizontal line represents the extent of the sequenced genomic DNA. Vertical hash marks are spaced 1 kb apart. CBF coding sequences are indicated by directional horizontal arrows. The partial coding sequences and those with multiple nonsense codons in all three reading frames are indicated with the Greek psi, ψ. The two coding sequences encoding non-CBF AP2 domain-containing proteins are indicated simply as *HvAP2*.

on our λ genomic clones. *HvCBF3* actually occupies a central location on an approximately 30 kb segment of contiguous sequenced genomic DNA (Fig. 4.3) before there is another *CBF*. However, this is not *HvCBF6* or *HvCBF10A*; rather an additional and novel *HvCBF*, *HvCBF13*, not pulled out in any of the cDNA library screens that we have deemed (Fig. 4.3). A second additional *HvCBF* not present in our cDNA probe was *HvCBF12*; and also existed as an isolated *HvCBF* gene (Fig. 4.3).

The barley *CBF* locus structure also differed from the *Arabidopsis* and tomato loci in a number of other attributes. For starters, the barley *CBF* genes were interspersed with other coding sequences. For example, upstream of *HvCBF2B* and downstream of *HvCBF12* were two different, non-CBF AP2 domain-containing proteins with similarity to the *Arabidopsis* sequences AtRAP2.3 and At4g344410 (Fig. 4.3). In addition, there were numerous *CBF* pseudogenes linked to these bona fide *CBF* coding sequences (Fig. 4.3). The pseudogenes either had multiple nonsense codons throughout all three reading frames or fragments of a *CBF* coding sequence. Also present were numerous repeat elements that flanked and separated many of the *CBF*s. Thus, the barley *CBF* genomic regions showed a much higher level of complexity than did the two sequenced dicot *CBF* loci.

Although we have covered nearly 100 kb of genomic DNA, we have still not isolated the genomic region encompassing every 5H *CBF* cDNA clone that we currently have in hand. *HvCBF6* and *HvCBF4A* are two that have so far eluded us and assembly of the *HvCBF9* genomic region is at present incomplete. Perhaps when all *HvCBF*s are in hand, including any additional novel *CBF*s, this 100 kb will double in size to 200 kb.

Concluding Remarks

So why are there so many *CBF*s in a genomic context which appears to be a veritable gene jumble? Are they all redundant or do they have specialized functions; i.e. do some play a critical role in winter crown survival whereas others play key roles in protecting reproductive structures during sudden spring frosts? We are now beginning detailed expression analyses of the *HvCBFs* and so it is hoped that answers to these questions will be revealed. But then there is the question of why are there so many highly identical sister *CBF* genes and a high number of pseudogenes? Are these the result of successful and failed gene duplications that occurred during chromosome pairing and crossing over, and is this then perpetuated and increased by the presence of even more *CBF*s? And has the selection pressure placed upon plants for winter survival in the plant breeder's nursery played a role in all of this; what does the wild species *Hordeum vulgare* subsp. *spontaneum CBF* locus look like? However, the fundamental question of whether differences between genotype-specific locus structure forms the molecular basis of the NT *QLtt-2* is still unresolved. We hope to answer some of these questions through comparative sequence analyses of the four barley cultivars, and also hope that the structural determination of the *CBF* locus in the other *Triticeae*, including wheat and rye, will further illuminate the complex nature of this chromosomal region.

It is anticipated that addressing these questions will lead to a better understanding of the structure of the *Triticeae CBF* locus and how these genes themselves are regulated. Since the CBFs are master regulatory switches that control the expression of gene batteries effecting cold acclimation, it is hoped that knowing how they are regulated will lead to the development of spring types that can be planted earlier in the season and winter types that can endure colder winters.

Acknowledgements

This research was supported in part by a grant from the National Science Foundation Plant Genome Program (DBI 0110124), The Ohio State University/Ohio Agricultural and Research Development Centre and the North American Barley Genome Project.

References

Artus, N.N., Uemura, M., Steponkus, P.L., Gilmour, S.J., Lin, C. and Thomashow, M.F. (1996) Constitutive expression of the cold-regulated *Arabidopsis thaliana COR15a* gene affects both chloroplast and protoplast freezing tolerance. *Proceedings of the National Academy of Sciences USA* 93, 13404–13409.

Baker, S.S., Wilhelm, K.S. and Thomashow, M.F. (1994) The 5′-region of *Arabidopsis thaliana cor15a* has *cis*-acting elements that confer cold-, drought- and ABA-regulated gene expression. *Plant Molecular Biology* 24, 701–713.

Choi, D.W., Zhu, B. and Close, T.J. (1999) The barley (*Hordeum vulgare* L.) dehydrin multigene family: sequences, allele types, chromosome assignments, and expression characteristics of 11 *Dhn* genes of cv Dicktoo. *Theoretical and Applied Genetics* 98, 1234–1247.

Choi, D.W., Rodriguez, E.M. and Close, T.J. (2002) Barley *Cbf3* gene identification, expression pattern, and map location. *Plant Physiology* 129, 1781–1787.

Crosatti, C., Nevo, E., Stanca, A.M. and Cattivelli, L. (1996) Genetic analysis of the accumulation of COR14 proteins in wild (*Hordeum spontaneum*) and cultivated (*Hordeum vulgare*) barley. *Theoretical and Applied Genetics* 93, 975–981.

Dal Bosco, C., Busconi, M., Govoni, C., Baldi, P., Stanca, A.M., Crosatti, C., Bassi, R. and Cattivelli, L. (2003) *cor* gene expression in barley mutants affected in chloroplast development and photosynthetic electron transport. *Plant Physiology* 131, 793–802.

Danyluk, J., Houde, M., Rassart, E. and Sarhan, F. (1994) Differential expression of a gene encoding an acidic dehydrin in chilling sensitive and freezing tolerant Gramineae species. *FEBS Letters* 344, 20–24.

Danyluk, J., Perron, A., Houde, M., Limin, A., Fowler, B., Benhamou, N. and Sarhan, F. (1998) Accumulation of an acidic dehydrin in the vicinity of the plasma membrane during cold acclimation of wheat. *Plant Cell* 10, 623–638.

Fowler, D.B., Chauvin, L.P., Limin, A.E. and Sarhan, F. (1996) The regulatory role of vernalization in the expression of low-temperature-induced genes in wheat and rye. *Theoretical and Applied Genetics* 93, 554–559.

Francia, E., Rizza, F., Cattivelli, L., Stanca, A.M., Galiba, G., Toth, B., Hayes, P.M., Skinner, J.S. and Pecchioni, N. (2004) Two loci on chromosome 5H determine low-temperature tolerance in a 'Nure' (winter) × 'Tremois' (spring) barley map. *Theoretical and Applied Genetics* 108, 670–680.

Galiba, G., Quarrie, S.A., Sutka, J., Morounov, A. and Snape, J.W. (1995) RFLP mapping of the vernalization (*Vrn1*) and frost resistance (*Fr1*) genes on chromosome 5A of wheat. *Theoretical and Applied Genetics* 90, 1174–1179.

Gilmour, S.J., Zarka, D.G., Stockinger, E.J., Salazar, M.P., Houghton, J.M. and Thomashow, M.F. (1998) Low temperature regulation of the *Arabidopsis* CBF family of AP2 transcriptional activators as an early step in cold-induced *COR* gene expression. *Plant Journal* 16, 433–442.

Gilmour, S.J., Sebolt, A.M., Salazar, M.P., Everard, J.D. and Thomashow, M.F. (2000) Overexpression of the *Arabidopsis CBF3* transcriptional activator mimics multiple biochemical changes associated with cold acclimation. *Plant Physiology* 124, 1854–1865.

Grossi, M., Giorni, E., Rizza, F., Stanca, A.M. and Cattivelli, L. (1998) Wild and cultivated barleys show differences in the expression pattern of a cold-regulated gene family under different light and temperature conditions. *Plant Molecular Biology* 38, 1061–1069.

Haake, V., Cook, D., Riechmann, J.L., Pineda, O., Thomashow, M.F. and Zhang, J.Z. (2002) Transcription factor CBF4 is a regulator of drought adaptation in *Arabidopsis*. *Plant Physiology* 130, 639–648.

Hayes, P.M., Blake, T., Chen, T.H.H., Tragoonrung, S., Chen, F., Pan, A. and Liu, B. (1993) Quantitative trait loci on barley (*Hordeum vulgare* L.) chromosome 7 associated with components of winter hardiness. *Genome* 36, 66–71.

Houde, M., Dhindsa, R.S. and Sarhan, F. (1992) A molecular marker to select for freezing tolerance in Gramineae. *Molecular and General Genetics* 234, 43–48.

Iwasaki, T., Kiyosue, T., Yamaguchi-Shinozaki, K. and Shinozaki, K. (1997) The dehydration-inducible *Rd17* (*Cor47*) gene and its promoter region in *Arabidopsis thaliana* (accession No. AB004872). *Plant Physiology* 115, 1287–1289.

Jaglo-Ottosen, K.R., Gilmour, S.J., Zarka, D.G., Schabenberger, O. and Thomashow, M.F. (1998) *Arabidopsis CBF1* overexpression induces *COR* genes and enhances freezing tolerance. *Science* 280, 104–106.

Krause, G.H. and Weis, E. (1991) Chlorophyll fluorescence and photosynthesis: the basics. *Annual Review of Plant Physiology and Plant Molecular Biology* 42, 313–349.

Limin, A.E., Danyluk, J., Chauvin, L.P., Fowler, D.B. and Sarhan, F. (1997) Chromosome mapping of low-temperature induced *Wcs120* family genes and regulation of cold-tolerance expression in wheat. *Molecular and General Genetics* 253, 720–727.

Liu, Q., Kasuga, M., Sakuma, Y., Abe, H., Miura, S., Yamaguchi-Shinozaki, K. and Shinozaki, K. (1998) Two transcription factors, DREB1 and DREB2, with an EREBP/AP2 DNA binding domain separate two cellular signal transduction pathways in drought- and low-temperature-responsive gene expression, respectively, in *Arabidopsis*. *Plant Cell* 10, 1391–1406.

Livingston, D.P. III, Olien, C.R. and Freed, R.D. (1989) Sugar composition and freezing tolerance in barley crowns at varying carbohydrate levels. *Crop Science* 29, 1266–1270.

NDong, C., Danyluk, J., Wilson, K.E., Pocock, T., Huner, N.P. and Sarhan, F. (2002) Cold-regulated cereal chloroplast late embryogenesis abundant-like proteins. Molecular characterization and functional analyses. *Plant Physiology* 129, 1368–1381.

Pan, A., Hayes, P.M., Chen, F., Chen, T.H.H., Blake, T., Wright, S., Karsai, I. and Bedo, Z. (1994) Genetic analysis of the components of winter hardiness in barley (*Hordeum vulgare* L.). *Theoretical and Applied Genetics* 89, 900–910.

Roberts, D.W.A. (1990) Identification of loci on chromosome 5A of wheat involved in control of cold hardiness, vernalization, leaf length, rosette growth habit, and height of hardened plants. *Genome* 33, 247–259.

Sarhan, F., Ouellet, F. and VazquezTello, A. (1997) The wheat *wcs120* gene family. A useful model to understand the molecular genetics of freezing tolerance in cereals. *Physiologia Plantarum* 101, 439–445.

Snape, J.W., Semikhodskii, A., Fish, L., Sarma, R.N., Quarrie, S.A., Galiba, G. and Sutka, J. (1997) Mapping frost tolerance loci in wheat and comparative mapping with other cereals. *Acta Agricultural Hungary* 45, 265–270.

Stockinger, E.J., Gilmour, S.J. and Thomashow, M.F. (1997) *Arabidopsis thaliana CBF1* encodes an AP2 domain-containing transcriptional activator that binds to the C-repeat/DRE, a *cis*-acting DNA regulatory element that stimulates transcription in response to low temperature and water deficit. *Proceedings of the National Academy of Sciences USA* 94, 1035–1040.

Sukumaran, N.P. and Weiser, C.J. (1972) An excised leaflet test for evaluating potato frost tolerance. *HortScience* 7, 467–468.

Sutka, J. and Snape, J.W. (1989) Location of a gene for frost resistance on chromosome 5A of wheat. *Euphytica* 42, 41–44.

Tóth, B., Galiba, G., Feher, E., Sutka, J. and Snape, J.W. (2003) Mapping genes affecting flowering time and frost resistance on chromosome 5B of wheat. *Theoretical and Applied Genetics* 107, 509–514.

Vágújfalvi, A., Crosatti, C., Galiba, G., Dubcovsky, J. and Cattivelli, L. (2000) Two loci on wheat chromosome 5A regulate the differential cold-dependent expression of the *cor14b* gene in frost-tolerant and frost-sensitive genotypes. *Molecular and General Genetics* 263, 194–200.

Vágújfalvi, A., Galiba, G., Cattivelli, L. and Dubcovsky, J. (2003) The cold-regulated transcriptional activator *Cbf3* is linked to the frost-tolerance locus *Fr-A2* on wheat chromosome 5A. *Molecular Genetics and Genomics* 269, 60–67.

Vazquez-Tello, A., Ouellet, F. and Sarhan, F. (1998) Low temperature-stimulated phosphorylation regulates the binding of nuclear factors to the promoter of *Wcs120*, a cold-specific gene in wheat. *Molecular and General Genetics* 257, 157–166.

Wang, H., Datla, R., Georges, F., Loewen, M. and Cutler, A.J. (1995) Promoters from *kin1* and *cor6.6*, two homologous *Arabidopsis thaliana* genes: transcriptional regulation and gene expression induced by low temperature, ABA, osmoticum and dehydration. *Plant Molecular Biology* 28, 605–617.

Welin, B.V., Olson, A. and Palva, E.T. (1995) Structure and organization of two closely related low-temperature-induced *dhn/lea/rab*-like genes in *Arabidopsis thaliana* L. Heynh. *Plant Molecular Biology* 29, 391–395.

Wilhelm, K.S. and Thomashow, M.F. (1993) *Arabidopsis thaliana cor15b*, an apparent homologue of *cor15a*, is strongly responsive to cold and ABA, but not drought. *Plant Molecular Biology* 23, 1073–1077.

Xue, G.P. (2002) An AP2 domain transcription factor HvCBF1 activates expression of cold-responsive genes in barley through interaction with a (G/a)(C/t)CGAC motif. *Biochimica et Biophysica Acta* 1577, 63–72.

Xue, G.P. (2003) The DNA-binding activity of an AP2 transcriptional activator HvCBF2 involved in regulation of low-temperature responsive genes in barley is modulated by temperature. *Plant Journal* 33, 373–383.

Yamaguchi-Shinozaki, K. and Shinozaki, K. (1994) A novel *cis*-acting element in an *Arabidopsis* gene is involved in responsiveness to drought, low-temperature, or high-salt stress. *Plant Cell* 6, 251–264.

Zhang, X., Fowler, S.G., Cheng, H., Lou, Y., Rhee, S.Y., Stockinger, E.J. and Thomashow, M.F. (2004) Freezing-sensitive tomato has a functional CBF cold response pathway, but a CBF regulon that differs from that of freezing-tolerant *Arabidopsis*. *Plant Journal* 39, 905–919.

5 The Genetic Basis of Vernalization Responses in Barley

L.L.D. COOPER,[1] J. VON ZITZEWITZ,[1] J.S. SKINNER,[1,2] P. SZŰCS,[1,3] I. KARSAI,[3] E. FRANCIA,[4] A.M. STANCA,[4] N. PECCHIONI,[5] D.A. LAURIE,[6] T. H.H. CHEN[2] AND P.M. HAYES[1,*]

[1]Department of Crop and Soil Science, Oregon State University, Corvallis, OR 97331, USA; [2]Department of Horticulture, Oregon State University, Corvallis, OR 97331, USA; [3]Agricultural Research Institute of the Hungarian Academy of Sciences, H-2462 Martonvásár, Hungary; [4]Experimental Institute for Cereal Research – C.R.A., Section of Fiorenzuola, I-29017 Fiorenzuola d'Arda, Italy; [5]Facoltà di Agraria, Università di Modena e Reggio Emilia, I-42100 Reggio Emilia, Italy; [6]John Innes Centre, Norwich Research Park, Colney, Norwich NR4 7UH, UK

Introduction

Vernalization can be defined as the induction of flowering by exposure of a plant to extended periods of low temperature. The range of effective temperatures at which flowering is promoted by cold is usually between 1°C and 7°C; however, in some cases vernalizing temperatures can be as low as –6°C (Michaels and Amasino, 2000). Vernalization is characteristic of many temperate-zone plants, including 'winter growth habit' forms of the *Triticeae*. Vernalization requirement is of particular interest in the cereal crops due to its role in determining adaptation range and its association with winter hardiness, i.e. the capacity of a genotype to survive the winter (Hayes *et al*., 1997). Candidate genes for two *Vrn* loci were recently cloned in diploid wheat (*Triticum monococcum*) (Yan *et al*., 2003, 2004b); and *Vrn-1* loci candidates have been identified in hexaploid wheat (*Triticum aestivum*) (Danyluk *et al*., 2003; Murai *et al*., 2003; Trevaskis *et al*., 2003; Fu *et al*., 2005). This has considerably advanced our understanding of this trait in economically important crop species. We are interested in the genetics of vernalization in barley, and have been able to avail ourselves of the progress in

*Corresponding author (e-mail: Patrick.m.hayes@oregonstate.edu; Fax: +1-541-737-1589)

Triticum spp. to clone the barley *Vrn-H1* and *Vrn-H2* candidate genes and characterize their structure and function (von Zitzewitz, 2004). This work has been submitted for publication (von Zitzewitz *et al.*, 2005) and accordingly only pertinent components of the full report will be summarized in this chapter.

Vernalization: from *Arabidopsis* to Crop Plant Model

Progress in unravelling the genetic basis of vernalization in the *Triticeae* has been made possible, in part, by advances made by the *Arabidopsis* research community (recently reviewed by Henderson *et al.*, 2003; Amasino, 2004; Henderson and Dean, 2004). While our knowledge of the genetics of vernalization in the *Triticeae* is not as complete as that in *Arabidopsis*, it is already apparent that different pathways lead to the same end phenotype (Yan *et al.*, 2003, 2004b). At this juncture, it is worth briefly reviewing key genes and processes in the *Arabidopsis* vernalization pathway in order later to compare and contrast with our findings in barley.

In *Arabidopsis*, vernalization provides the competence to flower by downregulating the expression of a flowering repressor, *Flowering Locus C* (*FLC*), by a perception of the environment. The repressed state is maintained through mitotic divisions, but it is reset again in the next generation (Amasino, 2004). The other principal gene involved in the response and requirement for vernalization is *Frigida* (*FRI*). *FRI* is a dominant gene that acts in concert with other loci to determine the vernalization-responsive, delayed-flowering phenotype of winter annual forms (Napp-Zinn, 1987). The rapid-flowering, summer annual accessions are, for the most part, homozygous for the recessive *fri* allele that confers early flowering and many of these are loss-of-function mutations. *FRI* and *FLC* act together to block the ability of a non-vernalized shoot apex to flower and the requirement for vernalization results from the synergistic interaction of dominant alleles of *FRI* and *FLC*. There is evidence that *FRI* acts in combination with other proteins such as *FRI-LIKE1* (*FRL1*). *FRI* and *FRL1* encode plant-specific proteins, but their mode of action to upregulate expression of *FLC* is still unknown (Amasino, 2004).

FLC encodes a MADS-domain protein that acts as a potent repressor of flowering (Amasino, 2004); and was shown to bind to the regulatory region of the MADS-box gene *SOC1*, repressing its transcription before vernalization (Hepworth *et al.*, 2002). Although expression of *FLC* alone (from a heterologous promoter) is sufficient to block flowering, the role of *FRI* is to elevate the expression of *FLC* to levels that block flowering (Amasino, 2004). The expression of *FLC* is regulated by at least three independent pathways (Henderson and Dean, 2004). The effects of *FRI* are overridden and flowering is promoted through a permanent epigenetic suppression of *FLC* expression in the autonomous floral promotion pathway, independent of *FRI* (He *et al.*, 2003). Mutations in the autonomous floral promotion pathway in a summer annual background (*fri* null) cause elevated *FLC* expression, similar to dominant alleles of *FRI* (Amasino, 2004).

Vernalization also overrides the effects of *FRI*, promoting flowering independently of the autonomous floral promotion pathway. Three genes, *Vernalization*

Insensitive (*VIN3*), *Vernalization1* (*VRN1*) and *Vernalization2* (*VRN2*), have been characterized that are involved in the repression of *FLC* expression during vernalization by changes in chromatin structure as a result of histone modifications at the *FLC* locus (Amasino, 2004; Bastow *et al.*, 2004; Henderson and Dean, 2004). In *vin3* mutants, the vernalization-mediated histone modifications do not occur and *FLC* is not repressed by vernalization. Interestingly, *VIN3* expression increases with the plants' exposure to cold and *VIN3* protein significantly accumulates only after a period of cold sufficient to trigger vernalization (Sung and Amasino, 2004). Mutants of the other two genes, *vrn1* and *vrn2*, are distinct from those of *VIN3* in that initial repression of *FLC* still occurs, but *FLC* repression is not maintained on the plants' return to ambient temperatures. The proteins encoded by *VRN1* and *VRN2* differ in their structure and mode of action. *VRN1* is plant-specific and encodes a protein that binds DNA *in vitro* in a non-sequence-specific manner (Levy *et al.*, 2002), while *VRN2* encodes a nuclear-localized zinc finger protein with similarity to Polycomb-group proteins of plants and animals (Gendall *et al.*, 2001). *VRN1* may function either downstream or independently of *VRN2* during vernalization-mediated *FLC* repression (Henderson and Dean, 2004).

Differences in the genetics of vernalization between *Arabidopsis* and the *Triticeae* should be mentioned, because identical gene names such as *Vrn* do not necessarily reflect orthology nor does orthology necessarily predict function. For example, the *Triticeae AP1* candidate gene for *Vrn1* is similar to the *Arabidopsis* meristem identity gene *APETALA1* (*AP1*). *Arabidopsis AP1* is not directly involved in vernalization (Murai *et al.*, 2003; Trevaskis *et al.*, 2003; Yan *et al.*, 2003), but acts downstream from *SOC1*, the putative central component in all the flowering promotion pathways (Jack, 2004). In *Arabidopsis*, *VRN1* and *VRN2* are involved in the repression maintenance of *FLC*, clearly distinguishing their role from the *T. monococcum* genes with the same name. Likewise, *ZCCT1*, the candidate for *Vrn2*, is similar to *FLC* in its role as a repressor, but belongs to a different family of transcription factors. The *Triticeae* repressor, *ZCCT1*, contains a putative zinc finger in the first exon and a *Constans* (*CO*), *CO*-like and *TOC1* (CCT) domain in the second exon, in common with the *Arabidopsis CO* and *CO*-like proteins (Yan *et al.*, 2004b). In *Arabidopsis, CO* is the central gene in the day length-controlled flowering pathway and the *CO* protein contains a CCT domain, which localizes it in the nucleus (Robson *et al.*, 2001; Valverde *et al.*, 2004). However, *ZCCT1* lacks the B-box domain characteristic to all the *CO* and *CO*-like genes (Griffiths *et al.*, 2003), which may be indicative of a different controlling mechanism.

Arabidopsis is, of course, an ideal model plant for genetic analysis, but the economic importance of *Triticeae* crops demands validation and extension from the model to the crop. The *Triticeae* includes wheat, barley, rye and related grasses, and forms a homogeneous genetic system in which results from one species are frequently applicable to other members of the cereal tribe (Dubcovsky *et al.*, 1998). Within the *Triticeae*, barley is an economically important crop and an ideal genetic model. Barley is a simple genetic system due to its self-pollinating and diploid nature, and it displays genetic variation for the components of winter hardiness. In addition, there is an ever-expanding set of tools for genetic and molecular analyses (reviewed in Hayes *et al.*, 2003), including multiple mapping populations, quantitative trait loci (QTL) studies, libraries of bacterial artificial chromosome (BAC) clones (Yu *et al.*, 2000), several large expressed sequence tag (EST) collections and a microarray chip (Close *et al.*, 2004).

Vernalization in the Context of Barley Growth Habit and Winter Hardiness

While we use the term vernalization *requirement* for standardization with other cereal nomenclature, winter barley varieties are better described as vernalization-*responsive* rather than -*requiring*, as they will eventually flower; however, without vernalization, the difference in days to flowering between winter and spring forms can be as great as 140 days (Karsai *et al.*, 2004). Furthermore, the terms 'winter' and 'spring' require definition, along with the term 'facultative'. As mentioned earlier, vernalization is of economic importance in the *Triticeae*, because it is a component of winter hardiness. The principal components of cereal winter hardiness are low-temperature tolerance, vernalization requirement and photoperiod (day length) sensitivity. Relative to these trait combinations, three growth habit classes, i.e. winter, facultative and spring, are present in the available barley germplasm. Winter varieties display all three traits, being low-temperature tolerant, photoperiod sensitive and vernalization requiring. Facultative varieties are strongly photoperiod sensitive and as tolerant to low temperature as winter varieties, but lack the vernalization requirement (Karsai *et al.*, 2001). In contrast, spring varieties have minimal low-temperature tolerance capacity, do not require vernalization, and are typically insensitive to short-day photoperiod. The three phenotypes are usually associated; for example, winter varieties that are low-temperature tolerant usually require vernalization and are photoperiod sensitive. However, Karsai *et al.* (2001), in an extensive analysis of a diverse array of barley germplasm, observed that some of the genotypes most tolerant to low temperature are facultative. The variety Dicktoo, which is a model for winter hardiness research in the barley community (Pan *et al.*, 1994; van Zee *et al.*, 1995; Hayes *et al.*, 1997; Choi *et al.*, 2000; Mahfoozi *et al.*, 2000; Fowler *et al.*, 2001; Karsai *et al.*, 2001), is facultative in growth habit. Dicktoo lacks a vernalization requirement, is highly photoperiod-sensitive and has a level of cold tolerance comparable to other winter barley genotypes. The intriguing question, of course, is: 'Are trait associations due to linkage and/or pleiotropy?' We are still answering this question, building on the base of vernalization genetics established over 30 years ago.

The Genetics of Vernalization in Barley: Historical Perspectives

The foundations for the genetics of vernalization requirement in barley were established by Takahashi and Yasuda (1971), who proposed a three-locus epistatic model in which genotypes that require vernalization (winter types) have the allelic architecture *Sh_sh2sh2sh3sh3*. All other allelic configurations lead to facultative or spring growth habits. Based on wheat:barley orthology, the *Sh* loci are now described using standard *Triticeae* nomenclature, with an 'H' indicating the *Hordeum* genome as follows: *Sh2* = *Vrn-H1* (chromosome 5H), *Sh* = *Vrn-H2* (chromosome 4H) and *Sh3* = *Vrn-H3* (chromosome 1H). Allelic variants at the *Vrn-H3* locus are reported only in barley from extremely high or low latitudes (Takahashi and Yasuda, 1971). Subsequently, the advent of abundant, phenotypically neutral, DNA markers facilitated linkage map construction and the development of QTL analysis tools. These advances provided powerful tools for the identification of genome regions associated with the components of winter hardiness.

In barley, the use of doubled haploid progeny has simplified the dissection of the genetics of vernalization by QTL analysis. For example, in the Igri × Triumph (winter × spring) mapping population, Laurie *et al.* (1995) reported vernalization QTLs on chromosomes 4H and 5H that correspond to the predicted locations of the *Vrn-H2* and *Vrn-H1* loci, and the same location for *Vrn-H1* was reported in the Nure × Tremois (winter × spring) mapping population (summarized in Table 5.1; Francia *et al.*, 2004). Multiple low temperature and photoperiod winter hardiness-related traits were mapped by QTL in the Dicktoo × Morex (facultative × spring) population to chromosome 5H at the predicted position of *Vrn-H1* (Karsai *et al.*, 1997), although this population does not segregate for vernalization requirement. Additional information on other winter hardiness-related phenotypes is available at the Barley QTL summary website http://www.barleyworld.org/northamericanbarley/qtlsummary.php.

The Genetics of Vernalization in Barley and the *Triticeae*: Contemporary Perspectives

A molecular model explaining the QTL effects and the *Vrn2/Vrn1* epistatic interaction in the *Triticeae* was proposed by Yan *et al.* (2004b) based on the positional cloning of *Vrn-A1* and *Vrn-A2* in *T. monococcum*. According to the model, *Vrn2* encodes a dominant repressor of flowering (*ZCCT1*) that inhibits the expression of the *Vrn1* flowering gene (*TmAP1*). Yan *et al.* (2003) suggested that the promoter region of *Vrn-A1* was a possible *Vrn-A2* target site, as they identified a deletion in that region that correlated with spring versus winter growth habit. Vernalization of winter accessions downregulates the expression of the repressor *Vrn2* (*ZCCT1*) that allows expression of *TmAP1*. In accessions with a physical deletion of the *Vrn2* locus (*ZCCT* genes), no vernalization requirement is observed regardless of the allele at *Vrn1*. This is also true in genotypes that have *Vrn2,* but lack a target-binding site for the repressor in the *TmAP1* gene (i.e. dominant *Vrn1*).

We have extended this *Triticum* spp. work to barley, using the available QTL data as a framework (von Zitzewitz *et al.*, 2005). Our goal is to characterize the structure and function of *Vrn-H1* and *Vrn-H2* in representative germplasm that has served as the basis for QTL detection. Accordingly, we used a reference germplasm array of ten cultivated barley accessions, shown in Table 5.1, and one wild progenitor accession (subsp. *spontaneum*). The subsp. *spontaneum* is the donor parent of a fully characterized set of recombinant chromosome substitution

Table 5.1. Summary of *Vrn-H1* and *Vrn-H2* allele types compared to vernalization-requirement QTL reported in the literature. The presence or deletion of *HvZCCT* (*Vrn-H2*) is indicated by '+' or '–'. *HvBM5A* (*Vrn-H1*) alleles are classified as dominant '*Vrn-H1*' or recessive '*vrn-H*' based on promoter and intron features. Significant QTL effects are shown as '+'; '–' indicates no QTL effects reported.

			Chromosome 4H		Chromosome 5H		
Mapping populations	Parents	Growth habit	*HvZCCT*	*QTL*	*HvBm5A*	*QTL*	Citation
Dicktoo ×	Dicktoo	Facultative	–		*vrn-H1*		Hayes *et al.* (1997)
Morex	Morex	Spring	–	–	*Vrn-H1*	–	
Dicktoo ×	Dicktoo	Facultative	–		*vrn-H1*		
Kompolti korai	Kompolti	Winter	+	+	*vrn-H1*	–	Karsai *et al.* (2005)
Strider ×	Strider	Winter	+		*vrn-H1*		Hayes *et al.* (unpublished)
88Ab536	88Ab536	Facultative	–	+	*vrn-H1*	–	
Nure ×	Nure	Winter	+		*vrn-H1*		Francia *et al.* (2004)
Tremois	Tremois	Spring	–	+	undetermined	+	
Igri ×	Igri	Winter	+		*vrn-H1*		Laurie *et al.* (1995)
Triumph	Triumph	Spring	–	+	*Vrn-H1*	+	

lines (RCSLs) (Matus *et al.*, 2003). This array includes representatives of the three barley growth habits: winter, facultative and spring. The cultivated accessions are all parents of mapping populations that segregate for vernalization.

HvBM5, the barley orthologue of *TmAP1*, was cloned during a MADS-box screen by Schmitz *et al.* (2000), and its orthology with *TmAP1* (*Vrn-A1*) and *TaVRT-1* (*Vrn1*) has been established (Danyluk *et al.*, 2003; Trevaskis *et al.*, 2003; Yan *et al.*, 2003). We have generated multiple lines of evidence, including sequence, linkage map location and expression data, that suggest that *HvBM5A* is *Vrn-H1* (von Zitzewitz, 2004; von Zitzewitz *et al.*, 2005). Our data provide very strong evidence that differences in vernalization requirement are due to gene regulation and not to differences in the encoded polypeptide, as the coding regions of all 11 genotypes we surveyed are identical (von Zitzewitz *et al.*, 2005). Integration of these results with those from *Triticum* spp. (Danyluk *et al.*, 2003; Yan *et al.*, 2003) reveals a high degree of *Vrn1* gene sequence conservation, and presumably function, in the *Triticeae*. Based on the lack of polypeptide variation and the report of Yan *et al.* (2003) that promoter mutations in a CArG-like motif correlated with differences in *T. monococcum* growth habit, we searched for promoter differences in the 11 representative barley genotypes. We found that the CArG-like motif of each accession was invariant. Yan *et al.* (2004a) also found dominant alleles of *Vrn1* that do not have CArG-like motif mutations, and these findings prompted a more comprehensive analysis of the promoter and intron 1 regions in the A, B, D and H genomes (Fu *et al.*, 2005). There was, however, no consistent promoter polymorphism separating winter from spring *Vrn-H1* alleles that appeared to account for the difference in vernalization requirement. Combined with the lack of coding region variation, this implied that the vernalization-critical regulatory region might be intron-based.

We found complete association of the deleted *Vrn-H2* candidate *ZCCT-H* genes with a loss of vernalization requirement in the important North American and European genotypes used to generate key winter hardiness mapping populations. Only winter growth habit cultivars contained the *Vrn-H2* locus, while it had been deleted from the genomes of both the spring and facultative growth habit cultivars. The agronomic growth habit term 'facultative' has lacked a rigorous definition and is loosely used to define genotypes that, while lacking a vernalization requirement, have retained sufficient low-temperature tolerance to allow autumn sowing and over-winter survival in environments lethal to spring growth habits. Interestingly, both the facultative genotypes examined have a 'winter' *HvBM5A* (recessive *vrn-H1*) allele. This suggests that the deleted *Vrn-H2* locus, the *ZCCT-H* gene set, is responsible for converting an otherwise winter genotype into a facultative type that lacks a vernalization requirement. The physical deletion of the *Vrn-H2* candidate gene results in the loss of the *trans*-acting repressor of the winter *Vrn-H1* allele, yielding the 'pseudo-spring' facultative phenotype, from which a vernalization requirement has essentially been eliminated. We propose that the term 'facultative' growth habit may be appropriate to describe genotypes that are low-temperature tolerant and have 'winter' (i.e. recessive) alleles at *Vrn-H1*, but lack the repressor encoded by *Vrn-H2* (i.e. recessive *vrn-H2*). Additional confirmation in a broader range of facultative genotypes is required to support this as a general explanation. As spring genotypes typically flower more quickly than facultative ones (Karsai *et al.*, 2004), additional regulatory inputs, relative to a spring *Vrn-H1* allele, are probably affecting the winter *vrn-H1* allele present in facultative genotypes, which may account in part for the maturation difference. This interpretation is supported by the full characterization of the candidate genes in the Dicktoo × Kompolti korai mapping population (Karsai *et al.*, 2005).

Our characterization of the *Vrn-H1* and *Vrn-H2* loci genotypes in mapping population parents explains QTL effects in all reported mapping populations (Table 5.1). The Igri × Triumph (Laurie *et al.*, 1995) and Nure × Tremois (Francia *et al.*, 2004) winter × spring crosses are variable for both *Vrn-H1* and *Vrn-H2* loci, so QTLs to each are expected and observed in these populations. No QTL was reported on 4H in association with *Vrn-H2* in the initial characterization of the Nure × Tremois population due to an incomplete genetic linkage map for this region (Francia *et al.*, 2004). We therefore scored additional loci providing expanded coverage of the long arm of chromosome 4H and re-examined the phenotypic data, identifying the *Vrn-H2* QTL coincident with the *ZCCT-H* locus (LOD = 5.4, R^2 = 0.2).

Both parents of a facultative × spring cross (Dicktoo × Morex) lack the *Vrn-H2* candidate *ZCCT-H* repressor gene, so neither a *Vrn-H1* nor *Vrn-H2* QTL will be detected (Hayes *et al.*, 1997), despite the presence of a repressible *Vrn-H1* allele in the facultative parent. In contrast, the Strider × 88Ab536 (P.M. Hayes *et al.*, unpublished results) and Dicktoo × Kompolti korai (Karsai *et al.*, 2005) winter × facultative crosses contrast for *Vrn-H2* alleles, where only the winter parent contains the *ZCCT-H* genes, while both parents harbour winter *Vrn-H1* alleles, and only the *Vrn-H2* QTL is detected. This concept is further illustrated in Fig. 5.1, which shows relevant portions of the chromosome 4H and 5H linkage maps for the Dicktoo × Kompolti korai and Dicktoo × Morex populations together with the corresponding vernalization QTL. The maps are, as expected, collinear, although the *ZCCT-H* gene family is deleted in both Dicktoo and Morex, necessitating the use of the physically linked *HvSnf2* locus. As predicted, a QTL effect is observed only in the Dicktoo × Komploti korai population (LOD = 13.9, R^2 = 0.52), and it is coincident with the predicted site of *Vrn-H2*. The finding that *Vrn-H1* and *Vrn-H2* allele data are sufficient to predict and explain all reported vernalization requirement QTL data from domesticated barley crosses is additional evidence that *Vrn-H3* is a determinant of vernalization response only in a limited range of exotic barley germplasm, as reported by Takahashi and Yasuda (1971).

Conclusions and Future Prospects

In summary, significant progress has been made in the 34 years that have elapsed since Takahashi and Yasuda (1971) first proposed the three-locus epistatic model for vernalization response in barley. With QTL analysis tools, the *Vrn* loci were assigned genome coordinates and were mapped in conjunction with other winter hardiness-related phenotypes. More recently, we have characterized these loci at the structural and functional levels (von Zitzewitz, 2004; von Zitzewitz *et al.*, 2005) and are initiating experiments to fully define the mechanism of the epistatic

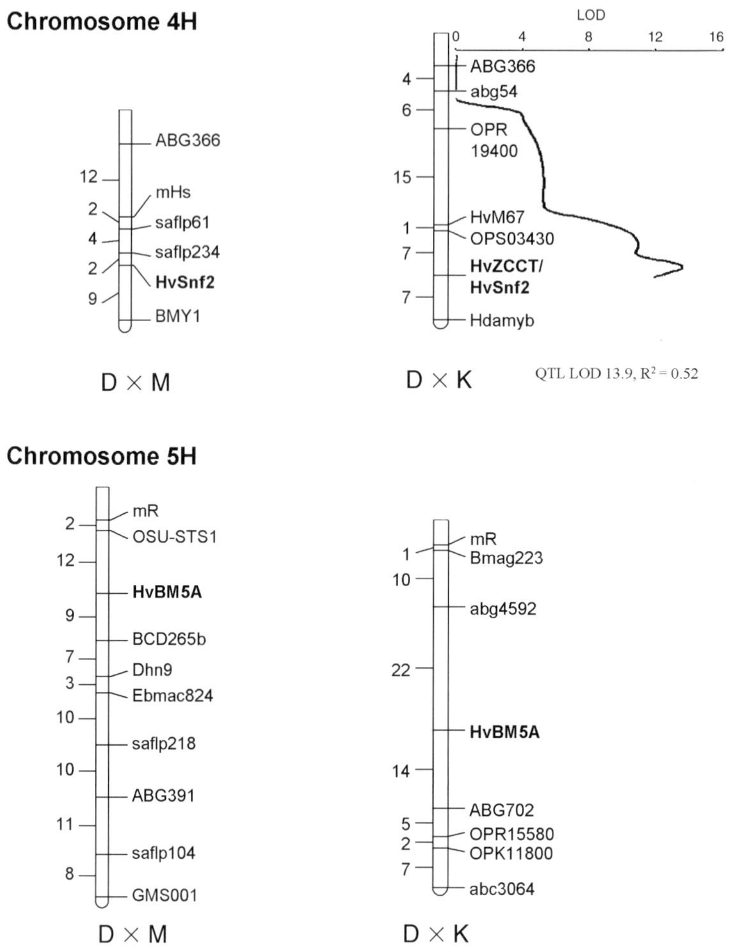

Fig. 5.1. Dicktoo × Morex and Dicktoo × Kompolti korai chromosomes 4H and 5H linkage maps (relevant portions only), showing the 4H QTL location and LOD score in the latter. Linkage maps were generated using JoinMap and QTL analysis was performed using WinQTL Cartographer V.2.0 by Composite Interval Mapping (LOD threshold set by permutation).

interaction. The barley and *Arabidopsis* vernalization pathways appear to be an excellent example of convergent evolution and are certainly evidence that model system genetics is a complement, not a substitute, for crop genetics. Experiments are under way in our lab, and elsewhere, to fully characterize the vernalization pathway in barley and to determine where and how it intersects with the photoperiod sensitivity and low-temperature tolerance pathways. The benefits will be of both scientific and agronomic importance.

Acknowledgements

This work was supported by NSF Functional Genomics grant 0110124 and the United States Barley Genome Project.

References

Amasino, R. (2004) Vernalization, competence, and the epigenetic memory of winter. *Plant Cell* 16, 2553–2559.

Bastow, R., Mylne, J.S., Lister, C., Lippman, Z., Martienssen, R.A. and Dean, C. (2004) Vernalization requires epigenetic silencing of *FLC* by histone methylation. *Nature* 427, 164–167.

Choi, D.-W., Koag, M.C. and Close, T.J. (2000) Map locations of barley *Dhn* genes determined by gene-specific PCR. *Theoretical and Applied Genetics* 101, 350–354.

Close, T., Wanamaker, S., Caldo, R., Turner, S., DA, A., Dickerson, J., Wing, R., Muehlbauer, G., Kleinhofs, A. and Wise, R. (2004) A new resource for cereal genomics: 22K barley GeneChip comes of age. *Plant Physiology* 134, 960–968.

Danyluk, J., Kane, N.A., Breton, G., Limin, A.E., Fowler, D.B. and Sarhan, F. (2003) *TaVRT-1*, a putative transcription factor associated with vegetative reproductive transition in cereals. *Plant Physiology* 132, 1–12.

Dubcovsky, J., Lijavetzky, D., Appendino, L. and Tranquilli, G. (1998) Comparative RFLP mapping of *Triticum monococcum* genes controlling vernalization requirement. *Theoretical and Applied Genetics* 97, 968–975.

Fowler, D.B., Breton, G., Limin, A.E., Mahfoozi, S. and Sarhan, F. (2001) Photoperiod and temperature interactions regulate low-temperature-induced gene expression in barley. *Plant Physiology* 127, 1676–1681.

Francia, E., Rizza, F., Cattivelli, L., Stanca, A.M., Galiba, G., Tóth, B., Hayes, P.M., Skinner, J.S. and Pecchioni, N. (2004) Two loci on chromosome 5H determine low-temperature tolerance in a Nure (winter) × Tremois (spring) barley map. *Theoretical and Applied Genetics* 108, 670–680.

Fu, D., Szűcs, P., Yan, L., Helguera, M., Skinner, J.S., von Zitzewitz, J., Hayes, P.M. and Dubcovsky, J. (2005) Large deletions within the *VRN-1* first intron are associated with spring growth habit in barley and wheat. *Molecular Genetics and Genomics* 273, 54–65.

Gendall, A.R., Levy, Y.Y., Wilson, A. and Dean, C. (2001) The *Vernalization 2* gene mediates the epigenetic regulation of vernalization in *Arabidopsis*. *Cell* 107, 525–535.

Griffiths, S., Dunford, R.P., Coupland, G. and Laurie, D.A. (2003) The evolution of CONSTANS-like gene families in barley, rice, and *Arabidopsis*. *Plant Physiology* 131, 1855–1867.

Hayes, P.M., Chen, F.Q., Corey, A., Pan, A., Chen, T.H.H., Baird, E., Powell, W., Thomas, W., Waugh, R., Bedo, Z., Karsai, I., Blake, T. and Oberthur, L. (1997) The Dicktoo × Morex population: a model for dissecting components of winter hardiness in barley. In: Li, P.H. and Chen, T.H.H. (eds) *Plant Cold Hardiness: Molecular Biology, Biochemistry and Physiology*. Plenum Press, New York, pp. 77–87.

Hayes, P.M., Castro, A., Marquez-Cedillo, L., Corey, A., Henson, C., Jones, B.L., Kling, J., Mather, D., Matus, I., Rossi, C. and Sato, K. (2003) Genetic diversity for quantitatively inherited agronomic and malting quality traits. In: von Bothmer, R., Knüpffer, H., van Hintum, T. and Sato, K. (eds) *Diversity in Barley (Hordeum vulgare)*. Elsevier Science Publishers, Amsterdam, pp. 201–226.

He, Y., Michaels, S. and Amasino, R. (2003) Regulation of flowering time by histone acetylation in *Arabidopsis*. *Science* 302, 1751–1754.

Henderson, I.R. and Dean, C. (2004) Control of *Arabidopsis* flowering: the chill before the bloom. *Development* 131, 3829–3838.

Henderson, I.R., Shindo, C. and Dean, C. (2003) The need for winter in the switch to flowering. *Annual Review of Genetics* 37, 371–392.

Hepworth, S., Valverde, F., Ravenscroft, D., Mouradov, A. and Coupland, G. (2002) Antagonistic regulation of flowering-time gene *SOC1* by *Constans* and *FLC* via separate promoter motifs. *EMBO Journal* 21, 4327–4337.

Jack, T. (2004) Molecular and genetic mechanisms of floral control. *The Plant Cell* 16, S1–17.

Karsai, I., Mészáros, K., Hayes, P. and Bedo, Z. (1997) Effects of loci on chromosome 2 (2H) and 7 (5H) on developmental patterns in barley (*Hordeum vulgare* L.) under different photoperiod regimes. *Theoretical and Applied Genetics* 94, 612–618.

Karsai, I., Mészáros, K., Lang, L., Hayes, P.M. and Bedo, Z. (2001) Multivariate analysis of traits determining adaptation in cultivated barley. *Plant Breeding* 120, 217–222.

Karsai, I., Hayes, P.M., Kling, J., Matus, I.A., Mészáros, K., Lang, L., Bedo, Z. and Sato, K. (2004) Genetic variation in component traits of heading date in *Hordeum vulgare* subsp. *spontaneum* accessions characterized in controlled environments. *Crop Science* 44, 1622–1632.

Karsai, I., Szücs, P., Mészáros, K., Filichkin, T., Hayes, P.M., Skinner, J.S., Lang, L. and Bedo, Z. (2005) The *Vrn-H2* locus is a major determinant of flowering time in a facultative × winter growth habit barley (*Hordeum vulgare* L.) mapping population. *Theoretical and Applied Genetics* 110, 1458–1466.

Laurie, D., Pratchett, N., Bezant, J. and Snape, J. (1995) RFLP mapping of five major genes and eight quantitative trait loci controlling flowering time in a winter × spring barley (*Hordeum vulgare* L.) cross. *Genome* 38, 575–585.

Levy, Y.Y., Mesnage, S., Mylne, J.S., Gendall, A.R. and Dean, C. (2002) Multiple roles of *Arabidopsis VRN1* in vernalization and flowering time control. *Science* 279, 243–246.

Mahfoozi, S., Limin, A.E., Hayes, P.M., Hucl, P. and Fowler, D.B. (2000) Influence of photoperiod response on the expression of cold hardiness in wheat and barley. *Canadian Journal of Plant Science* 80, 721–724.

Matus, I., Corey, A., Filichkin, T., Hayes, P.M., Vales, M.I., Kling, J., Riera-Lizarazu, O., Sato, K., Powell, W. and Waugh, R. (2003) Development and characterization of recombinant chromosome substitution lines (RCSLs) using *Hordeum vulgare* subsp. *spontaneum* as a source of donor alleles in a *Hordeum vulgare* subsp. *vulgare* background. *Genome* 46, 1010–1023.

Michaels, S.D. and Amasino, R.M. (2000) Memories of winter: vernalization and the competence to flower. *Plant, Cell and Environment* 23, 1145–1153.

Murai, K., Miyamae, M., Kato, H., Takumi, S. and Ogihara, Y. (2003) *WAP1*, a wheat *Apetala1* homolog, plays a central role in the phase transition from vegetative to reproductive growth. *Plant and Cell Physiology* 44, 1255–1265.

Napp-Zinn, K. (1987) Vernalization: environmental and genetic regulation. In: Atherton, J.G. (ed.) *Manipulation of Flowering*. Butterworths, London, pp. 123–132.

Pan, A., Hayes, P.M., Chen, T.H.H., Blake, T.K., Wright, S., Karsai, I. and Bedö, Z. (1994) Genetic analysis of the components of winter hardiness in barley (*Hordeum vulgare* L.). *Theoretical and Applied Genetics* 89, 900–910.

Robson, F., Costa, M., Hepworth, S., Vizir, I., Pineiro, M., Reeves, P., Putterill, J. and Coupland, G. (2001) Functional importance of conserved domains in the flowering-time gene *Constans* demonstrated by analysis of mutant alleles and transgenic plants. *Plant Journal* 28, 619–632.

Schmitz, J., Franzen, R., Ngyuen, T., Garcia-Maroto, F., Pozzi, C., Salamini, F. and Rohde, W. (2000) Cloning, mapping and expression analysis of barley MADS-box genes. *Plant Molecular Biology* 42, 899–913.

Sung, S. and Amasino, R.M. (2004) Vernalization in *Arabidopsis thaliana* is mediated by the PHD finger protein *VIN3*. *Nature* 427, 159–164.

Takahashi, R. and Yasuda, S. (1971) Genetics of earliness and growth habit in barley. In: Nilan, R.A. (ed.) *Barley Genetics III; Proceedings of Second International Barley Genetics Symposium.* Washington State University Press, Pullman, Washington, pp. 388–408.

Trevaskis, B., Bagnall, D.J., Ellis, M.H., Peacock, W.J. and Dennis, E.S. (2003) MADS box genes control vernalization-induced flowering in cereals. *Proceedings of the National Academy of Sciences USA* 100, 13099–13104.

Valverde, F., Mouradov, A., Soppe, W., Ravenscroft, D., Samach, A. and Coupland, G. (2004) Photoreceptor regulation of *Constans* protein in photoperiodic flowering. *Science* 303, 1003–1006.

van Zee, K., Chen, F.Q., Hayes, P.M., Close, T.J. and Chen, T. (1995) Cold-specific induction of a dehydrin gene family member in barley. *Plant Physiology* 108, 1233–1239.

von Zitzewitz, J. (2004) Structural and functional characterization of barley vernalization genes. MS thesis, Oregon State University, Corvallis, Oregon.

von Zitzewitz, J., Szűcs, P., Dubcovsky, J., Yan, L., Francia, E., Pecchioni, N., Casas, A., Chen, T.H.H., Hayes, P.M. and Skinner, J.S. (2005) Molecular and structural characterization of barley vernalization genes. *Plant Molecular Biology* (in press).

Yan, L., Loukoianov, A., Tranquilli, G., Helguera, M., Fahima, T. and Dubcovsky, J. (2003) Positional cloning of the wheat vernalization gene *VRN1*. *Proceedings of the National Academy of Science USA* 100, 6263–6268.

Yan, L., Helguera, M., Kato, K., Fukuyama, S., Sherman, J. and Dubcovsky, J. (2004a) Allelic variation at the *VRN-1* promoter region in polyploid wheat. *Theoretical Applied Genetics* 109, 1677–1686.

Yan, L., Loukoianov, A., Blechl, A., Tranquilli, G., Ramakrishna, W., SanMiguel, P., Bennetzen, J.L., Echenique, V. and Dubcovsky, J. (2004b) The wheat *VRN2* gene is a flowering repressor down-regulated by vernalization. *Science* 303, 1640–1644.

Yu, Y., Tomkins, J.P., Waugh, R., Frisch, D.A., Kudrna, D., Kleinhofs, A., Brueggeman, R.S., Muehlbauer, G.J., Wise, R.P. and Wing, R. (2000) A bacterial artificial chromosome library for barley (*Hordeum vulgare* L.) and the identification of clones containing putative resistance genes. *Theoretical and Applied Genetics* 101, 1093–1099.

6 Vernalization Genes in Winter Cereals

N.A. Kane, J. Danyluk and F. Sarhan*

Département des Sciences Biologiques, Université du Québec à Montréal, Case Postale 8888, Succursale Centre-ville, Montréal, Québec, Canada H3C 3P8

Introduction

Plants native to regions with a winter season have evolved the ability to time the transition from vegetative growth to flowering. This evolutionary adaptation allows plants to flower under favourable environmental conditions for sexual reproduction. The initiation and timing of this transition is determined by seasonal changes of the two main environmental cues, temperature and photoperiod. The term vernalization describes the period of low-temperature (LT) exposure that is necessary for these plants to acquire the capacity or to accelerate their ability to enter the reproductive phase (Chouard, 1960). Once vernalization or flowering competency is achieved, plants will initiate reproductive development only under appropriate photoperiod conditions in the spring.

During the period of growth at LT many physiological, biochemical and metabolic functions are altered in the plant (reviewed in Guy, 1999; Thomashow, 1999; Breton *et al.*, 2000). This cold acclimation (CA) period allows plants to develop efficient freezing tolerance (FT) mechanisms required to survive the winter months. One of these mechanisms is the accumulation of cold-regulated/late embryogenesis abundant (COR/LEA) proteins that are believed to promote the development of FT by protecting cellular components (Thomashow, 1999). In cereals, the development of maximum FT is known to be associated with COR/LEA protein accumulation and the vernalization saturation point (Fowler *et al.*, 1996a,b; Danyluk *et al.*, 2003). The transition from vegetative to reproductive growth phase can be perceived as a critical switch that initiates the down-regulation of LT-induced genes (Fowler *et al.*, 1996a,b, 2001; Mahfoozi *et al.*, 2001a,b; Danyluk *et al.*, 2003). As a result, full expression of cold-hardiness genes

*Corresponding author (e-mail: sarhan.fathey@uqam.ca; Fax: +1-514-987-4647)

 Cold Hardiness in Plants: Molecular Genetics, Cell Biology and Physiology (eds T.H.H. Chen *et al.*)

only occurs in the vegetative phase and plants in the reproductive phase have a limited ability to cold acclimate. These studies indicate that there is a close genetic linkage between vernalization and CA; and suggest that developmental genes also act to control genes affecting the expression of LT-induced genes associated with the acquisition of FT. Understanding the genetics mechanisms controlling the duration of vernalization and the timing of the switch to the reproductive phase will help to identify additional candidate genes for breeding new varieties with delayed transition to the reproductive stage. This will in turn increase the duration of stress resistance.

Plant species display a wide range of responses to vernalization. Some, like the biennial cabbage (*Brassica oleracea*), have an obligate vernalization requirement while others, like winter wheat, display a quantitative vernalization requirement. Studies on the growth habit of wheat cultivars have shown that spring genotypes have a very mild response or no response at all to vernalization compared to winter genotypes that continue reducing their final leaf number up to the point of vernalization saturation (49 days of LT exposure) (Fowler *et al.*, 1996a,b). Vernalization is a trait that evolved independently in flowering plants as they radiated into regions with a winter season (Amasino, 2004). Recent work in plants is revealing that differences in vernalization requirement do not necessarily imply fundamental modifications in the mechanisms that control flowering. Before reviewing the advances in our understanding of vernalization in wheat, we will briefly summarize the knowledge gained by studying flowering control in *Arabidopsis*.

The Model System *Arabidopsis*

The genetic and molecular analyses of *Arabidopsis* have revealed the existence of several interdependent pathways controlling flowering (reviewed in Mouradov *et al.*, 2002; Simpson and Dean, 2002; Henderson *et al.*, 2003). The photoperiod and vernalization pathways respond to environmental signals, while the autonomous and gibberellin (GA)-dependant pathways integrate the endogenous developmental state of the plant. Both the autonomous and vernalization pathways promote flowering by repressing the *FLOWERING LOCUS C* (*FLC*) gene that acts as a repressor of flowering. Subsequently, the photoperiod and GA pathways in combination with *FLC* levels (from autonomous/vernalization pathway integration) converge on the common flowering pathway integrators *FT*, *SUPPRESOR OF OVEREXPRESSION OF CO 1* (*SOC1/AGL20*) and *LEAFY* (*LFY*) that regulate floral initiation genes such as *APETALA1* (*AP1*) (Simpson and Dean, 2002).

Ecotypes of *Arabidopsis* display a natural variation in vernalization requirement (Burn *et al.*, 1993; Clarke and Dean, 1994; Koornneef *et al.*, 1994). The two main genes that were shown to display allelic variation in *Arabidopsis* ecotypes are *FRIGIDA* (*FRI*) and *FLC* (Micheals and Amasino, 1999; Johanson *et al.*, 2000). Winter annual ecotypes contain dominant alleles of *FRI* and *FLC* that delay flowering while spring ecotypes carry null and weak alleles of *FRI* and *FLC*, respectively (Michaels *et al.*, 2003). Molecular cloning of *FRI*, which encodes a plant-specific gene of unknown biochemical activity, revealed that *FRI* strongly enhanced *FLC* transcript levels (Johanson *et al.*, 2000). The *FLC* gene encodes a transcription factor of the MADS-box family whose expression alone is sufficient to repress flowering (Michaels and Amasino, 1999; Sheldon *et al.*, 1999). The effect of vernalization is to promote flowering through an epigenetic downregulation of *FLC* mRNA levels, thereby antagonizing *FRI* function (Michaels and Amasino, 1999; Sheldon *et al.*, 1999). The extent of this reduction is proportional to the duration of vernalization and is closely correlated with flowering time (Gendall *et al.*, 2001).

With the advancement of expressed sequence tags (EST) and genomic sequencing projects in plants, it became apparent that *FLC*-like genes were specifically found in the *Brassicaceae*. This indicated that other vernalization-responsive plants would at least contain some differences in the signalling pathways controlling the transition to the reproductive stage. The use of mutagenic approaches in *Arabidopsis* has led to the identification of many genes involved in the flowering pathway in plants. Information from such studies was invaluable in choosing a more targeted approach for identifying genes associated with the floral switch in wheat, a species less amenable to transformation studies.

Genetics of Vernalization in Cereals

Wheat is one of the leading food crops grown worldwide and is of great economic and nutritional importance. In an effort to maximize yield potential of wheat cultivars in diverse environments, extensive physiological and genetic studies have been undertaken since the beginning of the 20th century with the goal of uncovering the allelic variation that fine-tunes the life cycle for an appropriate flowering-time control (Chouard, 1960; Lang, 1965; Law and Wolfe, 1966; Pugsley, 1972; Klaimi and Qualset, 1974; Maistrenko, 1980; Stelmakh, 1993; Bezant *et al.*, 1996; Law and Worland, 1997; Dubcovsky *et al.*, 1998; Sourdille *et al.*, 2000). Chromosome substitution line analysis in wheat has shown that the genetic control of flowering time is complex, with nearly all homoeologous groups involved (Worland *et al.*, 1987; Law *et al.*, 1993; Laurie *et al.*, 1995; Bullrich *et al.*, 2002). These studies have revealed that distinct sets of genes regulate the adaptive mechanisms allowing cereals to properly time flowering: (i) the vernalization-LT responsive genes (*Vrn*) that regulate flowering; (ii) the photoperiod-responsive genes (*Ppd*) that regulate flowering using day length; and (iii) the earliness *per se* genes that influence the rate of development. Vernalization and photoperiod genes play important regulatory roles in the growth and development of winter cereals. In contrast to vernalization requirement and photoperiod sensitivity, earliness *per se* genes is environment-independent. In addition, genes from all three systems have shown pleiotropic effects on other aspects of plant growth and development. The variety of genes involved in these traits in plants builds in an added plasticity that ensures that the energy-demanding processes of flowering and setting seed is well timed to ensure the survival of the species.

During the last two decades, much of the efforts have been directed to precise QTL mapping and identification of the corresponding genes. The most studied alleles conferring the vernalization requirement are the *Vrn* loci found on chromosome 5 of wheat (Law *et al.*, 1993; Law and Worland 1997). In hexaploid wheat (*Triticum aestivum* L. 2n × 6 = 42), *Vrn-1* loci occur as a set of homoeologous genes named *Vrn-A1*, *Vrn-B1* and *Vrn-D1* that map to identical regions of the group 5 chromosomes (Galiba *et al.*, 1995; Nelson *et al.*, 1995; Dubcovsky *et al.*, 1998; Iwaki *et al.*, 2001; Danyluk *et al.*, 2003; McIntosh *et al.*, 2003; Yan *et al.*, 2003). Orthologous genes have been mapped in diploid wheat (*T. monococcum*, *Vrn-A^{m}1*), barley (*Hordeum vulgare*, *Vrn-H1*) and rye (*Secale cereale Vrn-R1*) (Plaschke *et al.*, 1993; Galiba *et al.*, 1995; Laurie *et al.*, 1995; Dubcovsky *et al.*, 1998). Alleles at the *Vrn-A1* locus appear to have a predominant effect in reducing the vernalization requirement compared to those on other *Vrn* loci (Snape *et al.*, 1985; Shindo *et al.*, 2003). For example, cultivars with *Vrn-A1* do not require vernalization at all, whereas those with *Vrn-B1* and *Vrn-D1* require vernalization for 15–30 days, and cultivars recessive for all of these genes require 45–60 days of vernalization (Shindo *et al.*, 2003). The *Vrn-1* loci have also been the ones most associated with FT. We recently reported an isogenic line containing recessive copies of all *vrn-1* loci in a spring Manitou background showed a higher accumulation of *cor/lea* transcripts and development of FT than the parental spring line (Limin and Fowler, 2002; Danyluk *et al.*, 2003), suggesting a genetic link between vernalization requirement and a higher capacity for developing freezing tolerance. In addition, these studies revealed that the *vrn-A1* locus does not act alone in determining the transition from the vegetative to the reproductive phase, confirming the importance of continuing the identification of other loci controlling the duration of the vegetative phase (Snape *et al.*, 2001).

A locus showing an epistatic interaction with *Vrn-A^{m}1* in diploid wheat was mapped to the distal region of chromosome *5A^{m}L* (Dubcovsky *et al.*, 1998; Tranquilli and Dubcovsky, 2000). The dominant winter allele, designated *Vrn-A^{m}2*, was shown to delay flowering only in the presence of *vrn-A^{m}1*. Allelic variation at the *Vrn-2* loci has been observed in barley but not in hexaploid wheat. In the latter, *Vrn-2* would only be able to determine a spring habit, if the three *Vrn-2* loci are simultaneously homozygous for the recessive spring alleles, which certainly can be predicted to occur at a lower probability.

Other loci affecting the transition to flowering have been reported in the literature but they remain less characterized. Genes sensitive to vernalization in wheat have been identified on chromosome 3B (Miura and Worland, 1994), and on group 1 and 6 chromosomes (Islam-Faridi *et al.*, 1996; Law *et al.*, 1998). Furthermore, markers associated with the QTL for ear emergence on chromosome 7HS from barley have been located to the physical consensus map of wheat (Boyko *et al.*, 2002) along with a possible vernalization locus. A QTL for heading time was located on chromosome 7BS in a doubled haploid population of wheat (Sourdille *et al.*, 2000), a region that may correspond to a QTL for earliness *per se*. Restriction fragment length polymorphism (RFLP) markers with significant linkage to heading date and photoperiod sensitivity have also been located to all group-7 chromosomes of wheat (Shindo *et al.*, 2003). In barley, RFLP markers with highly significant additive effects for ear emergence time were found on both the long and short arms of chromosome 7H in a spring habit cross (Bezant *et al.*, 1996). Baum *et al.* (2003) also reported finding a major QTL for days to heading on chromosome 7H. The extensive allelic variation that has been reported in cereal species is probably at the basis of their adaptive capabilities.

Identification of Vernalization Genes in Wheat

The identification and characterization of genes underlying adaptive traits in cereals is a prerequisite for understanding how they may be used in manipulating growth habit and/or vernalization response to achieve a better timing of flowering in wheat and other temperate cereals. Natural variation, mutation analyses and molecular characterization have identified the MADS-box family members as important floral regulators in *Arabidopsis* and other plants. To evaluate the possibility that this group is involved in the floral pathways of cereal plants, our group initially concentrated on determining if any were associated with the vegetative to reproductive transition in wheat. The following is a brief review of the recent developments on the characterization of these and related genes.

TaVRT-1/VRN-1

TaVRT-1 encodes a MADS-box protein and belongs to the AP1/SQUA-like clade of transcriptional regulators (Danyluk *et al.*, 2003), whose members have been implicated in functions such as determining inflorescence meristem identity and flower development. Mapping studies localized this gene to the *Vrn-1* regions of homoeologous group 5 chromosomes, regions that are associated with vernalization and FT in wheat. The level of expression of *TaVRT-1* is positively associated with the vernalization response and transition from vegetative to reproductive phase. Comparisons among different wheat genotypes, near-isogenic lines and cereal species that differ in their vernalization response, indicated that the gene is inducible only in those species that require vernalization, whereas it is constitutively expressed in spring habit genotypes. These expression studies suggested that *TaVRT-1* is a key developmental gene in the regulatory pathway that controls the transition from the vegetative to reproductive phase in cereals. The genetic evidence that *TaVRT-1* is a *VRN-1* gene from hexaploid wheat was provided by the positional cloning of the *Vrn-A^m1* gene from *T. monococcum* (Yan *et al.*, 2003). Allelic variation was found between the spring and winter accessions of tested diploids. Analysis of promoter sequences revealed that spring accessions contained a deletion encompassing a putative MADS-box protein-binding site (CArG-box) near the transcription start site. Subsequently, allelic variation was investigated in the promoter and first intron of *VRN-1* in polyploid wheat (Yan *et al.*, 2004b; Fu *et al.*, 2005). These studies have indicated that the regulation of *VRN-1* expression is complex and may involve multiple regulatory elements. Most of the polyploid wheat varieties containing a dominant spring *Vrn-A1* allele were characterized by an insertion or deletion in the vicinity of the CArG-box in the promoter, suggesting that these could interfere with the regulation mediated by this motif. The remaining *Vrn-A1* and spring *Vrn-B1* and *Vrn-D1* alleles did not show any polymorphism in the promoter region when compared with their corresponding recessive alleles. However, analysis of these spring alleles revealed important deletions in the first intron. These alleles were all missing a 440 bp region showing a high conservation between wheat and barley. This suggests that the DNA segment contains putative regulatory elements important for the vernalization requirement (Yan *et al.*, 2004b). Further experiments will be necessary to identify the putative vernalization-responsive regulatory regions in these alleles.

These molecular studies have revealed that the spring/winter growth habit of temperate cereals is being controlled mainly by genetic variability of the *VRN-1* locus that encodes an orthologue of *Arabidopsis AP1*. Thus, these relatives appear to function in promoting flowering in both species. However, the genetic variability in vernalization requirement in these species is being mediated by different genes. In addition, because the winter allele in wheat is believed to be the ancestral type, these studies revealed that spring alleles have undergone independent mutational changes affecting their responsiveness to vernalization (Yan *et al.*, 2004b; Beales *et al.*, 2005; Fu *et al.*, 2005). This created different variants that may, alone or in different combinations, provide an adaptive advantage in certain conditions.

VRN-2

The dominant repressor of flowering, *Vrn-A^m2*, was recently identified by positional cloning in *T. monococcum* (Yan *et al.*, 2004a). The gene encodes a substantially diverged zinc finger protein that is specifically found in temperate cereals, indicating that it evolved fairly recently. The *Vrn-A^m2* gene is down-regulated by vernalization and shows an opposite expression profile to *Vrn-A^m1*. The *vrn-A^m2* spring alleles were found to contain either deletions of the gene or a mutated version that is hypothesized to hinder its capacity for protein–protein interaction and function (Yan *et al.*, 2004a). Therefore, loss of function through either the presence of spring alleles or the reduction of transcript levels by RNA interference resulted in the promotion of flowering in wheat (Yan *et al.*, 2004a). With the cloning of this gene, allelic variation at the *VRN-1* and *VRN-2* loci are able to explain 46 of the 49 spring habits tested in *T. monococcum*. This suggests that other less-known loci conferring spring habit will be found in the coming years. The identification of these central genes in the vernalization pathway of wheat provides an opportunity to begin deciphering the molecular basis of the floral inductive pathway in cereals. These studies will eventually reveal how differences in controlling vernalization requirement in plants impact other floral signalling pathways.

TaVRT-2

In preliminary expression profiling of several wheat MADS-box genes, one showed an inverse pattern of expression relative to that of *TaVRT-1*, in that the level of expression was high during the early stage of vernalization but started to decline towards the vegetative/reproductive transition point (Kane *et al.*, 2005). This gene was named *TaVRT-2* to underline this close association with *TaVRT-1*. Molecular and phylogenetic analyses indicate that *TaVRT-2* encodes a member of the *St*MADS11-like clade of genes responsible for flowering repression in several species. Expression profiling of this gene using near-isogenic lines and different

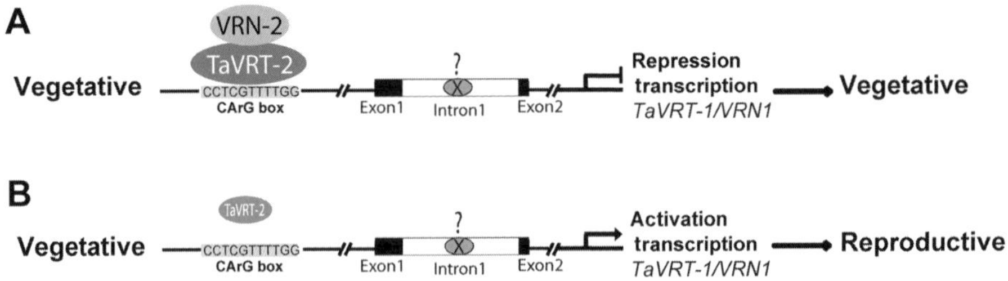

Fig. 6.1. Proposed model for the activation of *TaVRT-1/VRN-1* in winter wheat. (A) During vernalization TaVRT-2, a MADS-box protein, binds to the CArG box in the *TaVRT-1/VRN-1* promoter and interacts with VRN-2, a zinc finger protein. In addition, VRN-2 or another factor (?) binds conserved sequences in the first intron. This complex at the *TaVRT-1/VRN-1* gene leads to repression of expression and maintains plants in the vegetative phase. (B) After vernalization, VRN-2 and TaVRT-2 levels are repressed and none of these factors is bound to the promoter. This leads to expression of *TaVRT-1/VRN-1* and transition to the reproductive stage.

genotypes with natural variation in their response to vernalization and photoperiod showed a strong relationship with floral transition and photoperiod. In *Arabidopsis*, the *St*MADS11-like member, short vegetative phase (SVP), was shown to interact with the photoperiod pathway (Hartmann *et al.*, 2000).

Protein–protein interactions studies revealed that TaVRT-2 can form homodimers and interact with proteins encoded by the vernalization QTLs (TaVRT-1/VRN-1 and VRN-2) of wheat. The *St*MADS11-like genes have been found to interact with proteins of the AP1/SQUA clade in several species (Immink *et al.*, 2003; Fornara *et al.*, 2004; Masiero *et al.*, 2004). Therefore, the interaction between members of these two clades may represent an evolutionary conserved property that is important for their function. In addition, a shift between homo- and heterodimer formation was proposed to mediate a change in function for a member of this clade (Masiero *et al.*, 2004). Overall these results support the notion that TaVRT-2 is part of the flowering pathway in wheat, integrating developmental and environmental signals resulting in the vegetative/reproductive transition. In the proposed model (Fig. 6.1), TaVRT-2 binds to the CArG box present in winter *TaVRT-1/VRN-1* promoters, while VRN-2 binds unidentified elements in the first intron and/or interacts with TaVRT-2. Within this context, the expression of the *TaVRT-1/VRN-1* gene is repressed and the result is vegetative growth. After vernalization, the expression of *VRN-2* is repressed and levels of *TaVRT-2* diminished. This leads to the expression of *TaVRT-1/VRN-1* and the switch to the reproductive phase. Increased quantities of TaVRT-1/VRN-1 would then favour heterodimer formation with TaVRT-2, and a shift in function. The physical association of TaVRT-1, TaVRT-2 and VRN-2 may then be at the basis of the regulation of flowering transition in temperate cereals.

Future Prospects

An appropriate flowering time of cereals exposed to a wide range of environments is achieved by complex interactions of quantitative traits controlled by many genes. To date, only a limited number of traits have been identified at the molecular level from those reported to contain some genetic variability in temperate cereals. It is essential to increase our knowledge of these traits in the coming years through: (i) their detailed genetic mapping and positional cloning; and (ii) prioritize the characterization of cereal orthologous genes associated with floral pathways based on their similarities to other species. This will increase our knowledge of the full complement of variations that plants may have evolved for fine-tuning the life cycle to a target environment. Breeding of cereal varieties better adapted to specific environments would maximize yield potential and minimize risks from stress exposure.

An interesting question that can be asked is: how is vernalization genetically linked to the development of FT? Vernalization requirement enables winter wheat cultivars to maintain the expression of *cor/lea* genes at higher levels and for a longer period than in spring cultivars (Fowler *et al.*, 1996a,b; Danyluk *et al.*, 2003). The delay in the transition from the vegetative to the reproductive phase produces increased FT that is sustained for a longer period of time. This observation also explains why a high level of FT has not been observed in spring habit cultivars. Because LT gene expression is only upregulated when the plant is in the vegetative phase, the genetic potentials of spring habit cultivars are not given an opportunity to be fully expressed, leaving the impression that the spring habit *Vrn-A1* allele has a dominant pleiotropic effect for frost susceptibility (Fowler *et al.*, 1999). Although the two developmental pathways, vernalization and CA, respond to the same environmental cue, no molecular information is yet available to indicate how they are linked. Evidence from mutant analyses in the model system *Arabidopsis* indicates that some genes may be playing a role in both pathways. The first gene, *HOS1*, functions as a negative regulator of flowering and LT-responsive gene transcription (Lee *et al.*, 2001). It encodes a novel plant protein with a RING finger motif that was never previously associated with these developmental pathways. The second is the autonomous pathway gene *FVE* (Kim *et al.*, 2004) that negatively regulates the CBF/DREB pathway and positively flowering. Whether these genes play a similar role in temperate cereals will depend on how and when these functions evolved. Understanding this link may allow a more effective manipulation of both FT and vernalization to produce cereal varieties with higher and more sustained tolerance to LT stress.

Acknowledgements

This work was supported by a Natural Sciences and Engineering Research Council of Canada discovery grant, and by Genome Canada, Genome Québec, and Genome Prairie grants to FS.

References

Amasino, R. (2004) Vernalization, competence, and the epigenetic memory of winter. *Plant Cell* 16, 2553–2559.

Baum, M., Grando, S., Backes, G., Jahoor, A., Sabbagh, A. and Ceccarelli, S. (2003) QTLs for agronomic traits in the Mediterranean environment identified in recombinant inbred lines of the cross 'Arta' × *H. spontaneum* 41–1. *Theoretical and Applied Genetics* 107, 1215–1225.

Beales, J., Laurie, D.A. and Devos, K.M. (2005) Allelic variation at the linked AP1 and PhyC loci in hexaploid wheat is associated but not perfectly correlated with vernalization response. *Theoretical and Applied Genetics* 110, 1099–1107.

Bezant, J., Laurie, D., Pratchett, N., Chojecki, J. and Kearsey, M. (1996) Marker regression mapping of QTL controlling flowering time and plant height in a spring barley (*Hordeum vulgare* L.) cross. *Heredity* 77, 64–73.

Boyko, E., Kalendar, R., Korzun, V., Fellers, J., Korol, A., Schulman, A.H. and Gill, B.S. (2002) A high-density cytogenetic map of the *Aegilops tauschii* genome incorporating retrotransposons and defense-related genes: insights into cereal chromosome structure and function. *Plant Molecular Biology* 48, 767–790.

Breton, G., Danyluk, J., Ouellet, F. and Sarhan, F. (2000) Biotechnological applications of plant freezing associated proteins. *Biotechnology Annual Review* 6, 59–101.

Bullrich, L., Appendino, L., Tranquilli, G., Lewis, S. and Dubcovsky, J. (2002) Mapping of a thermo-sensitive earliness per se gene on *Triticum monococcum* chromosome 1A(m). *Theoretical and Applied Genetics* 105, 585–593.

Burn, J.E., Bagnall, D.J., Metzger, J.D., Dennis, E.S. and Peacock, W.J. (1993) DNA methylation, vernalization, and the initiation of flowering. *Proceedings of the National Academy of Sciences USA* 90, 287–291.

Chouard, P. (1960) Vernalization and its relations to dormancy. *Annual Review of Plant Physiology* 11, 238–238.

Clarke, J.H. and Dean, C. (1994) Mapping FRI, a locus controlling flowering time and vernalization response in *Arabidopsis thaliana*. *Molecular and General Genetics* 242, 81–89.

Danyluk, J., Kane, N.A., Breton, G., Limin, A.E., Fowler, D.B. and Sarhan, F. (2003) TaVRT-1, a putative transcription factor associated with vegetative to reproductive transition in cereals. *Plant Physiology* 132, 1849–1860.

Dubcovsky, J., Lijavetzky, D., Appendino, L. and Tranquilli, L. (1998) Comparative RFLP mapping of *Triticum monococcum* genes controlling vernalization requirement. *Theoretical and Applied Genetics* 97, 968–975.

Fornara, F., Parenicova, L., Falasca, G., Pelucchi, N., Masiero, S., Ciannamea, S., Lopez-Dee, Z., Altamura, M.M., Colombo, L. and Kater, M.M. (2004) Functional characterization of OsMADS18, a member of the AP1/SQUA subfamily of MADS box genes. *Plant Physiology* 135, 2207–2219.

Fowler, D.B., Chauvin, L.P., Limin, A.E. and Sarhan, F. (1996a) The regulatory role of vernalization in the expression of low-temperature-induced genes in wheat and rye. *Theoretical and Applied Genetics* 93, 554–559.

Fowler, D.B., Limin, A.E., Wang, S.Y. and Ward, S.W. (1996b) Relationship between low-temperature tolerance and vernalization response in wheat and rye. *Canadian Journal of Plant Science* 76, 37–42.

Fowler, D.B., Limin, A.E. and Ritchie, J.T. (1999) Low-temperature tolerance in cereals: model and genetic interpretation. *Crop Science* 39, 626–633.

Fu, D., Szűcs, P., Yan, L., Helguera, M., Skinner, J.S., von Zitzewitz, J., Hayes, P.M. and Dubcovsky, J. (2005) Large deletions within the first intron in VRN-1 are associated with spring growth habit in barley and wheat. *Molecular Genetics and Genomics* 273, 54–65.

Galiba, G., Quarrie, S.A., Sutka, J., Morgounov, A. and Snape, J.W. (1995) RFLP mapping of vernalization (*Vrn1*) and frost resistance (*Fr1*) genes on chromosome 5A of wheat. *Theoretical and Applied Genetics* 90, 1174–1179.

Gendall, A.R., Levy, Y.Y., Wilson, A. and Dean, C. (2001) The VERNALIZATION 2 gene mediates the epigenetic regulation of vernalization in *Arabidopsis*. *Cell* 107, 525–535.

Guy, C. (1999) Molecular responses of plants to cold shock and cold acclimation. *Journal of Molecular and Microbiological Biotechnology* 1, 231–242.

Hartmann, U., Hohmann, S., Nettesheim, K., Wisman, E., Saedler, H. and Huijser, P. (2000) Molecular cloning of SVP: a negative regulator of the floral transition in *Arabidopsis*. *Plant Journal* 21, 351–360.

Henderson, I.R., Shindo, C. and Dean, C. (2003) The need for winter in the switch to flowering. *Annual Review of Genetics* 37, 371–392.

Immink, R.G., Ferrario, S., Busscher-Lange, J., Kooiker, M., Busscher, M. and Angenent, G.C. (2003) Analysis of the petunia MADS-box transcription factor family. *Molecular Genetics and Genomics* 268, 598–606.

Islam-Faridi, M.N., Worland, A.J. and Law, C.N. (1996) Inhibition of ear-emergence time and sensitivity to day-length determined by the group 6 chromosomes of wheat. *Heredity* 77, 572–580.

Iwaki, K., Nishida, J., Yanagisawa, T., Yoshida, H. and Kato, K. (2002) Genetic analysis of Vrn-B1 for vernalization requirement by using linked dCAPS markers in bread wheat (*Triticum aestivum* L.). *Theoretical and Applied Genetics* 104, 571–576.

Johanson, U., West, J., Lister, C., Michaels, S., Amasino, R. and Dean, C. (2000) Molecular analysis of FRIGIDA, a major determinant of natural variation in *Arabidopsis* flowering time. *Science* 290, 344–347.

Kane, N.A., Danyluk, J., Tardif, G., Ouellet, F., Laliberté, J.-F., Limin, A.E., Fowler, D.B. and Sarhan, F. (2005) *TaVRT-Z*, a member of the *St*MADS-11 clade of flowering repressors, is regulated by vernalization and photoperiod in wheat. *Plant Physiology* 138, 2354–2363.

Kim, H.J., Hyun, Y., Park, J.Y., Park, M.J., Park, M.K., Kim, M.D., Kim, H.J., Lee, M.H., Moon, J., Lee, I. and Kim, J. (2004) A genetic link between cold responses and flowering time through FVE in *Arabidopsis thaliana*. *Nature Genetics* 36, 167–171.

Klaimi, Y.Y. and Qualset, C.O. (1974) Genetics of heading time in wheat (*Triticum aestivum* L.) II. The inheritance of vernalization response. *Genetics* 76, 119–133.

Koornneef, M., Blankenstijn-de Vries, H., Hanhart, C., Soppe, W. and Peters, T. (1994) The phenotype of some late-flowering mutants is enhanced by a locus on chromosome 5 that is not effective in the Landsberg erecta wild type. *Plant Journal* 6, 911–919.

Lang, A. (1965) Physiology of flower initiation. In: Ruhland, W. (ed.) *Encyclopedia of Plant Physiology*. Springer-Verlag, Berlin, pp. 1371–1536.

Laurie, D.A., Pratchett, N., Bezant, J.H. and Snape, J.W. (1995) RFLP mapping of five major genes and eight quantitative trait loci controlling flowering time in a winter × spring barley (*Hordeum vulgare* L.) cross. *Genome* 38, 575–585.

Law, C.N. and Wolfe, M.S. (1966) Location of genetic factors for mildew resistance and ear emergence time on chromosome 7B of wheat. *Canadian Journal of Genetics and Cytology* 8, 462–472.

Law, C.N. and Worland, A.J. (1997) Genetic analysis of some flowering time and adaptive traits in wheat. *New Phytologist* 137, 19–28.

Law, C.N., Dean, C. and Coupland, G. (1993) Genes controlling flowering and strategies for their isolation and characterization. In: Jordan, B.R. (ed.) *The Molecular Biology of Flowering*. CAB International, Wallingford, UK, pp. 47–68.

Law, C.N., Suarez, E., Miller, T.E. and Worland, A.J. (1998) The influence of the group 1 chromosomes of wheat on ear-emergence times and their involvement with vernalization and day length. *Heredity* 80, 83–91.

Lee, H., Xiong, L., Gong, Z., Ishitani, H., Stevenson, B. and Zhu, J.K. (2001) The *Arabidopsis HOS1* gene negatively regulates cold signal transduction and encodes a RING finger protein that displays cold-regulated nucleo-cytoplasmic partitioning. *Genes and Development* 15, 912–924.

Limin, A.E. and Fowler, D.B. (2002) Developmental traits affecting low-temperature tolerance response in near-isogenic lines for the vernalization locus Vrn-A1 in wheat (*Triticum aestivum* L. em Thell). *Annals of Botany* 89, 579–585.

Mahfoozi, S., Limin, A.E. and Fowler, D.B. (2001a) Developmental regulation of low-temperature tolerance in winter wheat. *Annals of Botany* 87, 751–757.

Mahfoozi, S., Limin, A.E. and Fowler, D.B. (2001b) Influence of vernalization and photoperiod responses on cold hardiness in winter cereals. *Crop Science* 41, 1006–1011.

Maistrenko, O.I. (1980) Cytogenetic study of the growth habit and ear-emergence time in wheat (*Triticum aestivum* L.). *Proceedings of the 14th International Congress on*

Genetics, Vol. 1, Bk 2. MIR Publishers, Moscow, pp. 267–282.

Masiero, S., Li, M.A., Will, I., Hartmann, U., Saedler, H., Huijser, P., Schwarz-Sommer, Z. and Sommer, H. (2004) INCOMPOSITA: a MADS-box gene controlling prophyll development and floral meristem identity in Antirrhinum. *Development* 131, 5981–5990.

McIntosh, R.A., Devos, K.M., Dubcovsky, J., Morris, C.F. and Rogers, W.J. (2003) Catalogue of gene symbols for wheat: 2003 Suppl. GrainGenes. Available at: http://wheat.pw.usda.gov/ggpages/wgc/2003upd.html

Michaels, S.D. and Amasino, R.M. (1999) FLOWERING LOCUS C encodes a novel MADS domain protein that acts as a repressor of flowering. *Plant Cell* 11, 949–956.

Michaels, S.D., He, Y., Scortecci, K.C. and Amasino, R.M. (2003) Attenuation of *FLOWERING LOCUS C* activity as a mechanism for the evolution of summer-annual flowering behavior in *Arabidopsis. Proceedings of the National Academy of Sciences USA* 100, 10102–10107.

Miura, H. and Worland, A.J. (1994) Genetic control of vernalization, day-length response and earliness *per se* by homoeologous group-3 chromosomes in wheat. *Plant Breeding* 113, 160–169.

Mouradov, A., Cremer, F. and Coupland, G. (2002) Control of flowering time: interacting pathways as a basis for diversity. *Plant Cell* 14, S111–S130.

Nelson, J.C., Sorrells, M.E., Van Deyne, A.E., Lu, Y.H., Atkinson, M., Bernard, M., Leroy, P., Faris, J.D. and Anderson, J.A. (1995) Molecular mapping of wheat: major genes and rearrangements in homoeologous groups 4, 5 and 7. *Genetics* 141, 721–731.

Plaschke, J., Xie, D.X., Koebner, R.M.D. and Gale, M.D. (1993) RFLP mapping of genes affecting plant height and growth habit in rye. *Theoretical and Applied Genetics* 85, 1049–1054.

Pugsley, A.T. (1972) Additional genes inhibiting winter habit in wheat. *Euphytica* 21, 547–552.

Sheldon, C.C., Burn, J.E., Perez, P.P., Metzger, J., Edwards, J.A., Peacock, W.J. and Dennis, E.S. (1999) The FLF MADS box gene: a repressor of flowering in *Arabidopsis* regulated by vernalization and methylation. *Plant Cell* 11, 445–458.

Shindo, C., Tsujimoto, H. and Sasakuma, T. (2003) Segregation analysis of heading traits in hexaploid wheat utilizing recombinant inbred lines. *Heredity* 90, 56–63.

Simpson, G.G. and Dean, C. (2002) *Arabidopsis*, the rosetta stone of flowering time? *Science* 296, 285–289.

Snape, J.W., Law, C.N., Parker, B.B. and Worland, A.J. (1985) Genetical analysis of chromosome 5A of wheat and its influence on important agronomic characters. *Theoretical and Applied Genetics* 71, 518–526.

Snape, J.W., Butterworth, K., Whitechurch, E. and Worland, A.J. (2001) Waiting for fine times: genetics of flowering time in wheat. *Euphytica* 119, 185–190.

Sourdille, P., Snape, J.W., Cadalen, T., Charmet, G., Nakata, N., Bernard, S. and Bernard, M. (2000) Detection of QTLs for heading time and photoperiod response in wheat using a doubled-haploid population. *Genome* 43, 487–494.

Stelmakh, A.F. (1993) Genetic effect of *Vrn* genes on heading date and agronomic traits in bread wheat. *Euphytica* 65, 53–60.

Thomashow, M.F. (1999) Plant cold acclimation: freezing tolerance genes and regulatory mechanisms. *Annual Review of Plant Physiology and Plant Molecular Biology* 50, 571–599.

Tranquilli, G. and Dubcovsky, J. (2000) Epistatic interaction between vernalization genes Vrn-Am1 and Vrn-Am2 in diploid wheat. *Journal of Heredity* 91, 304–306.

Worland, A.J., Gale, M.D. and Law, C.N. (1987) Wheat genetics. In: Lupton, F.G.H. (ed.) *Wheat Breeding: Its Genetic Basis.* Chapman and Hall, London, pp. 129–171.

Yan, L., Loukoianov, A., Tranquilli, G., Helguera, M., Fahima, T. and Dubcovsky, J. (2003) Positional cloning of the wheat vernalization gene VRN1. *Proceedings of the National Academy of Sciences USA* 100, 6263–6268.

Yan, L., Helguera, M., Kato, K., Fukuyama, S., Sherman, J. and Dubcovsky, J. (2004a) Allelic variation at the VRN-1 promoter region in polyploid wheat. *Theoretical and Applied Genetics* 109, 1677–1686.

Yan, L., Loukoianov, A., Blechl, A., Tranquilli, G., Ramakrishna, W., SanMiguel, P., Bennetzen, J.L., Echenique, V. and Dubcovsky, J. (2004b) The wheat VRN2 gene is a flowering repressor down-regulated by vernalization. *Science* 303, 1640–1644.

7

A Bulk Segregant Approach to Identify Genetic Polymorphisms Associated with Cold Tolerance in Lucerne

Y. CASTONGUAY,* J. CLOUTIER, S. LABERGE, A. BERTRAND AND R. MICHAUD

Soils and Crops Research and Development Centre, Agriculture and Agri-Food Canada, 2560 Hochelaga Blvd, Sainte-Foy, QC, Canada G1V 2J3

Introduction

The enormous contribution of forage legumes to grassland agriculture in temperate parts of the world is bound to intensify in the years to come as a result of new economic and environmental challenges associated with the current reliance on non-renewable energy sources in agriculture (Frame *et al.*, 1998). The perennial growth habit of forage legumes combined with their capacity to fix atmospheric N_2 symbiotically markedly contributes to long-term sustainability of agriculture through improved soil fertility, reduced reliance on fossil fuel consumption and by providing a renewable source of dietary N_2 (Vance, 2001). Lucerne (*Medicago sativa* L.) is the highest-yielding temperate forage legume and the most widely grown, with over 32 million hectares worldwide (Michaud *et al.*, 1988). Lack of winter hardiness constitutes a major limitation to lucerne persistence in areas experiencing harsh winter conditions (McKenzie *et al.*, 1988). The need to improve winter hardiness in cultivars of high agronomic value grown in northern climates has been frequently pointed out (Beuselinck *et al.*, 1994). However, a recent assessment of progress in breeding for greater winter hardiness in the USA in the last 20 years has revealed that efforts towards that goal have not been effective (Volenec *et al.*, 2002). Progress has been slowed down by a number of factors including the quantitative nature of winter hardiness inheritance, lack of effective screening methodologies and insufficient knowledge of the molecular and genetic bases of adaptation. Consequently, the identification of the molecular and

*Corresponding author (e-mail: castonguayy@agr.gc.ca; Fax: +1-418-648-2402)

genetic bases of superior adaptation to winter conditions remains an important objective of breeding programmes in their efforts to develop new selection technologies.

Freezing Tolerance of Lucerne

Numerous environmental (low subfreezing temperatures, freeze–thaw cycles, excess soil moisture, ice sheeting, low-temperature pathogens) and management (autumn-cutting management, fertilization, drainage) factors can weaken overwintering plants and reduce their winter survival. Tolerance to subfreezing temperatures is undoubtedly the single most important factor that determines the survival of lucerne to severe winter conditions (McKenzie *et al.*, 1988). There are numerous indications that freezing tolerance is at the very core of the winter hardiness complex and that winter damage is often the result of indirect effects of other stresses on the plants' capacity to tolerate subsequent exposure to cold temperatures. Schwab *et al.* (1996) noted a close agreement between survival scores for field-hardened lucerne exposed to a freezing test under environmentally controlled conditions and winter injury data taken from field plots in the following spring. In that perspective, it is envisioned that the improvement of freezing tolerance of lucerne will translate into better winter survival, extended longevity of the stands and higher crop productivity.

A wide variability in winter hardiness potential and freezing tolerance among lucerne cultivars grown in Canada and the US has long been documented (Heinrichs, 1973; Schwab *et al.*, 1996). Determination of freezing tolerance for a limited number of genotypes randomly selected within the cultivar Apica developed and recommended for growth in eastern Canada (Michaud *et al.*, 1983) revealed that extensive variability is also present within lucerne populations (Fig. 7.1). The wide range of freezing tolerance (≈6°C) observed among the small group of genotypes shown in Fig. 7.1 constitutes a conservative estimate of the overall variability that exists within this cultivar. Kidwell *et al.* (1994) concluded that there existed greater genetic diversity within than between populations of lucerne for most traits. There is reported evidence that such variability might also occur for freezing tolerance within cultivars of a given category of autumn dormancy (McCaslin *et al.*, 1990). Therefore, it appears that a large reservoir of genetic material harbouring superior potential for cold tolerance remains to be exploited within *M. sativa*. Thus far, the use of this genetic diversity has been hampered by an insufficient knowledge of the molecular and genetic bases of adaptation and the absence of high throughput methodologies to reliably screen a large number of plants to identify superior genotypes.

Molecular Bases of Cold Tolerance in Lucerne

Extensive modifications in the biochemical composition of perennial organs have been documented during cold acclimation of lucerne. Reports on the release of secondary messengers, changes in gene expression and modifications in the composition of soluble proteins, phospholipids, amino acids and soluble sugars have

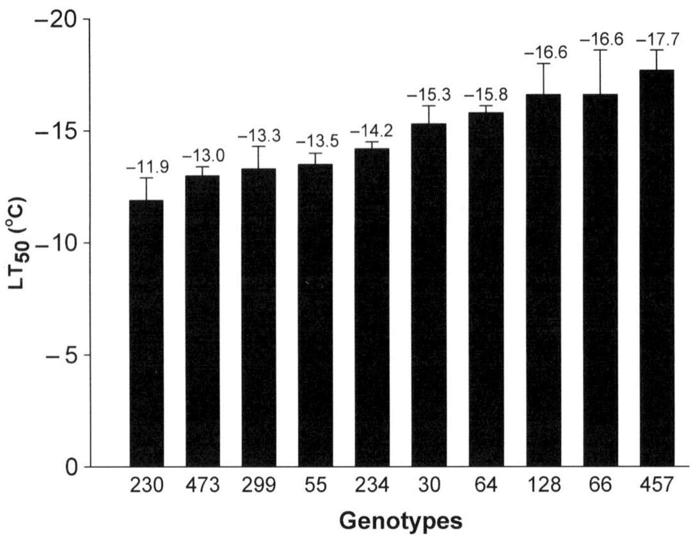

Fig. 7.1. Comparative assessment of the freezing tolerance (LT_{50}) of ten genotypes randomly selected within the lucerne cultivar Apica. Genotypes were vegetatively propagated and allowed to establish under environmentally controlled conditions prior to the evaluation of their freezing tolerance. Plants were cold acclimated under a constant temperature of 2°C, an 8-h photoperiod and 150 μmol photons/m^2/s PPFD for 2 weeks, followed by an additional incubation period of 2 weeks at −2°C in the dark to promote the second stage of hardening. Freezing tests were performed in programmed freezers as described in Castonguay *et al.* (1993). Means and standard deviations from three separate freezing tests that were conducted simultaneously are presented. (Results from Y. Castonguay, P. Nadeau and R. Michaud, unpublished manuscript).

been reviewed by Castonguay *et al.* (1997a). Among these, cold-induced accumulation of soluble sugars is thought to be particularly critical to winter survival of lucerne and has been the focus of research on cold tolerance mechanisms for many decades (see references in Cunningham *et al.*, 2003). Field studies by Paquin and Lechasseur (1982) have shown that maximum accumulation of total soluble sugars in crowns of lucerne coincides with the occurrence of maximum winter hardiness. Although the cryoprotectant sucrose accounts for nearly 90% of the total pool of soluble sugars in crowns and taproots of lucerne, differences in freezing tolerance between non-hardy and winter hardy cultivars were shown to be more closely related to the accumulation of the lower abundant stachyose and raffinose (Fig. 7.2). These members of the raffinose family of oligosaccharides (RFO) have also been closely related to the acquisition of freezing tolerance in many plant species and are thought to have important adaptive value in the tolerance to freeze-induced desiccation (see review in Castonguay *et al.*, 1998).

The enzyme galactinol synthase (GaS) is an important regulating point in the synthesis of RFO. A GaS cDNA isolated from cold-acclimated crowns of lucerne was recently shown to be cold-inducible (Cunningham *et al.*, 2003). Three

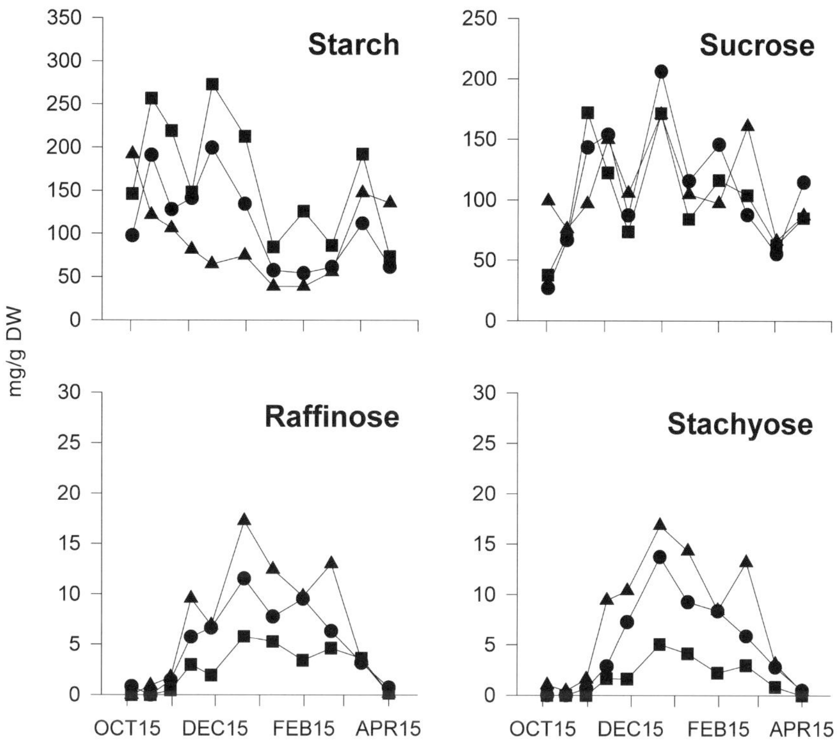

Fig. 7.2. Evolution of starch and soluble sugars (sucrose, raffinose and stachyose) levels in crowns of the very hardy cultivar Rambler (▲), moderately hardy cultivar Apica (●) and non-hardy cultivar CUF 101 (■) during their acclimation to simulated winter conditions in an unheated greenhouse during the 1994–1995 winter. Adapted from Castonguay and Nadeau (1998).

Arabidopsis GaS genes were found to be stress-responsive (drought, salt or cold stress) with one of them being specifically induced by cold (Taji *et al.*, 2002). In that particular study, overexpression of a drought-induced GaS led to the accumulation of raffinose under well-watered conditions and to better water-stress tolerance than the wild type. In cold-acclimated lucerne, GaS activity increased earlier and reached higher levels in two hardy cultivars than in a non-hardy cultivar; this response was tightly coupled to RFO accumulation in the crowns (Castonguay and Nadeau, 1998). A close relationship between RFO accumulation in roots sampled in December and genetic differences in winter survival was recently observed under field conditions in Indiana (Cunningham *et al.*, 2003). Such observations led to the suggestion that the accumulation of RFO oligosaccharides might be potentially used to screen for cold tolerance in lucerne (Castonguay *et al.*, 1998). However, the lack of a non-destructive method to assess RFO levels in perennial organs along with large variations between years and sites due to genotype–environment interactions have thus far precluded the use of biochemical end products as markers in breeding programmes. Observations of higher levels of RFO in plants of lucerne exposed to treatments known to reduce their winter hardiness (Dhont *et al.*, 2002; Haagenson *et al.*, 2003) reinforce the necessity of getting at the genetic make-up of plants in order to establish the true adaptive value of these molecular changes. The identification of robust genetic markers linked to their accumulation would pave the way to the development of effective screening approaches.

Many cold-regulated cDNAs have been isolated from cold-acclimated lucerne (see review in Castonguay *et al.*, 1997b). These genes encode for proteins homologous to well-known classes of stress-related proteins including dehydrins, pathogenesis-related proteins and glycine-rich proteins. Although this initial gene discovery based on differential screening of cDNA libraries provided a strong basis of information on the nature of cold-induced changes in gene expression, a more exhaustive coverage of the gene complement involved is currently being achieved by the simultaneous hybridization of thousands of genes on high-density grids (Desgagnés *et al.*, 2004).

Genetic Improvement of Cold Hardiness in Lucerne

Success in breeding stress-tolerant plants using conventional plant-breeding methodologies has been limited in spite, as mentioned earlier, of the presence of large genetic variation for morphological and physiological traits. Assessment of genetic diversity for winter hardiness has often been a bottleneck in the development of conventional selection technologies. In the absence of direct measurements of cold tolerance, winter-survival potential has often been predicted from autumn dormancy responses. It has long been known that plants of lucerne with early growth reduction in the autumn are generally more winter hardy than those that maintain active growth later in the season (Schwab *et al.*, 1996). However, early autumn dormancy is an undesirable trait from the productivity standpoint and therefore breeding programmes are trying to reach a compromise between extended growing season and winter survival. Reduced autumn dormancy does not necessarily have to be achieved at the cost of reduced winter hardiness. There are indeed indications in the literature that both traits can be dissociated, and that it should be theoretically possible to identify genotypes harbouring superior genetic potential for both forage productivity and winter survival (Brummer *et al.*, 2000). Progress in that regard will, however, necessitate a deeper understanding of the genetic bases of these traits.

Field Selection

Improvement of winter hardiness of lucerne has historically been based on field selection of genotypes that survived test winters. However, the unpredictability of test winters due to large variations between and within locations and between years in the environmental conditions to which the plants are exposed severely limits the usefulness of this approach (Schwab *et al.*, 1996). Field tests that allow adequate selection are a rare occurrence and most often result in either complete winterkill or lack of winterkill (Limin and Fowler, 1991). As a result, expensive assessment of the genetic material at multiple locations over many years is often used to increase the probability of discriminating between plants with regard to their winter hardiness potential. New approaches are clearly needed for a less costly and more reliable assessment of the genetic material currently used in breeding programmes.

Phenotypic Selection Under Environmentally Controlled Conditions

The success of gene discovery programmes is intimately related to the availability of a high-quality phenotyping assay to identify or produce biological material harbouring contrasted phenotypes. Our group has recently devised and tested a new method of selection performed under environmentally controlled conditions for the development of populations of improved cold tolerance and winter hardiness (Nadeau *et al.*, 2002). Large number of genotypes from commercial cultivars of lucerne are subjected to three successive freezing stresses consisting of cold hardening (2 weeks at 2°C and 2 weeks at −2°C) followed by a stepwise decrease in temperature down to the expected killing temperature for 50% of the plants (LT_{50}). After the third stress, the ≈10% most vigorous genotypes are selected and intercrossed to generate populations potentially more tolerant to freezing temperatures (TF). Using this approach, several cycles of selection (between three and six cycles depending on the cultivar) have been successively performed in order to generate populations selectively improved for their freezing tolerance through recurrent selection.

The evaluation of freezing tolerance, using plants acclimated to natural winter conditions in an unheated greenhouse, revealed striking improvement in freezing tolerance after six cycles of selection in the cultivar Apica (Fig. 7.3). Although significant damage was noticeable at around −17°C for the cultivar Apica, plants of A-TF6 remained unaffected until much lower temperatures (−26°C) were reached. A similar response that varied in intensity depending on the number of cycles of selection and the genetic background was observed with other cultivars. Increase in freezing tolerance in A-TF6 was associated with higher accumulation of cold-induced metabolites, including the cryoprotective RFOs (Fig. 7.4). Differences in levels of RFO between the original cultivar and the selected population were apparent during the acquisition of cold tolerance, and disappeared later in winter at the time when plants had started to deharden. Field evaluation of winter hardiness and spring yield revealed a marked increase in persistence of Apica-TF populations when exposed to stressful conditions (data not shown).

Genetic variation observed between lucerne populations selectively improved for their freezing tolerance (TF populations) could be attributable to an increase in the frequency of several adaptive alleles that have been selectively enriched in response to selection pressure. Such contrasted sources of biological material derived within a closed genetic background provide unique sources of information when used in combination with genome-wide analytical tools.

Fig. 7.3. Freezing tolerance of the lucerne cultivar Apica and the population A-TF6 derived from this cultivar through six cycles of recurrent selection for superior freezing tolerance. Plants were acclimated to natural winter conditions in an unheated greenhouse at an experimental research farm located near Québec City, Canada during the 2002–2003 overwintering season. Freezing tolerance was assessed in mid-January 2003 in programmed freezers as described in Castonguay *et al.* (1993) using a stepwise decline of temperatures between –16°C and –28°C. A group of plants used as controls for hardening survival (S) were not exposed to the freezing stress and were immediately transferred to the greenhouse for regrowth. Survival counts were taken after 3 weeks of regrowth in a greenhouse and the 50% killing temperature (LT_{50}) was computed using the SAS Probit procedure. (Results from Y. Castonguay, P. Nadeau, A. Bertrand and R. Michaud, unpublished manuscript).

Marker-assisted Selection

The development of molecular approaches for the improvement of stress tolerance relies on the effective identification of genetic markers associated with superior adaptation and their broad application in a wide array of genetic backgrounds. Many variants used in plant breeding programmes bear mutated forms of 'wild-type' alleles (Leung and An, 2004). Marker-assisted selection (MAS) offers many potential benefits to plant-breeding programmes, including the selection of genotypes that carry mutated alleles, the selective improvement of the trait of interest with limited impact on other agronomic traits and, in many cases, an acceleration of the selection process. A number of MAS approaches have been successfully applied in recent years. Their development relies on the availability of:

1. Genetic material harbouring variability for the trait of interest;
2. An accurate phenotypic-screening methodology for a sensitive discrimination of biological responses; and

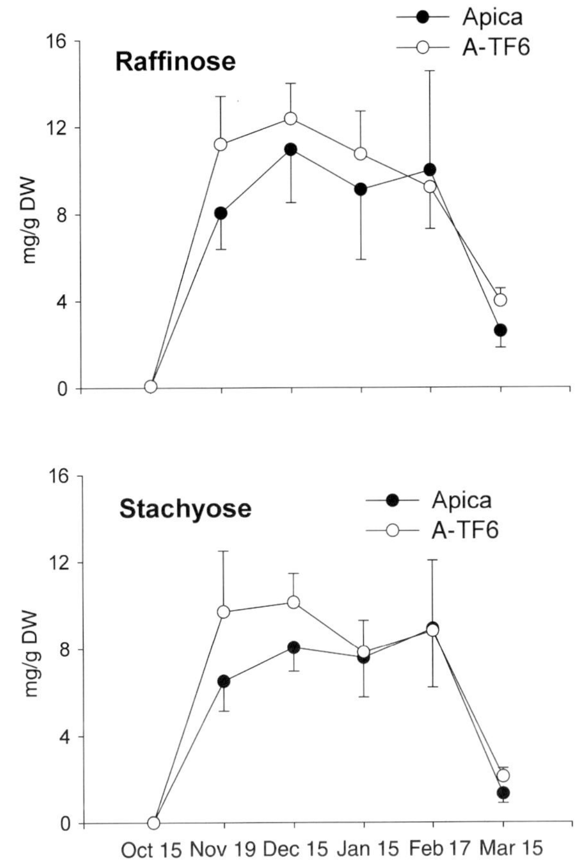

Fig. 7.4. Comparative accumulation of RFO (stachyose and raffinose) in crowns of the cultivar Apica and the population A-TF6 derived from this cultivar through six cycles of recurrent selection for superior freezing tolerance. Plants were acclimated to natural winter conditions in an unheated greenhouse at an experimental research farm located near Québec City, QC, Canada during the 2002–2003 overwintering season. Soluble sugars were analysed by high-performance liquid chromatography as described in Castonguay and Nadeau (1998). Mean values (n = 4) and standard deviations are presented. (Results from Y. Castonguay, P. Nadeau, A. Bertrand and R. Michaud, unpublished manuscript).

3. A genome-probing methodology for the discovery and validation of allelic forms associated with phenotypic variation of the trait.

Searching for markers associated with traits that exhibit a continuous variation like freezing tolerance is most frequently pursued by quantitative trait loci (QTL) mapping. A number of studies have illustrated the feasibility of mapping QTLs controlling freezing tolerance in a number of plant species (Galiba *et al.*, 1995; Kole *et al.*, 2002; Tsarouhas *et al.*, 2004) including lucerne (Brouwer *et al.*, 2000). These studies have unequivocally confirmed the polygenic nature of the trait and the involvement of genes with major effects. They also helped to establish important relationships between QTLs controlling various cold-adaptive processes like autumn growth reduction, vernalization requirements, autumn frost resistance and mid-winter freezing tolerance (Tsarouhas *et al.*, 2004). However, up to now, mapping of QTLs has seldom led to the identification of genes responsible for variation in freezing tolerance or the development of DNA markers that are diagnostic for freezing tolerance potential.

The QTL approach relies on the availability of high-density genetic linkage maps to study the relationship between the inheritance of phenotypic traits and that of genetic markers in segregating populations. Even though high-density linkage maps of tetraploid lucerne are becoming available (Julier *et al.*, 2003), their usefulness in the precise mapping of genes associated with agronomic traits is made difficult by characteristics of the lucerne genetics including autotetraploidy, allogamy and strong inbreeding depression (Osborn *et al.*, 1998). To palliate the inherent difficulties associated with genetic analysis of complex genomes, it is theoretically possible to make inferences on gene localization using comparative mapping of orthologous genes in model species (Devos and Gale, 2000). The effectiveness of this approach relies on the conservation of linkage (synteny) and gene order (collinearity) between related species (Yan *et al.*, 2004). Although macrosynteny (10–20 cM range) has been frequently observed, rearrangements, insertions and deletions often disrupt the microcollinearity (within the 100 kb range), sometimes making cross species–gene prediction and positional cloning a difficult, if not futile, pursuit (Leung and An, 2004). *Medicago truncatula*, an annual diploid species of the *Medicago* genus, with a relatively small genome (≈500 Mb), has been proposed as a model species to facilitate studies in legume genetics (Cook, 1999). High microsynteny between *M. truncatula* and *M. sativa* has been confirmed and recently exploited to clone a gene involved in nod-factor perception in lucerne (Endre *et al.*, 2002). However, such identification of allelic variants through genetic mapping of orthologous genes in *M. truncatula* remains a tedious task that can still be tempered by the occurrence of microstructural genome rearrangements (Choi *et al.*, 2004).

Thus far, the identification of DNA markers by QTL mapping has been mostly restricted to traits controlled by one or two genes having major effects on the trait of interest. This is due to the fact that in many cases there is a need to locate the gene of interest by chromosomal walking and to characterize the allelic forms associated with a particular phenotype. One additional drawback of QTL mapping as a gene discovery technique pertains to the narrow genetic basis under study. The typical use of two polymorphic parents that are intercrossed to generate F1 or F2 mapping populations could be particularly reductive when studying the inheritance of multigenic traits in an outcrossed polyploid. The likelihood that the two parents will be polymorphic at every locus linked to the trait of interest is low and therefore critical information is bound to be missed. Use of loosely linked markers can also be of limited value in molecular-breeding programmes as the linkage to a locus of interest may be lost as a result of different recombination patterns in other genetic backgrounds (Dwivedi *et al.*, 2003). Therefore, the development of robust markers that can effectively screen a wide array of germplasm could request the identification of the specific allelic forms (mutations) associated with a given phenotypic variation.

Bulk Segregant Analysis of Candidate Genes Polymorphisms

Combination of a bulk segregant approach (BSA) described by Michelmore *et al.* (1991) and candidate genes that play key roles in stress-induced pathways can be a very effective approach in discovering genes having major effects on quantitative traits. While QTL analysis of mapping populations is a more precise approach to locate genes associated with a given trait, BSA analysis of candidate genes offers a useful alternative for gene identification that does not require time-consuming and expensive genotyping of large segregating populations (Quarrie *et al.*, 1999). Germplasm, contrasted in their expression of the trait of interest, is at the very basis of this gene discovery strategy. It allows candidate genes potentially associated with phenotypic variation to be tested for differences in alleles frequency between populations derived from a given genetic background and selectively improved for that trait of interest. BSA has recently been successfully used in the development of markers tightly linked to disease-resistance genes (Caixeta *et al.*, 2003; Varshney *et al.*, 2004)

Lucerne populations selectively improved for freezing tolerance are particularly well suited to the use of BSA in the search for genetic markers associated with cold adaptation. Taking advantage of their availability, we pooled DNA extracts from ≈45 genotypes from each of the lucerne cultivar Apica and its derived TF populations selectively improved for superior freezing tolerance. Restricted-DNA was subsequently hybridized with candidate genes typically associated with cold acclimation. Clear polymorphisms that intensified with the number of selection cycles were uncovered for a number of genes including homologues of GaS (Fig. 7.5A) and dehydrins (data not shown). The expression of these genes has been previously related to the acquisition of cold tolerance (Bravo *et al.*, 2003; Cunningham *et al.*, 2003); therefore, they were good candidates for association with this trait. Three DNA segments identified as A, B and C in Fig. 7.5A showed progressive intensification in populations recurrently selected for superior freezing tolerance. Two bands were regrouped as polymorphism B, based on their co-inheritance in segregating genotypes. The likelihood that these polymorphisms are genetically linked to a QTL-controlling freezing tolerance is high, considering that selection was performed within a single genetic background and was solely targeted towards the improvement of that trait. Even though one polymorphic change was similarly observed in TF populations derived from another cultivar (Evolution), distinct polymorphisms were also observed between the two populations (data not shown). This observation hints at potential differences in the initial frequency of mutated alleles or in their neighbouring sequences between the two cultivars. It also highlights the need to characterize the specific mutations associated with a given polymorphism in order to be able to track the inheritance of alleles associated to contrasted phenotypes in progenies derived from crosses involving multiple genetic backgrounds. Interestingly, no polymorphism was observed for raffinose synthase (Fig. 7.5B), another gene involved in the RFO metabolic pathway, that has no reported link with stress adaptation. This observation reinforces the candidate gene approach and indicates that not all genes involved in key metabolic pathways are going to harbour genetic variability associated with superior adaptation to cold. Detection of the presence of polymorphic

GaS alleles in segregating genotypes randomly selected within each population confirms a progressive increase in their frequency with successive TF selections and, thus, the validity of observations made with pooled samples. For instance, the frequency of the GaS band 'A' in Fig. 7.5A, which is inferior to 2.5% in the Apica cultivar, increased to 20% after five cycles of recurrent selection (A-TF5).

Conclusions and Prospectives

Our results illustrate that a bulk segregant analysis of candidate genes polymorphism in populations selectively improved for freezing tolerance can be a very effective approach to identify genes that are associated with superior adaptation to cold in lucerne. In open-pollinated species with large genomes and complex inheritance systems like lucerne, this approach might prove to be a more efficient way to identify variant DNA sequences linked to stress tolerance than map-based searches relying on syntenic relationships with model species. This type of

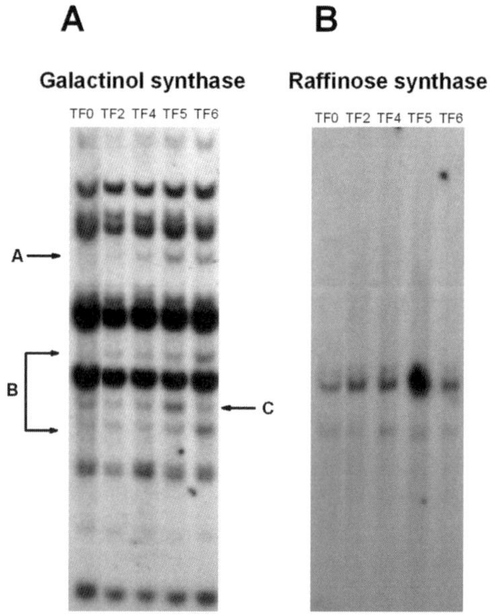

Fig. 7.5. Southern blot hybridization of pooled genomic DNA extracts from 40 genotypes from each of the cultivar Apica (TF0) and the populations (TF2, 4, 5 and 6) derived from this cultivar through recurrent cycles selection for superior freezing tolerance. Total genomic DNA was extracted using the CTAB procedure and 10 μg were digested with DraI, electrophoresed on 0.8% agarose gels and transferred onto nylon membranes. Homologues of galactinol synthase (A) (Cunningham *et al.*, 2003) and raffinose synthase (B) (unpublished) isolated from cold-acclimated lucerne were used for hybridization. Purified cDNA probes (25–50 ng) were labelled with alpha^{32}P dCTP using the oligolabelling procedure. Hybridizations were performed at 68°C using 2× SSC, 0.5% SDS and 0.25% (W/V) low-fat milk powder. (Results from Y. Castonguay, J. Cloutier, S. Laberge, A. Bertrand and R. Michaud, unpublished manuscript.)

approach necessitates the participation of a multidisciplinary group of experts working in close collaboration with breeding programmes for the rapid integration of the genes of interest into commercial cultivars.

A key element in the success of BSA in the identification of molecular markers for the improvement of cold hardiness of lucerne is the use of candidate genes with high probability of being polymorphic between contrasted phenotypes. Knowledge of the physiological and biochemical bases of adaptation is therefore a key element in the successful application of this marker discovery strategy. The choice of candidate genes relies mainly on information from the literature on the genetic and molecular bases associated with the trait of interest. Insufficient knowledge of the physiological and molecular bases of adaptation to cold certainly constitutes a limitation in that respect. However, information on the molecular bases of cold adaptation will rapidly expand since the identification of genes associated to phenotypic variation is a major endeavour of many genomic programmes worldwide. In the future, new high-throughput genotyping platforms will allow researchers to quickly genotype progeny at high resolution (hundreds to thousands of markers at once) and to accurately map the mutations responsible for candidate genes polymorphisms identified by BSA (Borevitz *et al.*, 2003).

The next steps in our BSA-assisted search of candidate genes associated with cold adaptation in lucerne will include:

1. The use of high-throughput gene-expression profiling of ≈10,000 expressed sequence tags (EST) from cold-acclimated lucerne (Desgagnés *et al.*, 2004) to identify genes that are more highly or uniquely expressed in germplasm selectively improved for cold tolerance in order to delineate a group of candidate genes to target for BSA.
2. The determination of freezing tolerance of genotypes randomly selected within the cultivar Apica and its TF-derived populations (≈45 genotypes in each population) following their acclimation to natural winter conditions.
3. Scoring for presence or absence of polymorphic genes in each genotype that will be tested for freezing tolerance.
4. The identification of the combination of polymorphic alleles that maximally discriminate TF populations using multivariate statistical analyses. Correlations between phenotypic variation and the polymorphism of candidate genes will provide strong circumstantial evidence for their association with the trait of interest.

Enormous efforts will need to be devoted in the future for the functional analysis of an ever-increasing number of stress-induced genes that are being identified by high-throughput screening of gene expression (Zhu and Provart, 2003). BSA coupled with physiological studies is bound to be a useful alternative to the time-consuming, and sometimes inconclusive, approach of plant transformation for the functional assessment of stress-induced genes (Quarrie *et al.*, 1999). The observation of the intensification of polymorphic DNA segments that hybridize with genes with unknown function, or that are not typically related to stress tolerance, will be a powerful means to unravel new gene functions. As pointed out by Volenec *et al.* (2002), function assignments based strictly on the basis of the degree of similarity between a gene of interest and sequences deposited in databases may sometimes be of limited value. While such information provides some insight into the potential role of the encoded protein, it does not determine its role(s) *in planta*.

The identification of candidate genes associated with improved freezing tolerance in lucerne is an initial, but nevertheless major, step towards the integration of DNA markers as selection tools into breeding programmes. We believe that marker-assisted selection will not supersede but will advantageously complement conventional plant-breeding programmes in the improvement of complex traits like tolerance to environmental stresses. There are indeed indications that DNA markers used in combination with conventional phenotypic selection might offer the best improvement gains per unit of time in applied breeding programmes (Lande and Thompson, 1990; Dwivedi *et al.*, 2003). Use of DNA markers would allow breeders to more rapidly increase the frequency of favourable alleles; to pyramid genes for stress tolerance in a given genetic background; and eventually, to accelerate the breeding process. These important contributions would facilitate efforts towards a more systematic integration of cold adaptive traits in high-quality cultivars that are being released at an ever-increasing pace.

Acknowledgements

The authors would like to thank Josée Bourassa, Lucette Chouinard and Pierre Lechasseur for their excellent technical assistance in the realization of this project.

References

Beuselinck, P.R., Bouton, J.H., Lamp, W.O., Matches, A.G., McCaslin, M.H., Nelson, C.J., Rhodes, L.H., Sheaffer, C.C. and Volenec, J.J. (1994) Improving legume persistence in forage crop systems. *Journal of Production Agriculture* 7, 311–322.

Borevitz, J.O., Liang, D., Plouffe, D., Chang, H.S., Zhu, T., Weigel, D., Berry, C.C., Winzeler, E. and Chory, J. (2003) Large-scale identification of single-feature polymorphisms in complex genomes. *Genome Research* 13, 513–523.

Bravo, L.A., Gallarde, J., Navarrete, A., Olave, N., Martinez, J., Alberdi, M., Close, T.J. and Corcuera, L.J. (2003) Cryoprotective activity of a cold-induced dehydrin purified from barley. *Physiologia Plantarum* 118, 262–269.

Brouwer, D.J., Duke, S.H. and Osborn, T.C. (2000) Mapping genetic factors associated with winter hardiness, fall growth, and freezing injury in autotetraploid alfalfa. *Crop Science* 40, 1387–1396.

Brummer, E.C., Shah, S.M. and Luth, D. (2000) Reexamining the relationship between fall dormancy and winter hardiness of alfalfa. *Crop Science* 40, 971–977.

Caixeta, E.T., Borém, A., Fagundes, S.A., Nietestche, S., Barros, E.G. and Moreira, M.A. (2003) Inheritance of angular leaf spot resistance in common bean line BAT 332 and identification of RAPD markers linked to resistance gene. *Euphytica* 134, 297–303.

Castonguay, Y. and Nadeau, P. (1998) Enzymatic control of soluble carbohydrate accumulation in cold-acclimated crowns of alfalfa. *Crop Science* 38, 1183–1189.

Castonguay, Y., Nadeau, P. and Laberge, S. (1993) Freezing tolerance and alteration of translatable mRNAS in alfalfa (*Medicago sativa* L.) hardened at subzero temperatures. *Plant and Cell Physiology* 34, 31–38.

Castonguay Y., Laberge, S., Nadeau P. and Vézina L.-P. (1997a) Temperature and drought stress. In: McKersie, B.D. and Brown, D.W. (eds) *Biotechnology and the Improvement of Forage Legumes.* CAB International, Wallingford, UK, pp. 175–202.

Castonguay, Y., Nadeau, P. and Michaud, R. (1997b) Differential accumulation of oligosaccharides and freezing tolerance of alfalfa. In: Chen, H. and Li, P.H. (eds) *Plant Cold Hardiness*. Plenum Press, New York, pp. 293–299.

Choi, H.K., Mun, J.-H., Kim, D.J., Zhu, H., Baek, J.M., Mudge, J., Roe, B., Ellis, N., Doyle, J., Kiss, G.B., Young, N.D. and Cook, D.R. (2004) Estimating genome conservation between crop and model legume species. *Proceedings of the National Academy of Sciences USA* 101, 15289–15294.

Cook, D.R. (1999) *Medicago truncatula* – a model in the making. *Current Opinion in Plant Biology* 2, 301–304.

Cunningham, S.M., Nadeau, P., Castonguay, Y., Laberge, S. and Volenec, J.J. (2003) Raffinose and stachyose accumulation, galactinol synthase expression, and winter injury of contrasting *Medicago sativa* germplasms. *Crop Science* 43, 562–570.

Desgagnés, R., Gagné, D., Castonguay, Y. and Laberge, S. (2004) Macroarray analysis of cold tolerance in alfalfa. *39th North American Alfalfa Improvement Conference*, Quebec City, Canada (Online). Available at: http://www.naaic.org/Meetings/National/index.html (undated).

Devos, K.M. and Gale, M.D. (2000) Genome relationships: the grass model in current research. *Plant Cell* 12, 637–646.

Dhont, C., Castonguay, Y., Nadeau, P., Bélanger, G. and Chalifour, F.-P. (2002) Alfalfa root carbohydrates and regrowth potential in response to fall harvests. *Crop Science* 42, 754–765.

Dwivedi, S.L., Crouch, J.H., Nigam, S.N., Ferguson, M.E. and Paterson, A.H. (2003) Molecular breeding of groundnut for enhanced productivity and food security in the semi-arid tropics: opportunities and challenges. *Advances in Agronomy* 80, 153–221.

Endre, G., Kereszt, A., Kevei, Z., Mihaceas, S., Kaló, P. and Kiss, G.B. (2002) A receptor kinase gene regulating symbiotic nodule development. *Nature* 41, 962–966.

Frame, J., Charlton, J.F.L. and Laidlaw, A.S. (1998) Prospects for forage legumes. In: *Temperate Forage Legumes*. CAB International, Wallingford, UK, pp. 313–317.

Galiba, G., Quarie, S.A., Sutka, J., Morgounov, A. and Snape, J.W. (1995) RFLP mapping of vernalization (*Vrn1*) and frost resistance (*Fr1*) on chromosome 5A of wheat. *Theoretical and Applied Genetics* 90, 1174–1179.

Haagenson, D.M., Cunningham, S.M., Joern, B.C. and Volenec, J.J. (2003) Autumn defoliation effects on alfalfa winter survival, root physiology, and gene expression. *Crop Science* 43, 1340–1348.

Heinrichs, D.H. (1973) Winter hardiness of alfalfa cultivars in southern Saskatchewan. *Canadian Journal of Plant Science* 53, 773–777.

Julier, B., Flajoulot, S., Barre, P., Cardinet, G., Santoni, S., Huguet, T. and Huyghe, C. (2003) Construction of two genetic linkage maps in cultivated tetraploid alfalfa (*Medicago sativa*) using microsatellite and AFLP markers. *BMC Plant Biology* 3:9. Available at: http://www.biomedcentral.com/1471-229/3/9 (undated).

Kidwell, K.K., Austin, D.F. and Osborn, T.C. (1994) RFLP evaluation of nine *Medicago* accessions representing the original germplasm sources for North American alfalfa cultivars. *Crop Science* 34, 230–236.

Kole, C., Thormann, C.E., Karlsson, B.H., Palta, J.P., Gaffney, P., Yandell, B. and Osborn, T.C. (2002) Comparative mapping of loci controlling winter survival and related traits in oilseed *Brassica rapa* and *B. napus*. *Molecular Breeding* 9, 201–210.

Lande, R. and Thompson, R. (1990) Efficiency of marker-assisted selection in the improvement of quantitative traits. *Genetics* 124, 743–756.

Leung, H. and An, G. (2004) Rice functional genomics: large-scale gene discovery and applications to crop improvement. *Advances in Agronomy* 82, 55–111.

Limin, A.E. and Fowler, D.B. (1991) Breeding for cold hardiness in winter wheat: problems, progress and alien gene expression. *Field Crops Research* 27, 201–218.

McCaslin, M., Brown, H., Deery, H. and Miller, D. (1990) Fall dormancy and winter survival and alfalfa variation within and between populations. In: Bauchan, G.R. *et al.* (eds) *Reports of the 32nd North American Alfalfa Improvement Conference*, Pasco, Washington, DC, 19–23 August 1990. North American Alfalfa Improvement Conference, Beltsville, Maryland, p. 9.

McKenzie, J.S., Paquin, R. and Duke, S.H. (1988) Cold and heat tolerance. In: Hanson, A.A *et al.* (eds) *Alfalfa and Alfalfa Improvement. Agronomy Monograph* 29. ASA, CSSA, and SSSA, Madison, Wisconsin, pp. 259–302.

Michaud, R., Richard, C., Willemot, C. and Gasser, H. (1983) Apica alfalfa. *Canadian Journal of Plant Science* 63, 547–549.

Michaud, R., Lehman, W.F. and Rumbaugh, M.D. (1988) World distribution and historical development. In: Hanson, A.A. *et al.* (eds) *Alfalfa and Alfalfa Improvement.* Agronomy Monograph 29. ASA, CSSA, and SSSA, Madison, Wisconsin, pp. 25–91.

Michelmore, R.W., Paran, I. and Kesseli, R.V. (1991) Identification of markers linked to disease-resistance genes by bulked segregant analysis: a rapid method to detect markers in specific genomic regions by using segregating populations. *Proceedings of the National Academy of Sciences USA* 88, 9828–9832.

Nadeau, P., Castonguay, Y., Laberge, S. and Michaud R. (2002) Characteristics of alfalfa populations selected for improved freezing tolerance. *38th North American Alfalfa Improvement Conference*, Sacramento, California (Online). Available at: http://www.naaic.org/Meetings/National/index.html (undated).

Osborn, T.C., Brouwer, D.J., Kidwell, K.K., Tavolettti, S. and Bingham, E.T. (1998) Molecular marker applications to genetics and breeding of alfalfa. In: Brummer, C. (ed.) *Molecular and Cellular Technologies for Forage Improvement.* ASA-SSSA-CSSA, pp. 25–31.

Paquin, R. and Lechasseur, P. (1982) Acclimatation naturelle de la luzerne (*Medicago sativa* Pers.) au froid. II. Variations de la teneur en sucres totaux des feuilles et des collets. *Acta Oecologica/Oecologia Plantarum* 3, 27–38.

Quarrie, S.A., Vesna, L.-J., Kovacevic, D., Steed, A. and Pekic, S. (1999) Bulk segregant analysis with molecular markers and its use for improving drought resistance in maize. *Journal of Experimental Botany* 50, 1299–1306.

Schwab, P.M., Barnes, D.K., Sheaffer, C.C. and Li, P.H. (1996) Factors affecting a laboratory evaluation of alfalfa cold tolerance. *Crop Science* 36, 318–324.

Taji, T., Ohsumi, C., Luchi, S., Seki, M., Kasuga, M., Kobayashi, M., Yamaguchi-Shinozaki, K. and Shinozaki, K. (2002) Important roles of drought- and cold-inducible genes for galactinol synthase in stress tolerance in *Arabidopsis thaliana. Plant Journal* 29, 417–426.

Tsarouhas, V., Gullberg, U. and Lagercrantz, U. (2004) Mapping of quantitative trait loci (QTLs) affecting autumn freezing tolerance and phenology in *Salix. Theoretical and Applied Genetics* 108, 1335–1342.

Vance, C.P. (2001) Symbiotic nitrogen fixation and phosphorus acquisition. Plant nutrition in a world of declining renewable resources. *Plant Physiology* 127, 390–397.

Varshney, A., Mohapatra, T. and Sharma, R.P. (2004) Development and validation of CAPS and AFLP markers for white rust resistance gene of *Brassica juncea. Theoretical and Applied Genetics* 109, 153–159.

Volenec, J.J., Cunningham, S.M., Haagenson, D.M., Berg, W.K., Joern, B.C. and Wiersma, D.W. (2002) Physiological genetics of alfalfa improvement: past failures, future prospects. *Field Crops Research* 75, 97–110.

Yan, H.H., Mudge, D.-J., Kim, D.-J., Shoemaker, R.C., Cook, D.R. and Young, N.D. (2004) Comparative physical mapping reveals features of microsynteny between *Glycine max*, *Medicago truncatula*, and *Arabidopsis thaliana. Genome* 47, 141–155.

Zhu, T. and Provart, N.J. (2003) Transcriptional responses to low temperature and their regulation in *Arabidopsis. Canadian Journal of Botany* 81, 1168–1174.

8 Ectopic Overexpression of *AtCBF1* in Potato Enhances Freezing Tolerance

M.-T. Pino,[1] J.S. Skinner,[1,2] Z. Jeknić,[1,2] E.J. Park,[1]
P.M. Hayes[2] and T.H.H. Chen[1,*]

[1]*Department of Horticulture, ALS 4017, Oregon State University, Corvallis, OR 97331, USA;* [2]*Department of Crop and Soil Science, Oregon State University, Corvallis, OR 97331, USA*

Introduction

The incidence of frost has a significant impact on worldwide agricultural production, causing considerable crop-productivity loss and limiting the geographical distribution of many important crop species. While it is grown in a wide range of climates, cultivated potato (*Solanum tuberosum*) is a frost-sensitive species. In most potato production areas, frost can significantly reduce yield, and particularly hard frosts can even result in the complete destruction of an entire plantation. All *S. tuberosum* cultivars are frost-sensitive and lack the ability to cold acclimate. In contrast, many other wild tuber-bearing *Solanum* species (e.g. *S. acaule*, *S. commersonii*) can survive temperatures of about −5°C pre-acclimation, and tolerate temperatures as low as −10°C following cold acclimation (Chen and Li, 1980; Costa and Li, 1993).

An initial research approach to solve the frost injury problem of potato has involved the attempt to genetically transfer frost-tolerant genes from frost-hardy wild species to the frost-sensitive cultivated varieties. Such attempts to transfer frost-tolerant genes to cultivated potato by traditional breeding have proved to be time-consuming, and have not resulted in a significant increase in frost tolerance of cultivated potato (Cardi *et al.*, 1993; Estrada *et al.*, 1993; Iovene *et al.*, 2004). This slow progress in frost-tolerance improvement in potato and other crop plant species is mainly due to the fact that frost tolerance and cold acclimation are complex genetic traits that involve multiple genes at distinct chromosomal locations, and to the lack of knowledge of the key genes involved in the process.

*Corresponding author (e-mail: chent@science.oregonstate.edu; Fax: +1-541-737-3479)

Recently, new biotechnology-based approaches have shown promise for both the direct improvement of plant cold tolerance, and as a means of evaluating the roles of individual genes important to the process of cold acclimation. This has principally involved a concerted effort conducted in the model plant system *Arabidopsis* to understand the key genes involved in the cold signal transduction mechanism and the resultant genes that are regulated by low temperature. Cold acclimation involves the action of a large suite of cold-regulated (*cor*) genes (Fowler and Thomashow, 2002). The identification of CBF transcription factors (also called DREB1s) and their role in the coordinate regulation of *cor* genes has significantly advanced our understanding of the genes and the ways by which plants adapt to low temperature (Stockinger *et al.*, 1997; Thomashow, 1999; Shinozaki and Yamaguchi-Shinozaki, 2000; Thomashow, 2001; Fowler and Thomashow, 2002). The *Arabidopsis CBF* genes and their homologues from other species have been employed as a new approach towards improving cold tolerance in crop plants such as *Brassica napus* (Jaglo *et al.*, 2001), tomato (Hsieh *et al.*, 2002a,b) and tobacco (Kasuga *et al.*, 2004).

Wild *Solanum* species differ in frost-tolerance level and cold-acclimation potential, representing a rich genetic source for improvement of these traits in cultivated potato. The advances made in *Arabidopsis* in understanding the function of the genetic components of cold tolerance have led us to study whether an endogenous CBF pathway is involved in controlling the frost-tolerance and cold-acclimation potential of wild *Solanum* species. We are investigating this by: (i) determining if characterized *CBF*s from *Arabidopsis* affect frost-tolerance traits in wild and domesticated potato; (ii) isolating and characterizing the endogenous *CBF*s of potato; and (iii) determining the role they play in potato freezing-tolerance and cold-acclimation. The latest results on ectopic expression of *Arabidopsis CBF1* (*AtCBF1*), its effect on enhancing freezing tolerance in wild and domesticated potato species, and that ectopic *AtCBF1* overexpression in *Solanum* species mimics many of the multiple biochemical and structural changes that are observed in wild-type (WT) plants during cold acclimation will be reported in this chapter.

Frost has a Significant Impact on Potato Production

S. tuberosum, or cultivated potato, is widely grown around the world and represents the fourth most important food crop after rice, maize and wheat. In many of these regions, the risk of frost damage is prevalent and frequently results in reduced potato yield and tuber quality. In most temperate zones, frost risks for potato are mainly encountered in the early spring and late autumn. In contrast, in the Andean Highland of South America, for example, frost damage can occur any time during the growing season and seriously reduce potato production (Chen and Li, 1980; Barrientos *et al.*, 1994; Vega and Bamberg, 1995). Frost damage is also a serious problem for the highly profitable early potato crop in the Mediterranean, which is planted from late autumn to early spring, when the potential for frost damage is highest and can cause significant loss in production (Iovene *et al.*, 2004).

Frost injury of plants and tissues primarily results from the severe cellular dehydration that occurs upon formation of extracellular ice and the damage associated with it. Extracellular ice formation can lead to altered membrane properties, cellular membrane damage, protein denaturation and the generation of reactive oxygen species (ROS) that damage various cellular components (Palta and Li, 1980; Toivio-Kinnican *et al.*, 1981; Steponkus, 1984; Guy, 1990; McKersie, 1991; Thomashow, 1999). ROS accumulation is due to a failure in the electron transfer reaction, and has been connected to decreased photosynthetic efficiency, photosystem II damage and subsequent increased damage upon exposure to high intensity light (Steffen and Palta, 1989; McKersie, *et al.*, 1991; O'Kane *et al.*, 1996; McKersie *et al.*, 2000). Differences also exist in the capacity of wild versus domesticated potato to deal with freezing stress-related photosynthetic damage. While only a temporary reduction in photosynthetic capacity is observed in the wild potato *S. commersonii* following freezing stress, highly reduced and irreversible photosynthetic capacity occurs in *S. tuberosum* (Seppanen and Coleman, 2003).

Multiple Changes Occur in Potato Plants During Cold Acclimation

Multiple biochemical and morphological alterations in potato are associated with the increase in freezing tolerance that accompanies cold acclimation. These include changes in the level and composition of carbohydrates, proteins, nucleic acids, amino acids, growth regulators, phospholipids and fatty acids (Li, 1984). The process of cold acclimation is associated with both the synthesis of cryoprotective polypeptides (Artus *et al.*, 1996; Steponkus *et al.*, 1998) and the accumulation of compatible solutes with cryoprotective properties (Gilmour *et al.*, 2000; Taji *et al.*, 2002), which include free amino acids (e.g. proline), quaternary ammonium compounds (e.g. glycine betaine) and carbohydrates (e.g. sucrose). These substances are believed to play an important role by increasing the internal osmotic pressure and preventing the loss of water from the cell during freezing-induced dehydration (Nanjo *et al.*, 1999; Thomashow, 1999).

The commonly observed accumulation of free proline in the leaves, shoots and roots of angiosperms during cold stress is considered one of the most widespread stress-induced responses (Chu *et al.*, 1974). Cold acclimation-induced changes in proline content have been reported in multiple species including perennial rye grass, barley, lucerne, winter rape, winter wheat and annual bluegrass (Draper, 1972; Chu *et al.*, 1974; Paquin, 1977; Stefl *et al.*, 1978; Dionne, 2001b). In potato, an increase in leaf proline content of three- to tenfold was observed during cold hardening without changes in water status (Van Swaaij *et al.*, 1985). Also, the exogenous application of proline increased potato frost tolerance (Van Swaaij *et al.*, 1985). While some studies suggest that the increase in proline content is related more to drought stress than cold stress, this association has not been demonstrated in potato (Levy, 1983). Later studies in *Arabidopsis* have shown similar results, where photoperiod and cold acclimation were highly associated with proline-content accumulation and an enhancement in freezing tolerance (Wanner and Junttila, 1999). Similar to cold-acclimated plants, transgenic *Arabidopsis* plants ectopically expressing *CBF* genes have increased proline content and an improved freezing-tolerance capacity (Liu *et al.*, 1998; Gilmour *et al.*, 2000).

Accumulation of sugars is another commonly observed biochemical change associated with cold acclimation. Sugars act as effective cryoprotectants *in vitro* (Carpenter and Crowe, 1988) and impart a protective benefit to membranes, cells and plants during cold stress (Sanitarius, 1973; Strauss and Hausser, 1986; Livingston and Henson, 1998; Vijn and Smeekens, 1999; Taji *et al.*, 2002). Detailed descriptions of alterations in sugar content during cold acclimation have been reported for many plant species (Gilmour *et al.*, 2000; Hincha *et al.*, 2000; Dionne *et al.*, 2001a). A comparison of multiple *Solanum* species found that both free sugars and starch increase during cold acclimation, and *S. commersonii* in particular, one of the most cold-tolerant *Solanum* species, showed the highest increase in sugar content following cold acclimation (Chen and Li, 1980). However, this study also demonstrated that sugar accumulation alone cannot be the basis for the differences between *Solanum* species in cold-acclimation potential. *S. tuberosum*, which cannot cold acclimate, was also found to accumulate sugars when grown under low temperatures (Chen and Li, 1980).

Previous research has shown that a specific subset of proteins is synthesized during cold acclimation (Guy, 1990). Members of the dehydrin protein family accumulate to high levels in response to low temperature, abscisic acid (ABA) application and drought stress (Close, 1996, 1997). COR, late embryogenesis abundant (LEA) and similar soluble proteins also accumulate in cold-acclimated plants and are probably critical to the freezing-tolerance mechanism (Thomashow, 1998; Iba, 2002). An early study in tuber-bearing *Solanum* species found a high correlation between protein synthesis and freezing tolerance where the *Solanum* species capable of cold acclimation exhibited an increase in soluble proteins (Chen and Li, 1980). In *S. commersonii*, cold acclimation induces the synthesis of many new polypeptides (Tseng and Li, 1987, 1990; Ryu and Li, 1994), and similar results were reported for *S. commersonii* cell cultures in response to cold acclimation and ABA application (Lee *et al.*, 1992).

It is well documented that the membrane lipid composition changes during cold acclimation in a wide range of plants. In fact, changes in membrane fluidity and lipid composition are thought to be associated with freezing tolerance (Palta and Li, 1980; Williams *et al.*, 1988; Palta *et al.*, 1993; Steponkus *et al.*, 1993; Welti *et al.*, 2002). In a comparison between freezing-tolerant *S. commersonii* and freezing-sensitive *S. tuberosum*, of the membrane lipid changes that occur during cold acclimation, common changes noted include decreased palmitic acid and cerebroside, while increases occurred for the unsaturated to saturated fatty acid ratio, free sterols and sitosterol (Palta *et al.*, 1993). *S commersonii*-specific lipid changes included decreased linolenic acid content and the sterol to phospholipid ratio, while increases occurred for phosphatidylethanolamine, linoleic acid and the acylated steryl glycoside to steryl glycoside ratio. The results indicated that the changes in lipid composition are associated with increased freezing tolerance during cold acclimation.

The transition to low temperature is known to alter leaf growth patterns and leaf cell ultrastructure in some plant species, suggesting that these structural and morphological alterations may play an important role in freezing tolerance (Kaku, 1973; Palta and Li, 1979; Ristic and Ashworth, 1993). Changes in leaf structure and cell-wall thickness, such as formation of double to triple palisade layers, occur in frost-tolerant potato species when grown at low temperature, but are not observed in freezing-sensitive potatoes (Chen *et al.*, 1977; Palta and Li, 1979; Estrada, 1982). Similar results have been reported in other species as well, such as changes in cell ultrastructure and increased thickness of mesophyll cells leading to leaf cell enlargement in *Arabidopsis* (Ristic and Ashworth, 1993) and winter oilseed rape (Stefanowsna *et al.*, 1999, 2002).

The expression of many genes is affected during the development of freezing tolerance in cold-hardy wild potato, and has been studied most extensively in the cold-hardy wild potatoes *S. commersonii* and *S. sogarandinum*. Many of these genes encode homologues to proteins proposed to play roles in the stress-tolerance responses of other plant systems. Examples include potato homologues to osmotin, cyclophilin – an RNA-binding glycine-rich protein, dehydrins, SAM decarboxylase, glucosyl transferase and cell-cycle protein CD48, among others (Zhu *et al.*, 1993; Rorat *et al.*, 1997; Meza-Zepeda *et al.*, 1998; Rorat *et al.*, 1998; Baudo *et al.*, 1999). In addition to cold, the expression of many of these genes also responds to exogenous ABA application and other abiotic stresses, such as salt and drought. Together, these results suggest that many of these genes may be employed in multiple stress response pathways.

Attempts to Transfer Frost-hardiness Genes from Wild Potato Species to Cultivated Potato

Temperatures below −3°C are typically lethal to *S. tuberosum* cultivars (2n = 4× = 48), which are incapable of cold acclimation. In contrast, wild potato species are able to survive this temperature range, and can survive even colder temperatures after cold acclimation (Chen and Li, 1980; Costa and Li, 1993). Wild potato species, such as *S. acaule*, *S. commersonii*, *S. boliviense*, *S. chomatophium*, *S. multidissectum*, *S. megistacrolobum* and *S. sanctae-rosae*, can survive temperatures from −4°C to −6°C prior to acclimation. *S. commersonii* (2n = 2× = 24), endemic to Argentina, Paraguay and Uruguay, is one of the most cold-hardy potato species, tolerating temperatures as low as −10°C following cold acclimation (Chen and Li, 1980).

Attempts to transfer frost-hardiness genes by traditional breeding from wild genotypes to cultivated potato have been challenging and met with minimal success to date. Potato breeders have also employed somatic fusion, embryo rescue and bridging-cross strategies to overcome the natural barriers associated with interspecific crossing between wild and cultivated potato species. Unfortunately, linkage drag typically limits the use of wild potato species, with many exotic genes which impart undesirable traits (e.g. high alkaloid content, long stolons) being co-transferred with cold-hardiness genes. Under these potato breeding schemes, time-consuming backcrosses, evaluations and phenotypic selections are needed to restore the desired phenotype of an improved cultivar (Cardi *et al.*, 1993; Estrada *et al.*, 1993; Pavek and Corsini, 2001; Iovene *et al.*, 2004).

Inheritance studies on F1 and backcross progenies of *S. commersonii* × *S. cardiophylum* showed that freezing-tolerance and cold-acclimation traits are under independent genetic control and determined by a small number of genes

(Stone *et al.*, 1993). Characterization of somatic hybrids between frost-tolerant *S. commersonii* and frost-sensitive *S. tuberosum* has been carried out without any notable increase in cold hardiness, with most of the results showing a minor increase in freezing tolerance after cold acclimation, but not prior to (Cardi *et al.*, 1993; Nyman and Waara, 1997; Palta *et al.*, 1997). For example, the offspring of *S. commersonii* × *S. tuberosum* somatic hybrids, while freezing-sensitive like the *S. tuberosum* parent prior to cold acclimation, increase slightly in freezing tolerance following cold acclimation (Chen *et al.*, 1996). Similar results were reported in a later study characterizing the freezing tolerance of selfed and backcrossed progenies derived from *S. tuberosum* × *S. commersonii* somatic hybrids (Chen *et al.*, 1999). Aneuploid hybrids derived from 5X × 4X crosses of *S. commersonii* × *S. tuberosum* had lethal freezing temperatures (LT_{50}) that did not differ from those of cultivated potato prior to acclimation, while the LT_{50} of the hybrids following cold acclimation were distributed between the wild and cultivated parental values, with an observed maximum 3°C gain in freezing tolerance (Iovene *et al.*, 2004).

Transformation of Agronomic Crops with *Arabidopsis CBF* Genes Enhances Freezing Tolerance

The CBF (C-repeat-binding factor) transcription factors (also termed DREB1, dehydration-responsive element-binding factor) play an important role in abiotic stress response in *Arabidopsis*, including freezing tolerance and cold acclimation (Gilmour *et al.*, 1998; Jaglo-Ottosen *et al.*, 1998; Liu *et al.*, 1998; Medina *et al.*, 1999; Shinozaki and Yamaguchi-Shinozaki, 2000). A more detailed description of the role(s) of CBFs in plant cold acclimation can be found in several chapters in this volume (chapters 1–4).

CBF genes have been isolated from a wide range of plant species, indicating that the CBF response pathway is probably conserved among most plant species. Constitutive overexpression of *CBF1*, *CBF3* and *CBF4* in *Arabidopsis* induces both the expression of multiple *cor* genes and increase in freezing tolerance without a low-temperature stimulus (Jaglo-Ottosen *et al.*, 1998; Liu *et al.*, 1998; Gilmour *et al.*, 2000; Haake *et al.*, 2002). Overexpression of *CBF3* in *Arabidopsis* has been shown to induce the expression of *cor* genes and results in a concomitant increase in freezing, salt and drought tolerance under non-stressed conditions (Liu *et al.*, 1998; Kasuga *et al.*, 1999; Gilmour *et al.*, 2000; Maruyama *et al.*, 2004). Many of the biochemical changes associated with the cold acclimation process are also observed in *CBF3*-overexpressing plants, including an increase in the total soluble sugars glucose, fructose, sucrose and raffinose to levels similar to those of cold-acclimated *Arabidopsis* plants (Gilmour *et al.*, 2000).

Ectopic overexpression of the *Arabidopsis CBFs* has also been conducted in a number of crop plants and resulted in the successful improvement of stress-tolerance characteristics, including cold tolerance (Table 8.1). Due to its rapid and direct mode, in contrast to traditional breeding, genetic engineering with the *Arabidopsis CBF* genes, or *CBF* homologues from other species, has been suggested as an effective and general means of improving stress tolerance in crop plants (Holmberg and Bulow, 1998; Bajaj *et al.*, 1999; Zhang *et al.*, 2000). However, constitutive overexpression of the *Arabidopsis CBF* genes is often associated with undesirable phenotypic alterations of the transgenic plants. For example, while *Arabidopsis* plants constitutively overexpressing *CBF3* have enhanced cold tolerance, both severe growth retardation and a delay in flowering are also observed in unstressed conditions (Liu *et al.*, 1998; Gilmour *et al.*, 2000). Similar growth retardation results have also been observed in transgenic tobacco and tomato plants overexpressing *CBF1 and CBF3* (Hsieh *et al.*, 2002a,b; Kasuga *et al.*, 2004), side effects that are undesirable in crop plants. To address this problem, the stress-inducible *rd29A* promoter has been used to control *CBF* transgene expression. Studies in *Arabidopsis* and tobacco have shown that use of the *rd29A* promoter alleviates these negative effects on plant growth in transgenic plants for the most part (Kasuga *et al.*, 1999, 2004). Together, these results suggest that it is possible to use *CBF* genes to improve freezing tolerance in crops plants while minimizing negative pleiotropic effects.

Table 8.1. Examples of *Arabidopsis CBF*s imparting enhanced cold tolerance to transgenic plants.

Arabidopsis CBF	Promoter[a]	Transgenic species	Phenotypic expression	References
CBF1	35S	*Arabidopsis*	Increased freezing tolerance	Jaglo-Ottosen *et al.* (1998)
CBF1	rd29A, 35S	*Arabidopsis*	Increased cold, drought and salinity tolerance	Kasuga *et al.* (1999)
CBF3	35S	*Arabidopsis*	Increased freezing tolerance	Liu *et al.* (1998), Gilmour *et al.* (2000)
CBF1, CBF2, CBF3	35S	Canola	Increased freezing tolerance	Jaglo *et al.* (2001)
CBF1	35S	Strawberry	Increased freezing tolerance	Owens *et al.* (2002)
CBF1	35S	Tomato	Enhanced water deficit, oxidative and chilling stress tolerance	Hsieh *et al.* (2002a,b)
CBF3	rd29A, 35S	Tobacco	Increased drought and cold stress tolerance	Kasuga *et al.* (2004)
CBF4	35S	*Arabidopsis*	Increased freezing tolerance	Haake *et al.* (2002)

[a]35S: a constitutive CaMV 35S promoter; rd29A: a stress-inducible *Arabidopsis rd29A* promoter.

As a first step in dissecting the role of a CBF cold response pathway in potato, we transformed two species that contrast in cold tolerance, S. *commersonii* and *S. tuberosum*, with a 35S promoter-driven *AtCBF1* construct, and determined the effect on freezing-tolerance and cold-acclimation potential. We assessed whether *AtCBF1* overexpression leads to improvement of frost-tolerance and/or cold-acclimation potential in each species, if the level of improvement differed, and if any of the phenotypic changes observed in WT plants during cold acclimation were imparted.

Transformation of *S. commersonii* and *S. tuberosum* with *35S::AtCBF1*

We generated transgenic *S. tuberosum* and *S. commersonii* lines overexpressing *AtCBF1* under control of the constitutive CaMV 35S promoter. Explants of leaf and stem from *S. commersonii* PI 243503 (clone 13) and *S. tuberosum* (cv. Umatilla) plants were transformed via *Agrobacterium*-mediated transformation. Putative transgenic-potato shoots were selected *in vitro* based on Kanamycin resistance (50 mg/l), and polymerase chain reaction (PCR) was used to confirm the presence of the transgene in the potato genome for each line.

We identified 13 and 19 PCR-positive clones from *S. commersonii* and *S. tuberosum*, respectively. Northern blot analysis was used to determine which PCR-positive lines were expressing the transgene at a detectable level, and ten *S. commersonii* and three *S. tuberosum* lines expressing the *AtCBF1* transgene were identified. A much higher frequency (10/13) of the *S. commersonii* transgenics expressed detectable *AtCBF1*-transgene levels compared to *S. tuberosum*, where only 3 of 19 lines had detectable levels. Among the *S. commersonii* lines, most accumulated high levels of *AtCBF1* transcript. In contrast, *AtCBF1* transcript levels in the transgenic *S. tuberosum* lines were much lower than the level of the *S. commersonii* lines. The difference between the two species in recovery efficiency of *AtCBF1* transgene-expressing lines is currently unknown, but suggests that *S. commersonii* is better able to tolerate any negative side effects of increased *AtCBF1* expression than *S. tuberosum* cv. Umatilla; extending this to additional cultivars will determine if this observation is particular to the cv. Umatilla genotype.

Ectopic Overexpression of *AtCBF1* in *S. commersonii* Results in Altered Plant Morphology

Similar to previous studies evaluating *AtCBF* overexpression in foreign species (Jaglo *et al.*, 2001; Hsieh *et al.*, 2002a), the majority of the transgenic *S. commersonii* lines displayed visible phenotypic abnormalities relative to WT plants. Varying degrees of growth retardation were noted among the transgenic lines when grown in tissue culture. Following transplantation to soil and growth under greenhouse conditions, many of the lines recovered and grew to a similar height as those of WT plants (Fig. 8.1). This result suggests that some of the initial growth alterations may have been due to the effects of the tissue-culture growth conditions. Nevertheless, several independent transgenic lines continued to show severe growth retardation, even after a long growth period in the greenhouse, confirming that ectopic *AtCBF1* expression can also yield stable pleiotropic effects in *S. commersonii*; phenotypic alterations in flowers, leaf colour and leaf shape were also observed (data not shown). Significant phenotypic alterations or negative pleiotropic growth effects among the three transgenic *35S::AtCBF1 S. tuberosum* lines were not observed (data not shown). This may be due to the low *AtCBF1*-expression levels in these transgenic *S. tuberosum* lines, relative to the substantially higher levels seen in *S. commersonii*.

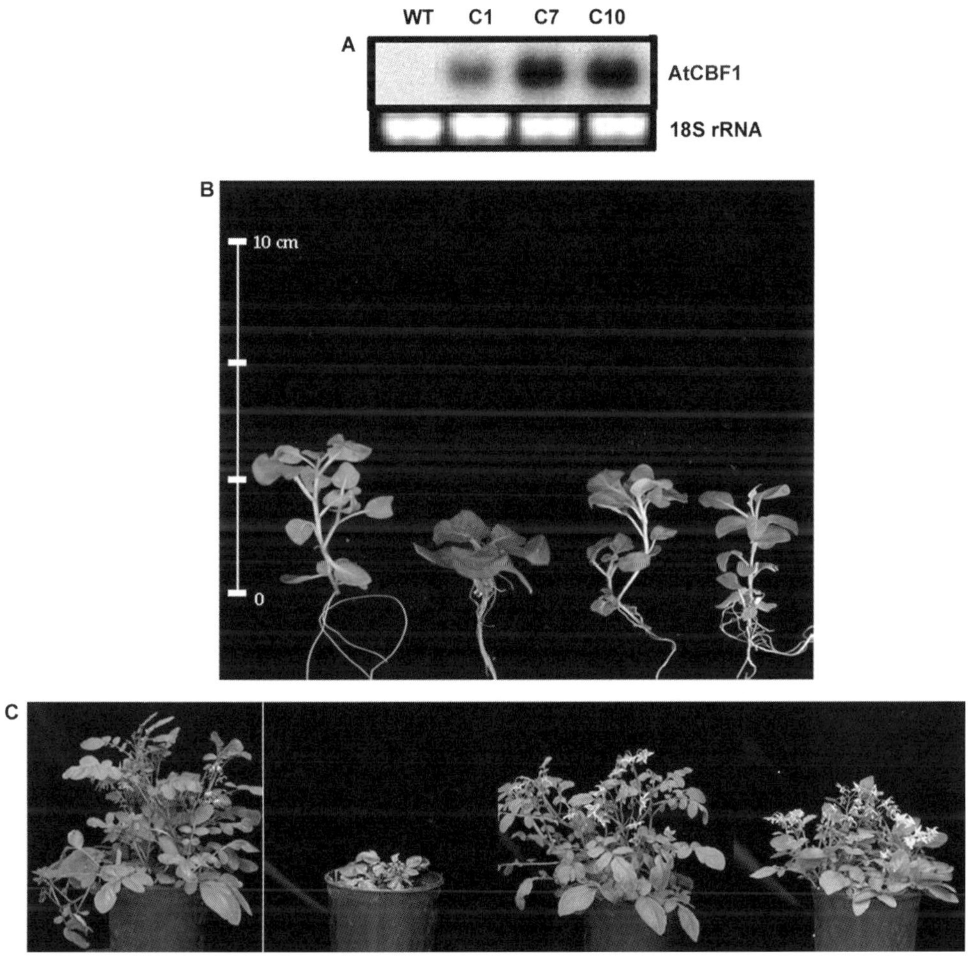

Fig. 8.1. Examples of phenotypic variations in wild-type (WT) and *AtCBF1*-overexpressing transgenic *S. commersonii*. (A) Northern blot analysis. (B) WT and *AtCBF1*-transgenic plants grown in tissue culture. (C) Twelve-week-old WT and *AtCBF1*-transgenic plants grown under greenhouse conditions. From left to right, WT and transgenic lines C1, C7 and C10.

Ectopic Overexpression of *AtCBF1* Enhances Freezing Tolerance

WT and transgenic *S. tuberosum* and *S. commersonii* plants were grown at either 25°C or cold acclimated for 2 weeks at 2°C, and freezing tolerance evaluated using a controlled freezing test (Jaglo-Ottosen *et al.*, 1998; Gilmour *et al.*, 2000). Freezing tolerance (expressed as LT_{50}) for WT *S. commersonii* plants was −6°C and −10°C, prior to and after 2 weeks of cold acclimation, respectively (Fig. 8.2). In contrast, WT plants of *S. tuberosum* had an LT_{50} of −3°C under both conditions, verifying that *S. tuberosum* is more sensitive to cold and fails to undergo cold acclimation (data not shown).

Under warm (non-acclimating) conditions, transgenic *S. commersonii* lines were from 2°C to 4°C cold hardier than WT plants (Fig. 8.2). Following 2 weeks of cold acclimation at 2°C, these transgenic lines were also hardier than the cold-acclimated WT plants. These results indicate that the constitutive overexpression of *AtCBF1* in *S. commersonii* increased both freezing-tolerance and cold-acclimation potential (Fig. 8.2). The results also suggest that an endogenous *CBF* pathway is involved in this process in *S. commersonii*, and that the constitutive *AtCBF1* activity is artificially stimulating and enhancing the pathway. We also determined LT_{50} values in WT and two of the transgenic lines cold acclimated at 2°C from 0 to 21 days, and found that both WT and transgenic plants displayed a linear increase in freezing tolerance that correlated with the length of cold acclimation at 2°C (r^2 = 0.85, $P \leq 0.0001$, from regression analysis). A similar exper-

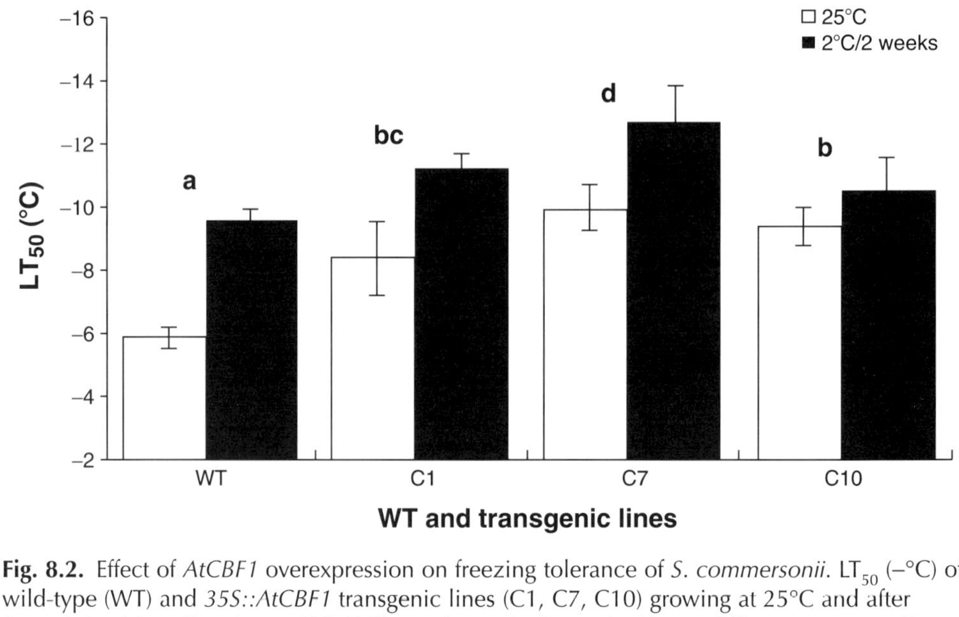

Fig. 8.2. Effect of *AtCBF1* overexpression on freezing tolerance of *S. commersonii*. LT_{50} (−°C) of wild-type (WT) and *35S::AtCBF1* transgenic lines (C1, C7, C10) growing at 25°C and after 2 weeks' cold acclimation at 2°C. Different letters indicate significant differences among lines ($P \leq 0.0001$) according to Duncan's multiple range test.

iment was conducted using WT and transgenic *S. tuberosum* plants, and transgenic lines were about 2°C cold hardier than WT plants both before and after cold acclimation (data not shown). This indicated that overexpression of *AtCBF1* increased freezing tolerance, but did not impart a cold-acclimation response to *S. tuberosum* plants. This also suggests that the freezing tolerance property of *S. tuberosum* may be responsive to enhanced levels of CBF activity.

Overexpression of *AtCBF1* in *S. commersonii* Results in Anatomical and Ultrastructural Changes in Leaves

Light microscopy cross-sectional analysis of transgenic *S. commersonii* leaves revealed both increased leaf thickness and an altered pattern of palisade and sponge cell distribution (Fig. 8.3). Transgenic plants showed significantly thickened leaves, enlarged intercellular spaces and elongated palisade cells when grown under both warm- and cold-acclimating conditions. Similar alterations were observed in WT leaves following cold acclimation, suggesting that the post-acclimation changes are related to an endogenous *CBF* activity and that constitutive *AtCBF1* expression under warm conditions in *S. commersonii* mimics the cold-acclimation effect (Fig. 8.3). In contrast, a comparative cross-sectional analysis of WT and transgenic *S. tuberosum* leaves did not show significant variations in the leaf traits mentioned above. While leaves of some *S. tuberosum* transgenic lines were thicker than WT leaves, the differences were not significant between the unacclimated and cold-acclimated plants (Fig. 8.4).

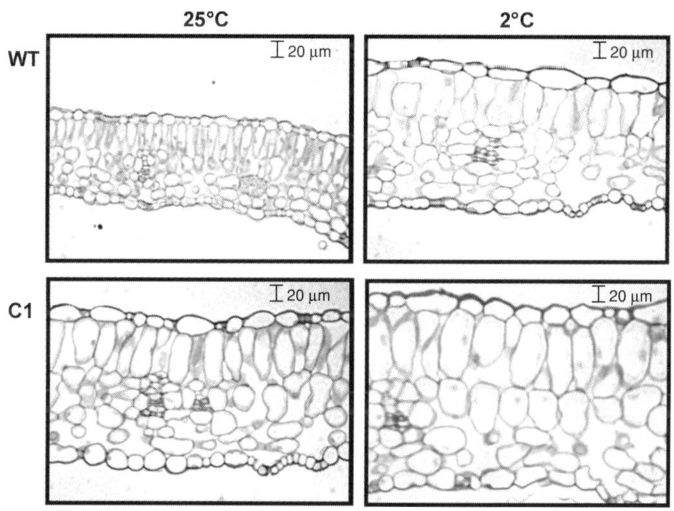

Fig. 8.3. Leaf cross sections from wild-type (WT) and *S. commersonii* transgenic-line C1 plants. Cross sections were prepared from leaves of WT and *AtCBF1*-overexpressing line C1 grown at 25°C and after 2 weeks of cold acclimation at 2°C. Light micrographs were taken at 400×.

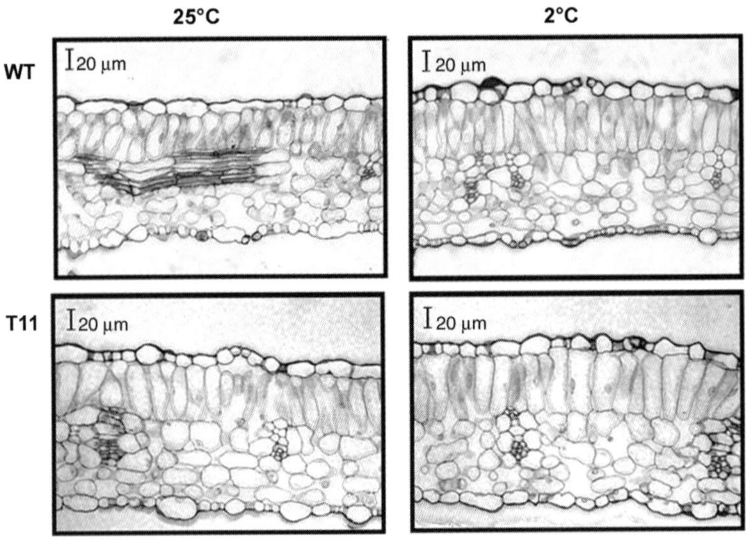

Fig. 8.4. Leaf cross sections from wild-type (WT) and *S. tuberosum* transgenic-line T11 plants. Cross sections were prepared from leaves of WT and *AtCBF1*-overexpressing line T11 grown at 25°C and after 2 weeks of cold acclimation at 2°C. Light micrographs were taken at 200×.

Transmission electron microscopy (TEM) conducted on the palisade and spongy mesophyll cells of transgenic lines revealed that, relative to WT, transgenics had an increase in chloroplast number, and decreases in the oil body number and both the size and number of starch grains (Fig. 8.5). Scanning EM analysis of the S. *commersonii* adaxial (top) and abaxial (bottom) leaf surfaces under non-acclimating conditions showed an increase in stomata per cm^2 and altered stomata morphology. This was especially true for the adaxial leaf surface, where an approximately threefold increase in stomata had occurred (data not shown).

Biochemical Changes Associated with Cold Acclimation are Observed in *AtCBF1*-Transgenic *S. commersonii*

In *Arabidopsis*, proline and sugar levels increase during cold acclimation (Wanner and Junttila, 1999). An increase in proline and sugar levels is also observed under non-acclimating conditions for *Arabidopsis* plants ectopically overexpressing *CBF3* (Gilmour *et al.*, 2000). We observed similar results in WT and *35S::AtCBF1* transgenic *S. commersonii* (Fig. 8.6). Both proline and sugar content increase in WT *S. commersonii* following 2 weeks of cold acclimation. In unacclimated transgenic *S. commersonii*, proline content increased more than twofold relative to WT plants. Following 2 weeks of cold acclimation, the proline content of transgenic *S. commersonii* increased above the pre-acclimation levels. Similar results were

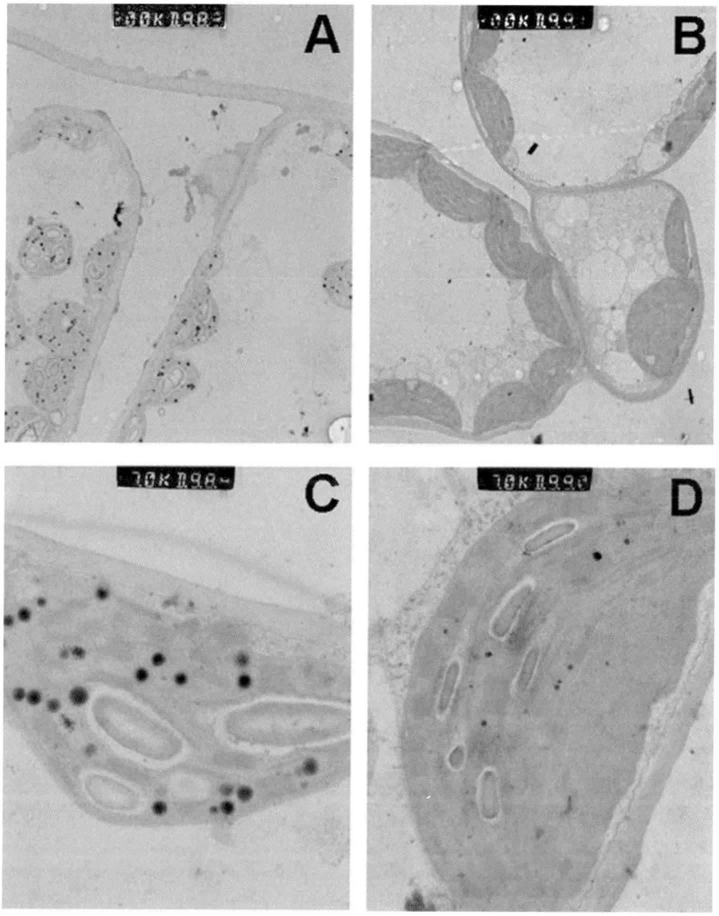

Fig. 8.5. Transmission electron micrographs of leaf cross sections of wild-type (A and C) and *AtCBF1*-transgenic *S. commersonii* line C7 (B and D).

observed for sugar content, which, relative to WT plants, was increased in transgenic plants under both unacclimated and acclimated conditions.

S. commersonii Plants Overexpressing *AtCBF1* Showed Higher Photosynthetic Capacity

We noticed that leaves of *S. commersonii AtCBF1*-overexpressing plants were generally greener than WT leaves when grown side by side, and that *AtCBF1*-overexpressing lines tend to have a higher chloroplast number per cell (Fig. 8.5). We thus compared total chlorophyll content of WT plants with one of the transgenic lines (C7) under both non-acclimated and acclimated conditions (Fig. 8.7B). While total chlorophyll (Chl *a* + *b*) content was indeed higher in the unacclimated leaves

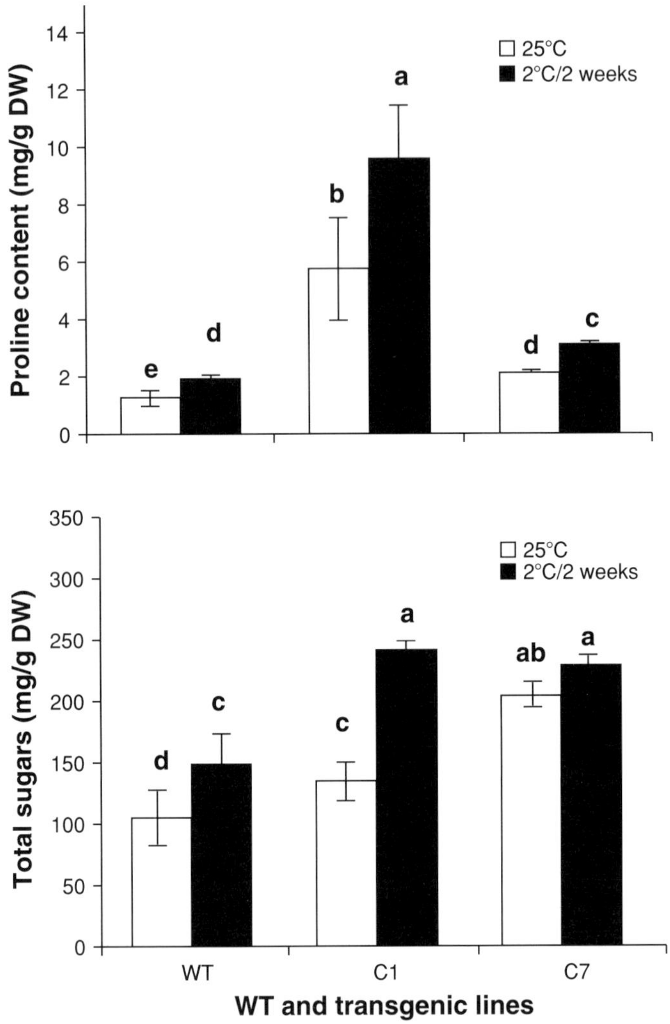

Fig. 8.6. Effect of *AtCBF1* overexpression on proline and total sugar content of *S. commersonii*. Proline and sugar content were determined from leaf tissue of wild-type (WT) and *AtCBF1*-overexpressing lines C1 and C7 grown at 25°C and after 2 weeks' cold acclimation at 2°C. Different letters indicate significant differences among lines and treatments ($P \leq 0.05$) according to Duncan's multiple range test.

of the *AtCBF1*-transgenic line relative to WT, the process of cold acclimation did not lead to a further alteration of the total chlorophyll content in either background.

Based on our observations, including the altered leaf anatomy and ultrastructural changes, we suspected that photosynthetic activities were also altered in *AtCBF1*-overexpressing lines. We compared a variety of photosynthetic parameters between WT and *AtCBF1*-overexpressing plants under both non-acclimated and acclimated conditions (Fig. 8.7). Transpiration rates were comparable between

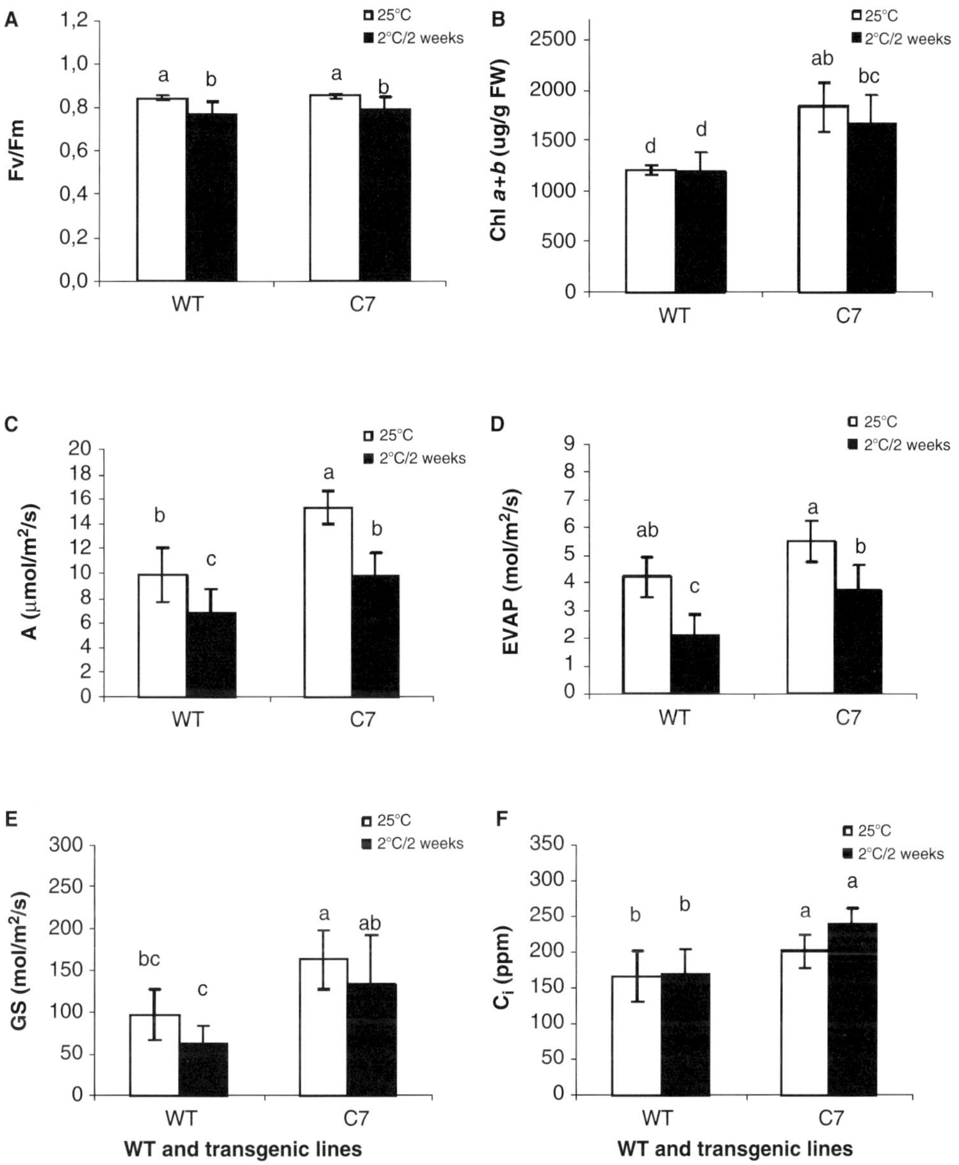

Fig. 8.7. Effect of *AtCBF1* overexpression on *S. commersonii* photosynthesis parameters. (A) Chlorophyll fluorescence [Fv/Fm]. (B) Total chlorophyll [Chl *a+b*]. (C) Photosynthesis rate [A]. (D) Transpiration rate [EVAP]. (E) Stomatal conductance [GS]. (F) Internal CO_2 concentration [C_i]. Assays were conducted on wild-type (WT) and *AtCBF1*-overexpressing line C7 grown at 25°C and after 2 weeks' cold acclimation at 2°C. Different letters note significant differences among lines and treatments ($P \leq 0.05$) according to Duncan's multiple range test.

WT and transgenic plants (Fig. 8.7D), and while both WT and *AtCBF1*-over-expressing lines are slightly photo-inhibited following cold acclimation, based on the decreased Fv/Fm ratio (Fig. 8.7A), the effect is not significantly different. Increases were found for photosynthetic rate (Fig. 8.7C), stomatal conductance (Fig. 8.7E) and internal CO_2 concentration (Fig. 8.7F) of transgenic plants relative to WT. These results suggest that *AtCBF1* overexpression results in an enhanced photosynthetic capacity of the transgenic plants.

Conclusions

Ectopic overexpression of *AtCBF1* significantly increased the freezing-tolerance and cold-acclimation potential of *S. commersonii*. In contrast, in *S. tuberosum*, while *AtCBF1* overexpression increased freezing tolerance, it failed to impart a cold-acclimation response to the plant. These results suggest that an endogenous *CBF* pathway is present in both plant systems and that enables the ectopic *AtCBF1* transgene expression to influence the plants' freezing-tolerance capacity. A *CBF*-based cold-acclimation pathway also appears to be present in *S. commersonii* and is affected by *AtCBF1*. This acclimation pathway appears to be disrupted in *S. tuberosum*, at some point past the *CBF* signal, as introduction of the constitutive presence of an active *CBF* failed to impart an acclimation capacity on this species.

Cold acclimation caused marked alterations in the leaf anatomy of *S. commersonii*. These included thickened leaves with smaller epidermal cells, disorganized palisade and spongy mesophyll cell layers, and increased intercellular spaces. Similar changes were not observed in *S. tuberosum* plants, which failed to cold-acclimate, suggesting that the *S. commersonii* changes are related to the acclimation process. Interestingly, we also observed these same anatomical changes in unacclimated *S. commersonii* plants overexpressing *AtCBF1*. This suggests that the cold-induced structural changes are an associated component of acclimation-based freezing-tolerance enhancement, and may be a product of the induction of endogenous *CBF* activity during the acclimation process. In addition to effects on freezing tolerance, *AtCBF1* overexpression also alters traits related to photosynthesis in *S. commersonii*. These included an increased chloroplast number and alterations of chloroplast structure and chlorophyll content, which collectively lead to an alteration of photosynthetic properties. Taken together, our results suggest that constitutive overexpression of *AtCBF1* in potato can enhance freezing tolerance. However, distinct morphological and physiological alterations are associated with this enhancement, which could have a negative contribution to the plant relative to desirable crop characteristics. We are currently investigating whether use of a stress-inducible promoter, as well as other *CBF* genes, can retain the beneficial gains in freezing tolerance, without the associated negative phenotypic alterations.

Acknowledgements

This study was supported in part by a grant from the NSF Plant Genome Project (DBI 0110124). M. Teresa Pino was supported by a fellowship from the National Research Institute for Agriculture of Chile (INIA).

References

Artus, N.N., Uemura, M., Steponkus, P.L., Gilmour, S.J., Lin, C.T. and Thomashow, M.F. (1996) Constitutive expression of the cold-regulated *Arabidopsis thaliana* COR15a gene affects both chloroplast and protoplast freezing tolerance. *Proceedings of the National Academy of Sciences USA* 93, 13404–13409.

Bajaj, S., Targolli, J., Liu, L.F., Ho, T.H.D. and Wu, R. (1999) Transgenic approaches to increase dehydration-stress tolerance in plants. *Molecular Breeding* 5, 493–503.

Barrientos, M., Mol, E., Peruzzo, A., Contreras, A. and Alberdi, M. (1994) Responses to cold of Chilean wild *Solanum* species. *Environmental and Experimental Botany* 34, 47–54.

Baudo, M., Meza-Zepeda, L., Palva, E. and Heino, P. (1999) Isolation of a cDNA corresponding to a low temperature- and ABA-responsive gene encoding a putative glycine-rich RNA-binding protein in *Solanum commersonii. Journal of Experimental Botany* 50, 1867–1868.

Cardi, T.K., Puite, K.S., Ramulu Dámbriosio, F.D. and Frusciante, L. (1993) Production of somatic hybrid between frost-tolerant *Solanum commersonii* and *Solanum tuberosum*: protoplast fusion, regeneration and isozyme analysis. *American Potato Journal* 70, 753–764.

Carpenter, J.F. and Crowe, J.H. (1988) The mechanism of cryoprotection of protein solutes. *Cryobiology* 25, 244–255.

Chen, P.M., Li, P.H. and Cunningham, W.P. (1977) Ultrastructure difference in leaf cell of *Solanum* species in relation to their frost resistance. *Botanical Gazette* 138, 267–285.

Chen, T.H.H. and Li, P.H. (1980) Characteristics of cold acclimation and deacclimation in tuber-bearing *Solanum* species. *Plant Physiology* 65, 1146–1148.

Chen, Y.K.H., Palta, J.P., Bamberg, J.B., Helgeson, J.B. and Haberlach, G.T. (1996) Expression of freezing tolerance in somatic hybrid hardy wild and cultivated potato species. *American Potato Journal* 73, 348.

Chen, Y.K.H., Palta, J.P. and Bamberg, J.B. (1999) Freezing tolerance and tuber production in selfed and backcross progenies derived from somatic hybrid between *Solanum tuberosum L.* and *Solanum commersonii Dun. Theoretical and Applied Genetics* 99, 100–107.

Chu, T.M., Aspinall, D. and Paleg, F.J. (1974) Stress metabolism VI. Temperature stress and the accumulation of proline in barley and radish. *Australian Journal of Plant Physiology* 1, 87–97.

Close, T.J. (1996) Dehydrins: emergence of a biochemical role of a family of plant dehydration proteins. *Physiologia Plantarum* 97, 795–803.

Close, T.J. (1997) Dehydrins: a commonality in the response of plants to dehydration and low temperature. *Physiologia Plantarum* 100, 291–296.

Costa, A. and Li, P.H. (1993) Development of cold hardiness in *Solanum tuberosum* by absisic acids and mefluidide. In: Li, P.H. and Christersson, L. (eds) *Advances in plant hardiness.* CRC Press, Boca Raton, Florida, pp. 139–140.

Dionne, J., Castonguay, Y., Nadeau, P. and Desjardins, Y. (2001a) Freezing tolerance and carbohydrate changes during cold acclimation of green-type annual bluegrass (*Poa annua L.)* ecotypes. *Crop Science* 41, 443–451.

Dionne, J., Castonguay, Y., Nadeau, P. and Desjardins, Y. (2001b) Amino acid and protein changes during cold acclimation of green-type annual bluegrass (*Poa annua L.*) ecotypes. *Crop Science* 41, 1862–1870.

Draper, S.R. (1972) Amino acids changes associated with low-temperature treatment of *Lolium perenne. Phytochemistry* 11, 639–641.

Estrada, N. (1982) Breeding wild and primitive potato species to obtain frost resistant cultivated varieties. In: Li, P.H. and Sakai, A. (eds) *Plant Hardiness and Freezing Stress: Mechanism and Crop Implications.* Academic Press, New York, pp. 615–633.

Estrada, N., Garcia, W., Carrasco, J. and Carrasco, E. (1993) Breeding potato for tolerance to frost and resistance to late blight. *American Potato Journal* 70, 809–810.

Fowler, S. and Thomashow, M.F. (2002) *Arabidopsis* transcriptome profiling indicates that multiple regulatory pathways are activated during cold acclimation in addition to the CBF cold response pathway. *Plant Cell* 14, 1675–1690.

Gilmour, S.J., Zarka, D.G., Stockinger, E.J., Salazar, M.P., Houghton, J.M. and Thomashow, M.F. (1998) Low temperature regulation of the *Arabidopsis* CBF family of AP2 transcriptional activators as an early step in cold-induced COR gene expression. *Plant Journal* 16, 433–442.

Gilmour, S.J., Sebolt, A.M., Salazar, M.P., Everard, J.D. and Thomashow, M.F. (2000) Overexpression of the *Arabidopsis CBF3* transcriptional activator mimics multiple biochemical changes associated with cold acclimation. *Plant Physiology* 124, 1854–1865.

Guy, C. (1990) Cold acclimation and freezing tolerance – role of protein metabolism. *Annual Review of Plant Physiology and Plant Molecular Biology* 41, 187–223.

Haake V., Cook, D., Riechmann, J.L., Pineda, O., Thomashow, M.F. and Zhang, J.F. (2002) Transcription factor CBF4 is a regulator of drought adaptation in *Arabidopsis*. *Plant Physiology* 130, 639–648.

Hincha, D.K., Hellwege, E.M., Heyer, A.G. and Crowe, J.H. (2000) Plant fructans stabilize phosphatidylcholine liposomes during freeze-drying. *European Journal of Biochemistry* 267, 535–540.

Holmberg, N. and Bulow, L. (1998) Improving stress tolerance in plants by gene transfer. *Trends in Plant Science* 3, 61–66.

Hsieh, T.H., Lee, J.T., Yang, P.T., Chiu, L.H., Charng, Y.Y., Wang, Y.C. and Chan, M.T. (2002a) Heterology expression of the *Arabidopsis* C-Repeat/dehydration response element binding factor 1 gene confers elevated tolerance to chilling and oxidative stress in transgenic tomato. *Plant Physiology* 129, 1086–1094.

Hsieh, T.H., Lee, J.T., Charng, Y.Y. and Chan, M.T. (2002b) Tomato plants ectopically expressing *Arabidopsis* CBF1 show enhanced resistance to water deficit stress. *Plant Physiology* 130, 618–626.

Iba, K. (2002) Acclimative response to temperature stress in higher plants: approaches of gene engineering for temperature tolerance. *Annual Review of Plant Physiology and Plant Molecular Biology* 53, 225–245.

Iovene, M., Barone, A., Frusciante, L. and Monti, L. (2004) Selection for aneuploid potato hybrids combining a low wild genome content and resistance traits from *S. commersonii*. *Theoretical and Applied Genetics* 109, 1139–1146.

Jaglo, K.R., Kleff, S., Amundsen, K.L., Zhang, X., Haake, V., Zhang, J.Z, Deits, T. and Thomashow, M.F. (2001) Components of *Arabidopsis* C-repeat/dehydration-responsive element binding factor cold-response pathway are conserved in *Brassica napus* and other plants species. *Plant Physiology* 127, 910–917.

Jaglo-Ottosen, K.R., Gilmour, S.J., Zarka, D.G., Schabenberger, O. and Thomashow, M.F. (1998) *Arabidopsis* CBF1 overexpression induces COR genes and enhances freezing tolerance. *Science* 280, 104–106.

Kaku, S. (1973) High ice nucleating ability in plants' leaves. *Plant and Cell Physiology* 14, 1035–1038.

Kasuga, M., Liu, Q., Miura, S., Yamaguchi-Shinozaki, K. and Shinozaki, K. (1999) Improving plant drought, salt, and freezing tolerance by gene transfer of a single stress-inducible transcription factor. *Nature Biotechnology* 17, 287–292.

Kasuga, M., Miura, S., Shinozaki, K. and Yamaguchi-Shinozaki, K. (2004) A combination of *Arabidopsis* DREB1A gene and stress inducible rd29A promoter improved drought and low temperature stress tolerance in tobacco by gene transfer. *Plant and Cell Physiology* 45, 346–350.

Lee, S.P., Zhu, B., Chen, T.H.H. and Li, P.H. (1992) Induction of freezing tolerance in potato (*Solanum commersonii*) suspension cultured cell. *Physiologia Plantarum* 84, 41–48.

Levy, D. (1983) Water deficit enhancement of proline and α-amino nitrogen accumulation in potato plants and its association with susceptibility to drought. *Physiologia Plantarum* 57, 169–173.

Li, P.H. (1984) Subzero temperature stress physiology of herbaceous plants. *Horticultural Review* 6, 373–416.

Liu, Q., Kasuga, M., Sakuma, Y., Abe, H. and Miura, S. *et al.* (1998) Two transcription factors, DREB1 and DREB2, with an EREBP/AP2 DNA binding domain separate two cellular signal transduction pathways in drought- and low-temperature-responsive gene expression, respectively, in *Arabidopsis*. *Plant Cell* 10, 1391–1406.

Livingston, D.P. and Henson, C.A. (1998) Apoplastic sugars, fructans, fructan exohydrolase, and invertase in winter oat: responses to second-phase cold hardening. *Plant Physiology* 116, 403–408.

Maruyama, K., Sakuma, Y., Kasuga, M., Ito, I., Seki, M., Goda, H., Shimada, Y., Yhoshida, S., Shinozaki, K. and Yamaguchi-Shinozaki, K. (2004) Identification of cold inducible downstream genes from *Arabidopsis* DREB1A/CBF3 transcriptional using two microarray system. *Plant Journal* 38, 982–993.

McKersie, B.D. (1991) The role of oxygen free radicals in mediating freezing and desiccation stress in plants. In: Pell, E. and Steffen, K. (eds) *Active Oxygen/Oxidative Stress and Plant Metabolism.* American Society of Plant Physiologists, Rockville, Maryland, pp. 107–118.

McKersie, B.D., Murnaghan, J., Jones, K.S. and Bowley, S.R. (2000) Iron-superoxide dismutase expression in transgenic alfalfa increase winter survival without a detectable increase in photosynthetic oxidative stress tolerance. *Plant Physiology* 122, 1427–1437.

Medina, J., Bargues, M., Terol, J., Pérez-Alonso, M. and Salinas, J. (1999) The *Arabidopsis* CBF gene family is composed of three genes encoding AP2 domain-containing proteins whose expression is regulated by low temperature but not by abscisic acid or dehydration. *Plant Physiology* 119, 463–470.

Meza-Zepeda, L., Baudoz, M., Palva, E. and Heino, P. (1998) Isolation and characterization of a cDNA corresponding to a stress-activated cyclophilin gene in *Solanum commersonii. Journal of Experimental Botany* 49, 1451–1455.

Nanjo, T., Kobayashi, M., Yoshiba, Y., Kakubari, Y., Yamaguchi-Shinozaki, K. and Shinozaki, K. (1999) Antisense suppression of proline degradation improves tolerance to freezing and salinity in *Arabidopsis thaliana. FEBS letters* 461, 205–210.

Nyman, M. and Waara, S. (1997) Characterization of somatic hybrids between *Solanum tuberosum* and its frost-tolerant relative *Solanum commersonii. Theoretical Applied Genetics* 95, 1127–1132.

O'Kane, D., Gill, V., Boyd, P. and Burdon, R. (1996) Chilling, oxidative stress and antioxidant responses in *Arabidopsis thaliana* callus. *Planta* 198, 371–377.

Owens, C.L., Thomashow, M.F., Hancock, J.F. and Iezzoni, A.F. (2002) CBF1 orthologs in sour cherry and strawberry and the heterologous expression of CBF1 in strawberry. *Journal of American Society for Horticultural Sciences* 127, 489–494.

Palta, J.P. and Li, P.H. (1979) Frost hardiness in relation to leaf anatomy and natural distribution of several *Solanum* species. *Crop Science* 19, 665–671.

Palta, J.P. and Li, P.H. (1980) Alterations in membrane transport properties by freezing injury in herbaceous plants, evidence against the rupture theory. *Physiologia Plantarum* 50, 169–175.

Palta, J.P., Whitaker, B.D. and Weiss, L.S. (1993) Plasma membrane lipids associated with genetic variability in freezing tolerance and cold acclimation in *Solanum* species. *Plant Physiology* 103, 793–803.

Palta, J.P., Bamberg, J.B., Chen, Y.K., Vega, S.E., Weiss, L.S. and Karlsson, B.H. (1997) Understanding genetic control of freezing resistance using potato species as a model system. In: Li, P.H. and Chen, T.H.H. (eds) *Plant Cold Hardiness: Molecular Biology, Biochemistry and Physiology*. Plenum Press, New York, pp. 67–75.

Paquin, R. (1977) Effet des bases temperatures sur la resistance au gel de la Luzerne (medicago nedia Pers) et son contenu en proline libre. *Physiologie Vegetative* 15, 657–665.

Pavek, J. and Corsini, D.L. (2001) Utilization of potato genetic resources in variety development. *American Journal of Potato Research* 78, 433–441.

Ristic, Z. and Ashworth, E.N. (1993) Changes in leaf ultrastructure and carbohydrates in *Arabidopsis thaliana* L (Heyn) cv. Columbia during rapid cold acclimation. *Protoplasma* 172, 111–123.

Rorat, T., Irzyykowski, W. and Grygorowizc, W.J. (1997) Identification and expression of cold specific genes in potato (*Solanum sogarandinum*). *Plant Science* 124, 69–78.

Rorat, T., Grygorowizc, W.J., Berbezy, P. and Irzyykowski, W. (1998) Isolation and expression of cold specific genes in potato (*Solanum sogarandinum*). *Plant Science* 133, 57–67.

Ryu, S.B. and Li, P.H. (1994) Potato cold hardiness development and abscisic acid. II. De novo synthesis of proteins is required for the increase in free abscisic acid during potato (*Solanum commersonii*) cold acclimation. *Physiologia Plantarum* 90, 21–26.

Sanitarius, K.A. (1973) The protective effect of sugars on chloroplast membranes during temperature and water stress and its relationship to frost, desiccation and heat resistance. *Planta* 113, 105–114.

Seppanen, M.M. and Coleman, G.D. (2003) Characterization of genotypic variation in stress gene expression and photosynthetic parameters in potato. *Plant, Cell and Environment* 26, 406–410.

Shinozaki, K. and Yamaguchi-Shinozaki, K. (2000) Molecular responses to dehydration and cold: differences and cross talk between two stress signal pathways. *Current Opinion in Plant Biology* 3, 217–223.

Stefanowsna, M., Kurás, M. and Kacperska, A. (1999) Low temperature affects pattern of leaf growth and structure of cell walls in winter oilseed rape (*Brassica napus L. var. oleifera L.*) *Annals of Botany* 84, 313–319.

Stefanowsna, M., Kurás, M. and Kacperska, A. (2002) Low temperature induced modifications in cell ultra structure and localization of phenolics in winter oilseed rape (*Brassica napus L. var. oleifera L.*) leaves. *Annals of Botany* 90, 637–645.

Steffen, K.L. and Palta, J.P. (1989) Light stress following a frost episode influences the frost tolerance of a wild potato species. *Journal of American Society for Horticultural Sciences* 114, 656–661.

Stefl, M., Treca, I. and Vratny P. (1978) Proline biosynthesis in winter plants due to exposure to low temperature. *Biology of Plants* 20, 119–128.

Steponkus, P.L. (1984) Role of the plasma membrane in freezing injury and cold acclimation. *Annual Review of Plant Physiology* 35, 543–584.

Steponkus, P.L., Uemura, M. and Webb, M.S. (1993) A contrast of the cryostability of the plasma of winter rye and spring oat: two species that widely differ in their freezing tolerance and plasma membrane lipid composition. In: Steponkus, P.L. (ed.) *Advances in Low-Temperature Biology*, Vol. 2, JAI Press, London, pp. 211–312.

Steponkus, P.L., Uemura, M., Joseph, S., Gilmour, J. and Thomashow, M.F. (1998) Mode of action of the COR15a gene on the freezing tolerance of *Arabidopsis thaliana*. *Proceedings of the National Academy of Sciences USA* 95, 14570–14575.

Stockinger, E.J., Gilmour, S.J. and Thomashow, M.F. (1997) *Arabidopsis thaliana CBF1* encodes an AP2 domain-containing transcriptional activator that binds to the C-repeat/DRE, a cis-acting DNA regulatory element that stimulates transcription in response to low temperature and water deficit. *Proceedings of the National Academy of Sciences USA* 94, 1035–1040.

Stone, J.M., Palta, J.P., Bamberg, J.B., Weiss, L.S. and Habage, J.F. (1993) Inheritance of freezing resistance in tuber-bearing *Solanum* species: evidence for independent genetic control of non-acclimated freezing tolerance and cold acclimation capacity. *Proceedings of the National Academy of Science USA* 90, 7869–7873.

Strauss, G. and Hauser, H. (1986) Stabilization of lipid bilayer vesicles by sucrose during freezing. *Proceedings of the National Academy of Sciences USA* 83, 2422–2426.

Taji, T., Ohsumi, C., Iuchi, S., Seki, M., Kasuga, M., Kobayashi, M., Yamaguchi-Shinozaki, K. and Shinosaki, K. (2002) Important role of drought and cold inducible genes for galactinol synthase in stress tolerance in *Arabidopsis thaliana*. *Plant Journal* 29, 417–426.

Thomashow, M.F. (1998) Role of cold-responsive genes in plant freezing tolerance. *Plant Physiology* 118, 1–7.

Thomashow, M.F. (1999) Plant cold acclimation, freezing tolerance genes and regulatory mechanisms. *Annual Review of Plant Physiology and Plant Molecular Biology* 50, 571–599.

Thomashow, M.F. (2001) So what's new in the field of plant cold acclimation? Lots! *Plant Physiology* 125, 89–93.

Toivio-Kinnican, M.A., Chen, H.H., Li, P.H. and Stushnoff, C. (1981) Plasma membrane alteration in callus tissues of tuber-bearing *Solanum* species during cold acclimation. *Plant Physiology* 67, 478–483.

Tseng, M.J. and Li, P.H. (1987) Changes in nucleic acids and proteins synthesis during the induction of cold hardiness. In: Li, P.H. (ed.) *Plant Cold Hardiness.* Alan R. Liss, New York, pp. 1–27.

Tseng, M.J. and Li, P.H. (1990) Alterations of gene expression in potato (*Solanum commersonii*) during cold acclimation. *Physiologia Plantarum* 78, 538–547.

Van Swaaij, A.C., Jacobsen, E. and Feenstra, W.J. (1985) Effect of cold hardening, wilting and exogenously applied proline on leaf proline content and frost tolerance of several genotypes of *Solanum. Physiologia Plantarum* 64, 230–236.

Vega, S.E. and Bamberg, J.B. (1995) Screening the US potato collection for frost hardiness. *American Potato Journal* 72, 13–21.

Vijn, I. and Smeekens, S. (1999) Fructan, more than a reserve carbohydrate? *Plant Physiology* 120, 351–359.

Wanner, L.A. and Junttila, O. (1999) Cold-induced freezing tolerance in *Arabidopsis. Plant Physiology* 120, 391–400.

Welti, R., Li, W., Li, M., Sang, Y., Biesiada, H., Zhou, H.E., Rajashekar, C.B., Williams, T.D. and Wang, X. (2002) Profiling membrane lipids in plant stress responses. Role of phospholipase D alpha in freezing-induced lipid changes in *Arabidopsis. Journal of Biological Chemistry* 277, 31994–32002.

Williams, J.P., Khan, M.U., Mitchell, K. and Johnson, G. (1988) The effect of temperature on the level and biosynthesis of unsaturated fatty acids in diacylglycerol of *Brassica napus* leaves. *Plant Physiology* 87, 904–910.

Zhang, J., Klueva, N., Wand, Z., Wu, R., Ho, T.H.D. and Nguyen, H.T. (2000) Genetic engineering for abiotic stress resistance in crop plants. *In Vitro Cellular and Development Biology – Plant* 36, 108–114.

Zhu, B., Chen, T.H.H. and Li, P.H. (1993) Expression of ABA-responsive osmotin-like gene during the induction of freezing tolerance in *Solanum commersonii. Plant Molecular Biology* 21, 729–735.

9 Overexpression of a Heat-inducible *apx* Gene Confers Chilling Tolerance to Rice Plants

Y. SATO[1,*] AND H. SARUYAMA[2]

[1]*National Agricultural Research Centre for Hokkaido Region, Hitsujigaoka 1, Toyohira-ku, Sapporo 062-8555, Japan;* [2]*Hokkaido Green-Bio Institute, Naganuma, Hokkaido 069-1301, Japan*

Introduction

Plants from temperate regions are in general chilling resistant. In contrast, plants indigenous to the tropics and subtropics suffer chilling injury upon exposure to non-freezing temperatures below 12°C (Lafuente *et al.*, 1991). Although some chilling-sensitive plants can be hardened, i.e. they exhibit increased resistance to chilling stress after exposure to intermediate temperatures (Moynihan *et al.*, 1995), a few chilling-sensitive plants, including rice seedlings, do not show hardening.

In chilling-sensitive plants, oxidative stress is a major component of chilling stress (Hodges *et al.*, 1997; Pinhero *et al.*, 1997). Active oxygen species (AOS) such as hydrogen peroxide, superoxide radicals and hydroxyl radicals can react very rapidly with DNA, lipids and proteins, which causes severe cellular damage (Van Breusegem *et al.*, 1999). Although AOS were thought to be involved in light-associated chilling stress (Prasad *et al.*, 1994a), molecular and biochemical evidence was presented to indicate that low temperature also imposes oxidative stress in dark-grown maize seedlings during chilling treatment.

Several enzymes can efficiently detoxify AOS; however, during prolonged stress conditions, such detoxification systems become saturated and damage occurs (Van Breusegem *et al.*, 1999). A major hydrogen peroxide-detoxifying system in plant chloroplasts and cytosol is the so-called ascorbate–glutathione cycle, in which ascorbate peroxidase (APX) is the key enzyme (Asada, 1992). APX, localized in the cytosol, peroxisomes, mitochondria and chloroplasts of plant cells, catalyses the reduction of H_2O_2 to water by using ascorbic acid as specific electron donor (Murgia *et al.*, 2004).

*Corresponding author (e-mail: yutaka@affrc.go.jp; Fax: +81-11-859-2178)

Prior exposure to heat-shock temperatures has been shown to increase the tolerance of sensitive tissue to subsequent chilling (Lurie and Klein, 1991; Saltveit, 1991). In the present study, we examined the effect of prior high-temperature exposure on the susceptibility of rice seedlings to chilling injury. We evaluated changes in APX activity and induction of transcriptions after exposure to high temperature. This was done to determine the possible role of APX in increasing the chilling tolerance to plants previously subjected to high temperature. We also characterized heat-inducible apx gene promoter sequences and examined chilling tolerance of transgenic rice with elevated levels of APX.

Materials and Methods

Plant material

Rice seeds (*Oryza sativa* L., cv. Kirara 397) were imbibed for 3 days in water at 27°C and sown in moist vermiculite in 9-cm diameter plastic Petri dishes, and germinated in the dark at 25°C and 80% RH. After 7 days, seedlings were exposed to 42°C at 100% RH for 1, 3, 9 and 24 h before chilling at 5°C and 80% RH in light or darkness for 7 days. Control seedlings were exposed to 25°C. After chilling, rice seedlings were transferred to the growth chamber, kept at 25°C, and grown for 7 days. The seedlings were evaluated for their per cent survival based on the observations that actively growing seedlings were determined to be survivors and the non-growing and wilted seedlings were determined to be non-survivors. All of the experiments were repeated at least twice.

Assays of APX, CAT and SOD

Leaves were homogenized with 50 mM phosphate buffer (pH 7.0) containing 1% Triton X-100. The homogenate was centrifuged at 12,000 × **g** for 20 min at 4°C and the supernatant was immediately used for the enzyme assay.

APX activity was determined according to Saruyama and Tanida (1995). The reaction mixture (1.0 ml) was composed of 50 mM potassium phosphate (pH 7.0) containing NaN_3, 0.5 mM ascorbate, 1.54 mM H_2O_2 and the enzyme fraction. The oxidation of ascorbate was started by addition of H_2O_2 and the decrease in the absorbance at 290 nm was monitored. Catalase (CAT) activity was measured by measuring the decrease in absorbance at 240 nm, as described by Tanida (1996). Superoxide dismutase (SOD) activity was assayed by using an assay kit (SOD-Test Wako) based on the NBT method (Beyer and Fridovich, 1987).

RNA analysis

The rice cytosolic *apx* gene probe was amplified by polymerase chain reaction (PCR) from cDNA prepared from rice seedlings heated at 42°C for 24 h using *Taq* DNA polymerase (GIBCO BRL) and the primers, 5′-ACCCGCAGCCATG-GCTAAGAACTAC-3′ and 5′-ACTAGAAACCTCTTAAGCATCAGCG-3′.

Primers were designed according to the known cDNA sequence of the rice *apxa* (Morita *et al.*, 1997). The cloned PCR fragment was sequenced and was identical to rice *apxa*. Total RNA was extracted from 1 g of shoots of rice seedlings by phenol/SDS method and purified by ethanol precipitation and LiCl precipitation. Purified total RNA (20 μg) was electrophoresed in 1.2% agarose gel and transferred onto nylon membrane, and allowed to hybridize with 32P-labelled *apxa* gene probe. Hybridization signals were visualized by an autoradiogram and a bio-imaging analyser (FUJIX, BAS 1000).

Isolation and analysis of rice *apxa* gene promoter

The upstream region of the *apxa* was amplified from rice genomic DNA by the method of thermal asymmetric interlaced polymerase chain reaction (TAIL-PCR) reported by Liu *et al.* (1995). The *apxa*-specific primers TR1 (5′-GTAGGA-GATGGTGGGTATCT-3′), TR2 (5′-TATCTCCTCCTTGATGGGCT-3′) and TR3 (5′-TCACGACGGGGTAGTTCTTA-3′) were synthesized according to the known cDNA sequence of the rice *apxa* (Morita *et al.*, 1997). In addition, an arbitrary degenerate primer, AD (GTNCGASWCANAWGTT), was synthesized according to Liu *et al.* (1995). Three PCR reactions were carried out sequentially to amplify target sequences using nested *apxa*-specific primers (TR1, TR2 and TR3) on one side and an AD primer on the other. Amplified products from tertiary reaction with TR3 and AD were electrophoresed in 1.2% agarose gel, and a 400-bp fragment was eluted from the gel and cloned into pCRII vector by using TA-cloning (Invitrogen). The insert DNA was sequenced using a DNA sequencer (LIC-4200LS-2, Aloka).

Rice transformation

A full-length *apxa* cDNA was introduced into a Ti-based vector, PMLH7133 (Mitsuhara *et al.*, 1996), in the sense orientation downstream of the CaMV 35S promoter. The construct was introduced into calli of rice (*Oryza sativa* L. cv., Yukihikari) by means of *Agrobacterium*-mediated transformation according to the previously reported protocol (Hiei *et al.*, 1994). Transformed calli were selected for hygromycin resistance, and then transgenic plants were regenerated. PCR screening of T0, T1 and T2 plants was carried out using primers specific for the CaMV 35S promoter and *apxa* cDNA to select homozygous lines.

Assay for chilling tolerance of transgenic plants overexpressing *apxa*

To examine chilling tolerance of transgenic plants, 10-day-old plants were exposed to a temperature of 5°C at 60% RH for 11 days. After chilling treatment, rice seedlings were transferred to a growth chamber and grown for 7 days at 25°C. Chilling tolerance was assessed by measuring survival rate. All of the experiments were repeated at least twice.

Results

Effect of prior high-temperature exposure on the susceptibility of rice seedlings to chilling injury

Rice seedlings exposed to chilling temperature (5°C) for 7 days developed severe symptoms of chilling injury (Fig. 9.1A). Heating seedlings for 24 h at 42°C before moving them to 5°C prevented the development of these symptoms (Fig. 9.1B). When heated seedlings were placed for 4 days at 25°C before being transferred to the low temperature, seedlings were no longer resistant to chilling (data not shown).

Figure 9.2 shows the effect of different periods of heat treatment on resistance to chilling stress in rice seedlings. After heating at 42°C for 6 h, more than 50% of seedlings could survive after chilling at 5°C for 7 days. No significant difference was observed in the protective effect of a heat-shock pretreatment between dark-grown seedlings and light-grown seedlings during chilling treatment. Chilling resistance was enhanced in parallel with the period of heat treatment. Twenty-four hours of exposure to 42°C resulted in a drastic increase in survival rate after chilling.

APX, CAT and SOD activities in response to heat and chilling stress

Changes in APX, CAT and SOD activity after exposure to 42°C were evaluated. The total protein amounts were decreased at the highest 10% after 24 h of heating (data not shown). The level of APX activity was higher in seedlings exposed to 42°C whereas CAT activity in heated seedlings declined to lower levels (Fig. 9.3). There was no significant difference in SOD activity between heated and unheated seedlings (Fig. 9.3). The effects of chilling stress on activity of these

Fig. 9.1. Comparison of the chilling damage between unheated (A) and preheated (B) rice seedlings. Seedlings were unheated or heated at 42°C for 24 h, and were chilled at 5°C for 7 days in light and transferred to the growth chamber, kept at 25°C, and grown for 7 days.

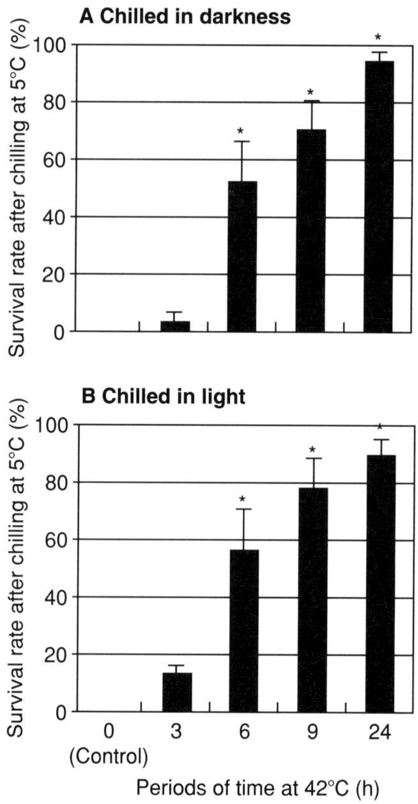

Fig. 9.2. Effect of prior high-temperature exposure on the resistance of rice seedlings to chilling injury. Rice seedlings were exposed to 42°C for 0, 3, 6, 9 and 24 h before chilling at 5°C in darkness (A) or light (B) for 7 days, and were transferred to the greenhouse and grown for 7 days. After a week of chilling treatment, the seedlings were evaluated for their per cent survival based on the observations that actively growing seedlings were determined to be survivors and the non-growing and wilted seedlings were determined to be non-survivors. The vertical bars represent the standard deviation of two individual experiments. *: Significant at the 1% level from the control.

enzymes in heated seedlings were examined (Fig. 9.4). The elevated activity of APX caused by heat stress slightly decreased but was still sustained at higher levels after chilling. In contrast, the activity of CAT drastically decreased with chilling stress. There was no significant difference in SOD activity between chilled and unchilled seedlings (Fig. 9.4). The increased and sustained activity of APX may be important in relation to the observed changes in the susceptibility of rice seedlings to chilling injury.

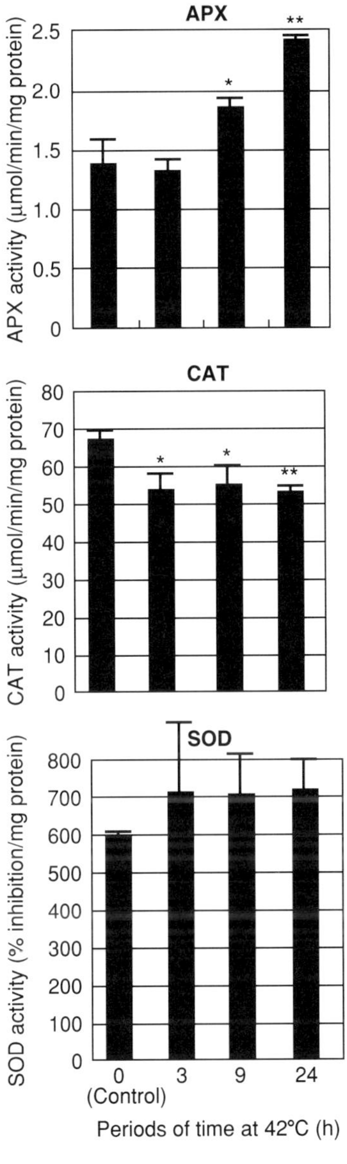

Fig. 9.3. Effect of heat stress on APX, CAT and SOD activities in rice seedlings. Rice seedlings were exposed to 42°C for 3, 9 and 24 h and samples were taken for determination of levels of each enzyme activity. The vertical bars represent the standard deviation of three individual experiments. *,**: Significant at the 5% and 1% levels from the control, respectively.

apxa mRNA levels in response to heat and chilling stress

A cDNA encoding the cytosolic APX was cloned from heated rice seedlings. The clone was identical in DNA sequence to rice *apxa* reported by Morita *et al.* (1997).

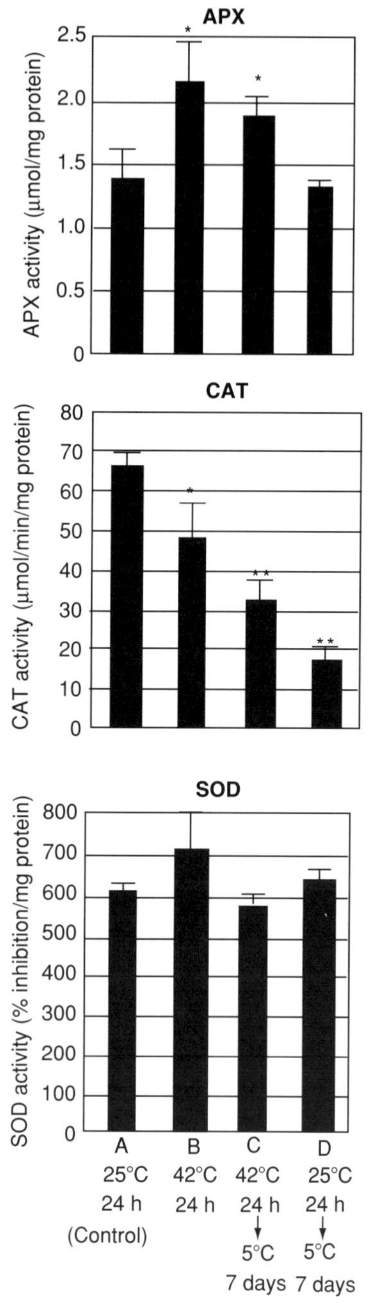

Fig. 9.4. Effect of prior high-temperature exposure on APX, CAT and SOD activities during chilling stress in rice seedlings. Each enzyme activity was examined from rice seedlings: unheated (A); heated at 42°C for 24 h (B); preheated at 42°C for 24 h and chilled at 5°C in the dark for 7 days (C); and unheated and chilled at 5°C in the dark for 7 days (D). The vertical bars represent the standard deviation of three individual experiments. *,**: Significant at the 5% and 1% levels from the control, respectively.

To determine whether heat and chilling stress has an effect on the cytosolic *apxa* mRNA levels, *apxa* gene expression in response to high and low temperature was analysed with *apxa* gene probe. This analysis revealed a 1.8-fold increase in a level of *apxa* mRNA after 1 h of heat stress (Fig. 9.5). Elevated *apxa* mRNA levels could also be detected after 6, 9, 12 and 24 h of heat stress. *apxa* mRNA levels declined after 7 days of cold stress both in preheated and control seedlings. However, the *apxa* mRNA level in preheated seedlings was higher than in control seedlings after 7 days of cold stress (Fig. 9.5).

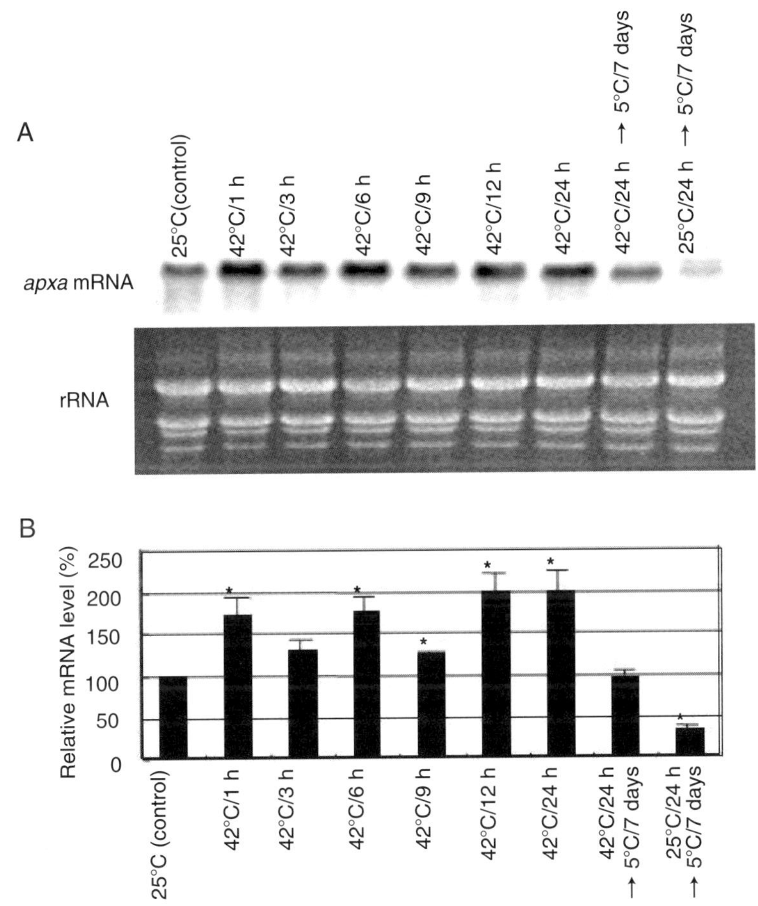

Fig. 9.5. The effects of heat stress and chilling stress on *apxa* mRNA levels of rice seedlings. (A) Rice seedlings were heated at 42°C for 0, 1, 3, 6, 9, 12 and 24 h. RNA was extracted from the seedlings at different times. The times in hours are indicated at the top of the lanes. In addition, RNA was extracted from seedlings that were either unheated or heated and then chilled at 5°C for 7 days. RNA was analysed by gel-blot hybridization using an *apxa* gene probe. The bottom panel (rRNA) shows the ethidium bromide-stained RNA gel as the loading control. (B) The results of the Northern analysis were quantitated, and are presented in the graph. The vertical bars represent the standard deviation of two individual experiments. *: Significant at the 5% level from the control.

Analysis of *apxa* promoter sequence

The upstream region including the promoter of *apxa* gene was cloned from rice genomic DNA by TAIL-PCR. A 400-bp fragment amplified by the third nested PCR was characterized by DNA sequencing. The region had very low similarity to *Arabidopsis apx1* gene promoter except for one region located 81-bp upstream of the TATA box, which contained putative heat-shock element (HSE) consisting of nGAAn and nTTCn (Fig. 9.6A). The HSE had a minimal heat-shock factor (HSF) binding motif 5′-nGAAnnTTCn-3′, the same as the HSE of *Arabidopsis apx1* promoter (Fig. 9.6B). A CAAT motif was also found at position −215.

Chilling tolerance of transgenic plants overexpressing *apxa*

A total of 15 transgenic hygromycin-resistant rice plants were generated. All the transgenic plants were phenotypically normal and were fertile. Out of these, three transformants (5-2, 5-6 and 6-4) were further analysed at the molecular level. Southern blot analysis on the T3 progeny of homozygous lines demonstrated the presence of a single copy of the transgene in all three transgenic lines. Northern blot analysis indicated that lines 5-2 and 5-6 expressed high levels of the *apxa* transcripts whereas the line 6-4 expressed the transcript at a low level similar to that of non-transformed control plants (data not shown). The level of APX activity was

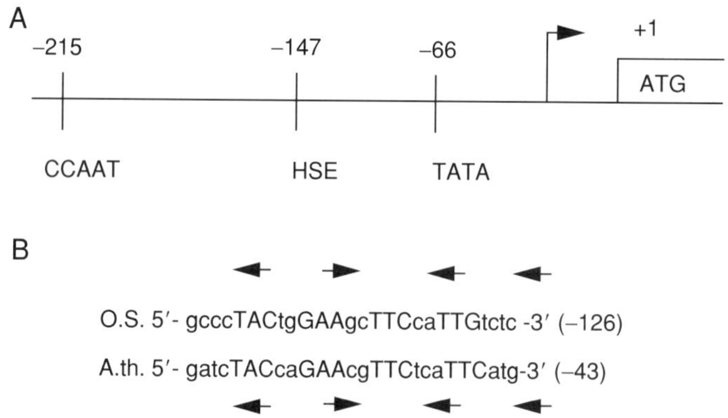

Fig. 9.6. (A) Schematic representation of *apxa* promoter organization. ATG indicates the initiating translation codon. The 5-foot end of the *apxa* cDNA is marked by an arrowhead and indicated as +1. The positions (in bp) of the putative *cis* elements as related to the transcription start are indicated above the line. (B) Comparison of the heat-shock elements from *Arabidopsis* (A. th.) *apx1* and rice (O.S.) *apxa*. Sequences matching the nGAAn, the basic 5-bp HSE motif, are indicated in uppercase letters. The two central motifs are in reverse orientation and perfectly match requirements for the minimal HSF-binding motif nGAAnnTTCn. They are flanked by two other motifs that have one tolerated substitution in rice *apxa*. Orientations of the nGAAn-like motifs are indicated by arrows.

1.8-fold higher in lines 5-2 and 5-6 than in the 6-4 and non-transformed control plants (data not shown). After chilling at 5°C for 11 days, more than 40% of seedlings of lines 5-2 and 5-6 survived whereas only 5% of seedlings of the line 6-4 and non-transformed control plants survived (data not shown).

Discussion

Exposure of plants to one stress can elicit responses similar to exposure to other stress and sometimes protect the plant against another stress (Lurie and Klein, 1991). Different stress conditions, such as drought and cold, can also result in the activation of similar stress response pathways (Seki *et al.*, 2001). Thus, a high degree of overlap may exist between gene clusters activated by different stresses. This overlap may explain the well-documented phenomenon of 'cross tolerance', in which a particular stress can induce resistance in plants to a subsequent stress that is different from the initial one (Bowler and Fluhr, 2000). Since it has been shown that heat shock induces cross-resistance in plant cells to salt shock (Harrington and Alm, 1988), toxic metals (Wollgiehn and Neumann, 1995), chilling injury (Sebehat *et al.*, 1996; Sato *et al.*, 2001) and oxidative injury (Banzet *et al.*, 1998), it is thought that small heat-shock proteins (smHSPs) must play an important role in plant adaptation to various environmental stress conditions. Heat shock can result in the oxidative stress, which induces genes involved in the oxidative-stress defence (Morgan *et al.*, 1986). In fact, pea and *Arabidopsis apx1* gene expressions are induced by heat stress as well as by oxidative stress and both have heat-shock *cis* elements in their promoters (Mittler and Zilinskas, 1992, 1994; Storozhenko *et al.*, 1998).

The objective of this study was to determine whether heat stress could confer tolerance to low-temperature stress in rice seedlings. The data presented here clearly demonstrate that a treatment of rice seedlings at 42°C before chilling at 5°C prevented chilling injury. Similar effects have been found in a number of fruits such as avocado (Woolf *et al.*, 1995), cucumber (Lafuente *et al.*, 1991; Jennings and Saltveit, 1994; McCollum *et al.*, 1995), pepper (Mencarelli *et al.*, 1993) and tomato (Lurie and Klein, 1991; Sabehat *et al.*, 1996).

Studies with heat-stressed tomato fruits have shown a correlation between the accumulation of smHSPs and the acquisition of chilling tolerance (Sabehat *et al.*, 1996, 1998; Kadyrzhanova *et al.*, 1998). Some smHSPs may protect plants against chilling injury by preventing denaturation of proteins from chilling (Sabehat *et al.*, 1998; Ukaji *et al.*, 1999). Thus, smHSPs are assumed to be involved in the ability to increase chilling tolerance.

However, the primary cause of chilling injury is believed to be lipid peroxidation caused by increases in oxygen-radical generation induced by chilling stress (Prasad *et al.*, 1994a,b). Saruyama and Tanida (1995) concluded that the tolerance of rice cultivars to chilling injury is closely linked to the cold stability of APX and CAT. Therefore, we addressed the question of whether heat stress enhances active oxygen-scavenging systems, such as APX, CAT and SOD, which contribute to the survival of chilled rice seedlings. The measurement of changes in activity of these enzymes after exposure to 42°C confirmed that APX was upregulated by heat stress. Furthermore, increased APX activity was sustained after 7 days of chilling at 5°C. This finding strongly suggests that APX is involved in the ability to increase chilling tolerance by heat stress. The important role of APX in relation to the increase in oxidative tolerance has been reported for many plants (Tanaka *et al.*, 1985; Schoner and Krause, 1990; Ushimaru *et al.*, 1992; Willekens *et al.*, 1994; Conklin and Last, 1995; Kubo *et al.*, 1995; Rao *et al.*, 1996; Orvar and Ellis, 1997; Morita *et al.*, 1999). Furthermore, the higher affinity for H_2O_2 (Scandalios *et al.*, 1972; Chen and Asada, 1989; Elia *et al.*, 1992) suggests that APX can work at low temperatures (Saruyama and Tanida, 1995).

We have demonstrated that exposure of seedlings to 42°C enhanced the *apxa* mRNA level within 1 h. The heat-shock induction of the gene suggested that *apxa* gene promoter might have HSE, which are binding sites for the regulatory HSF (Chen and Pederson, 1993). We cloned the promoter of *apxa* gene and characterized by DNA sequencing. *apxa* promoter had sequence motifs characteristic of HSE identified in promoters of all heat shock–inducible genes. The putative HSE is the sole conserved sequence among promoters of *Arabidopsis apx1*, *apx2*, pea *apx1* and rice *apxa* (Storozhenko *et al.*, 1998; our data). Recently, *in vivo* analysis of the interaction between recombinant tomato HSF and the *Arabidopsis apx1* promoter confirmed that the apx1 HSE represents a functional HSF-binding site (Storozhenko *et al.*, 1998). Furthermore, the *apx1* promoter with a mutated HSE loses inducibility and even becomes repressed under the heat-shock treatment (Storozhenko *et al.*, 1998). It is, therefore, possible that the HSE in the rice *apxa* promoter contributes to the heat-shock induction of *apxa* gene, although further experiments are necessary to show whether the HSE in the rice *apxa* promoter is recognized by the HSF.

Different protein isoforms of cytosolic APX are known in *Arabidopsis* (Storozhenko *et al.*, 1998). It is considered that rice also has a few different isoforms of cytosolic APX. The contribution of the *apxa* to total extractable foliar APX activity remains unknown and should be addressed in the future.

To test the hypothesis that plants endowed with higher levels of APX enzyme will be better protected against chilling stress, transgenic rice plants were generated via *Agrobacterium tumefaciens*-mediated transformation. The level of APX activity was higher in high-expression transgenic lines than the low-expression transgenic line and non-transformed control plants. Relative to the low-expression line and non-transformed control plants, high-expression lines showed significantly less injury from the chilling stress. The results clearly demonstrated that overexpression of the *apxa* is effective in enhancing chilling tolerance of rice plants.

References

Asada, K. (1992) Ascorbate peroxidase – a hydrogen peroxidase-scavenging enzyme in plants. *Physiologia Plantarum* 85, 235–241.

Banzet, N., Richaud, C., Deveaux, Y., Kazmaier, M., Gagnon, J. and Triantaphylides, C. (1998) Accumulation of small heat shock proteins, including mitochondrial *HSP22*, induced by oxidative stress and adaptive response in tomato cells. *Plant Journal* 13, 519–527.

Beyer, W.F. and Fridovich, I. (1987) Assaying for superoxide dismutase activity: some large consequences of minor changes in conditions. *Analytical Biochemistry* 161, 559–566.

Bowler, C. and Fluhr, R. (2000) The role of calcium and activated oxygens as signals for controlling cross-tolerance. *Trends in Plant Science* 5, 241–246.

Chen, G.-X. and Asada, K. (1989) Ascorbate peroxidase in tea leaves: occurrence of two isozymes and the differences in their enzymatic and molecular properties. *Plant and Cell Physiology* 30, 987–998.

Chen, J. and Pederson, D.S. (1993) A distal heat shock element promotes the rapid response to heat shock of the *HSP26* gene in the yeast *Saccharomyces cerevisiae*. *Journal of Biological Chemistry* 268, 7442–7448.

Conklin, P.L. and Last, R.L. (1995) Differential accumulation of antioxidant mRNAs in *Arabidopsis thaliana* exposed to ozone. *Plant Physiology* 109, 203–212.

Elia, M.R., Borraccino, G. and Dipierro, S. (1992) Soluble ascorbate peroxidase from potato tubers. *Plant Science* 85, 17–21.

Harrington, H.M. and Alm, D.M. (1988) Interaction of heat and salt shock in cultured tobacco cells. *Plant Physiology* 88, 618–625.

Hiei, Y., Ohta, S., Komari, T. and Kumashiro, T. (1994) Efficient transformation of rice (*Oryza sativa* L.) mediated by *Agrobacterium* and sequence analysis of the boundaries of the T-DNA. *Plant Journal* 6, 271–282.

Hodges, D.M., Andrews, C.J, Johnson, D.A. and Hamilton, R.I. (1997) Antioxidant enzyme and compound responses to chilling stress and their combining abilities in differentially sensitive maize hybrids. *Crop Science* 37, 857–863.

Jennings, P. and Saltveit, M.E. (1994) Temperature and chemical shocks induce chilling tolerance in germinating *Cucumis sativus* (cv. Poinsett 76) seeds. *Physiologia Plantarum* 91, 703–707.

Kadyrzhanova, D.K., Vlachonasios, K.E., Ververidis, P. and Dilley, D.R. (1998) Molecular cloning of a novel heat induced/chilling tolerance related cDNA in tomato fruit by use of mRNA differential display. *Plant Molecular Biology* 36, 885–895.

Kubo, A., Saji, H., Tanaka, K. and Kondo, N. (1995) Expression of *Arabidopsis* cytosolic ascorbate peroxidase gene in response to ozone or sulfur dioxide. *Plant Molecular Biology* 29, 479–489.

Lafuente, M.T., Belver, A., Guye, M.G. and Saltveit, M.E. (1991) Effect of temperature conditioning on chilling injury of cucumber cotyledons. *Plant Physiology* 95, 443–449.

Liu, Y.G., Mitsukawa, N., Oosumi, T. and Whittier, R.F. (1995) Efficient isolation and mapping of *Arabidopsis thaliana* T-DNA insert junctions by thermal asymmetric interlaced PCR. *Plant Journal* 8, 457–463.

Lurie, S. and Klein, J.D. (1991) Acquisition of low-temperature tolerance in tomatoes by exposure to high-temperature stress. *Journal of the American Society for Horticultural Science* 116, 1007–1012.

McCollum, T.G., Doostdar, H., Mayer, R.T. and McDonald, R.E. (1995) Immersion of cucumber fruit in heated water alters chilling-induced physiological changes. *Postharvest Biology Technology* 6, 55–64.

Mencarelli, F., Ceccantoni, B., Bolini, A. and Anelli, G. (1993) Influence of heat treatment on the physiological response of sweet pepper kept at chilling temperature. *Acta Horticulturae* 343, 238–243.

Mitsuhara, I., Ugaki, M., Hirochika, H., Ohshima, M., Murakami, T., Gotoh, Y., Katayose, Y., Nakamura, S., Honkura, R., Nishimiya, S., Ueno, K., Mochizuki, A., Tanimoto, H., Tsugawa, H., Otsuki, Y. and Ohashi, Y. (1996) Efficient promoter cassettes for enhanced expression of foreign genes in dicotyledonous and monocotyledonous plants. *Plant and Cell Physiology* 37, 49–59.

Mittler, R. and Zilinskas, B.A. (1992) Molecular cloning and characterization of a gene encoding pea cytosolic ascorbate peroxidase. *Journal of Biological Chemistry* 267, 21802–21807.

Mittler, R. and Zilinskas, B.A. (1994) Regulation of pea cytosolic ascorbate peroxidase and other antioxidant enzymes during the progression of drought stress and following recovery from drought. *Plant Journal* 5, 397–405.

Morgan, R.W., Christman, M.F., Jacobson, F.S., Storz, G. and Ames, B.N. (1986) Hydrogen peroxide-inducible proteins in *Salmonella typhimurium* overlap with heat shock and other stress proteins. *Proceedings of the National Academy of Sciences USA* 83, 8059–8063.

Morita, S., Kaminaka, H., Yokoi, H., Masumura, T. and Tanaka, K. (1997) Cloning and characterization of cytosolic ascorbate peroxidase cDNA from rice (Accession No. D45434). *Plant Physiology* 113, 306.

Morita, S., Kaminaka, H., Masumura, T. and Tanaka, K. (1999) Induction of rice cytosolic ascorbate peroxidase mRNA by oxidative stress; involvement of hydrogen peroxide in oxidative stress signalling. *Plant and Cell Physiology* 40, 417–422.

Moynihan, M.R., Ordentlich, A. and Raskin, I. (1995) Chilling-induced heat evolution in plants. *Plant Physiology* 108, 995–999.

Murgia, I., Tarantino, D., Vannini, C., Bracale, M., Carravieri, S. and Soave, C. (2004) *Arabidopsis thaliana* plants overexpressing thylakoidal ascorbate peroxidase show increased resistance to paraquat-induced photooxidative stress and to nitric oxide-induced cell death. *Plant Journal* 38, 940–953.

Orvar, B.L. and Ellis, B.E. (1997) Transgenic tobacco plants expressing antisense RNA for cytosolic ascorbate peroxidase show increased susceptibility to ozone injury. *Plant Journal* 11, 1297–1305.

Pinhero, R.G., Rao, M.V., Paliyath, G., Murr, D.P. and Fletcher, R.A. (1997) Changes in activities of antioxidant enzymes and their relationship to genetic and paclobutrazol-induced chilling tolerance of maize seedlings. *Plant Physiology* 114, 695–704.

Prasad, T.K., Anderson, M.D., Martin, B.A. and Stewart, C.R. (1994a) Evidence for chilling-induced oxidative stress in maize seedlings and a regulatory role for hydrogen peroxide. *Plant Cell* 6, 65–74.

Prasad, T.K., Anderson, M.D. and Stewart, C.R. (1994b) Acclimation, hydrogen peroxide, and abscisic acid protect mitochondria against irreversible chilling injury in maize seedlings. *Plant Physiology* 105, 619–627.

Rao, M.V., Paliyath, G. and Ormrod, D.P. (1996) Ultraviolet-B- and ozone-induced biochemical changes in antioxidant enzymes of *Arabidopsis thaliana. Plant Physiology* 110, 125–136.

Sabehat, A., Weiss, D. and Lurie, S. (1996) The correlation between heat-shock protein accumulation and persistence and chilling tolerance in tomato fruit. *Plant Physiology* 110, 531–537.

Sabehat, A., Lurie, S. and Weiss, D. (1998) Expression of small heat-shock proteins at low temperatures. *Plant Physiology* 117, 651–658.

Saltveit, M.E. (1991) Prior temperature exposure affects subsequent chilling sensitivity. *Physiologia Plantarum* 82, 529–536.

Saruyama, H. and Tanida, M. (1995) Effect of chilling on activated oxygen-scavenging enzymes in low temperature-sensitive and -tolerant cultivars of rice (*Oryza sativa* L.). *Plant Science* 109, 105–113.

Sato, Y., Murakami, T., Funatsuki, H., Matsuba, S., Saruyama, H. and Tanida, M. (2001) Heat shock-mediated APX gene expression and protection against chilling injury in rice seedlings. *Journal of Experimental Botany* 52, 145–151.

Scandalios, J.G., Liu, E.H. and Campeau, M.A. (1972) The effects of intragenic and intergenic complementation on catalase structure and function in maize: a molecular approach to heterosis. *Archives Biochemistry and Biophysics* 153, 695–705.

Schoner, S. and Krause, G.H. (1990) Protective systems against active oxygen species in spinach: response to cold acclimation in excess light. *Planta* 180, 383–389.

Sebehat, A., Weiss, D. and Lurie, S. (1996) The correlation between heat shock protein accumulation and persistence and chilling tolerance in tomato fruits. *Plant Physiology* 110, 531–537.

Seki, M., Narusaka, M., Abe, H., Kasuga, M., Yamaguchi-Shinozaki, K., Carninci, P., Hayashizaki, Y. and Shinozaki, K. (2001) Monitoring the expression pattern of 1300 *Arabidopsis* genes under drought and cold stresses by using a full-length cDNA microarray. *Plant Cell* 13, 61–72.

Storozhenko, S., Pauw, P.D., Van Montagu, M., Inze, D. and Kushnir, S. (1998) The heat-shock element is a functional component of the *Arabidopsis apx1* gene promoter. *Plant Physiology* 118, 1005–1014.

Tanaka, K., Suda, Y., Kondo, N. and Sugahara, K. (1985) O_3 tolerance and the ascorbate-dependent H_2O_2 decomposing system in chloroplasts. *Plant and Cell Physiology* 26, 1425–1431.

Tanida, M. (1996) Catalase activity of rice seed embryo and its relation to germination rate at a low temperature. *Breeding Science* 46, 23–27.

Ukaji, N., Kuwabara, C., Takezawa, D., Arakawa, K., Yoshida, S. and Fujikawa, S. (1999) Accumulation of small heat-shock protein homologs in the endoplasmic reticulum of cortical parenchyma cells in mulberry in association with seasonal cold acclimation. *Plant Physiology* 120, 481–489.

Ushimaru, T., Shibasaka, M. and Tsuji, H. (1992) Development of the O-2-detoxification system during adaptation to air of submerged rice seedlings. *Plant and Cell Physiology* 33, 1065–1071.

Van Breusegem, F., Slooten, L., Stassart, J.-M., Moens, T., Botterman, J., Van Montagu, M. and Inze, D. (1999) Overproduction of *Arabidopsis thaliana* FeSOD confers oxidative stress tolerance to transgenic maize. *Plant and Cell Physiology* 40, 515–523.

Willekens, H., Van Camp, W., Van Montagu, M., Inze, D., Langebartels, C. and Sandermann, H. (1994) Ozone, sulfur dioxide, and ultraviolet B have similar effects on mRNA accumulation of antioxidant genes in *Nicotiana plumbaginifolia* L. *Plant Physiology* 106, 1007–1014.

Wollgiehn, R. and Neumann, D. (1995) Stress response of tomato cell cultures to toxic metals and heat shock: differences and similarities. *Journal of Plant Physiology* 146, 736–742.

Woolf, A.B., Watkins, C.B., Bowen, J.H., Lay-Yee, M., Maindonald, J.H. and Ferguson, I.B. (1995) Reducing external chilling injury in stored 'Hass' avocados with dry heat treatments. *Journal of the American Society for Horticultural Science* 120, 1050–1056.

10 Physiological and Morphological Alterations Associated with Development of Freezing Tolerance in the Moss *Physcomitrella patens*

A. Minami,[1] M. Nagao,[1] K. Arakawa,[2] S. Fujikawa[2] and D. Takezawa[3,*]

[1]*Cryobiosystem Research Centre, Faculty of Agriculture, Iwate University, Morioka 020-8550, Japan;* [2]*Department of Environmental Resources, Graduate School of Agriculture, Hokkaido University, Sapporo 060-8589, Japan;* [3]*Department of Regulation Biology, Faculty of Science, Saitama University, Saitama 338-8570, Japan*

Introduction

It is known that some higher plants can acquire freezing tolerance through a cold-acclimation process during the period from late autumn to early winter. In this process, plant cells undergo drastic physiological and morphological changes leading to the development of freezing tolerance at the cellular level. Enhancement of freezing tolerance often accompanies transient increases in levels of endogenous abscisic acid (ABA); thus, ABA-induced physiological changes within the cells are thought to accelerate the process required for freezing tolerance (Chen *et al.*, 1983; Lalk and Dörfflung, 1985; Heino *et al.*, 1990; Lång *et al.*, 1994). A number of genes of which expression is associated with freezing tolerance have been identified in higher plant species, and it has been shown that many of these genes are also responsive to ABA or stress that causes water loss, such as desiccation (Seki *et al.*, 2001). Recent studies on *Arabidopsis* have revealed the existence of specific transcription factors and signalling molecules that control stress responses in plants (Thomashow, 1999; Fowler and Thomashow, 2002; Xiong *et al.*, 2002). From these studies, it has been suggested that cold, desiccation and ABA-triggered

*Corresponding author (e-mail: takezawa@post.saitama-u.ac.jp; Fax: +81-48-858-3698)

signalling processes converge on expression of a set of genes encoding stress proteins and enzymes that catalyse synthesis of cellular metabolites, including compatible solutes that protect cells from stress caused by freeze- and desiccation-induced dehydration (Shinozaki and Yamaguchi-Shinozaki, 2000).

Mosses, which are non-vascular plants, are known to have appeared in terrestrial regions on the earth more than 400 million years ago and they are now distributed throughout the world, from tropics to frigid regions. Mosses that have adapted to an environment with cold seasons are likely to possess a mechanism for tolerating freezing temperatures. Considering both mosses and higher plants have a common green algal ancestor that had successfully invaded a terrestrial environment, it is possible that mosses have cellular devices necessary for responding to environmental stress signals, including cold and desiccation. Several studies have provided evidence that moss cells are freezing tolerant, although little is known about their freezing behaviour and physiological changes associated with freezing tolerance. Some moss species are known to dominate in regions with severe winters such as continental Antarctica. These mosses are thought to have high degrees of freezing tolerance and some species can even be directly cryopreserved without going through a prior cryoprotective pretreatment (Burch, 2003). Not only these mosses, but also mosses of temperate regions, are likely to experience subzero temperatures in their life cycles, and they need to withstand such freezing stress. It has been shown that some species of naturally grown mosses have certain degrees of freezing tolerance (Dilks and Proctor, 1975), and freezing tolerance of some species changes seasonally (Rütten and Santarius, 1992a). These studies suggested that mosses have the ability to develop freezing tolerance in response to changes in environmental conditions such as decrease in temperature.

The moss *Physcomitrella patens*, a temperate species, has been used as an experimental plant for various physiological and genetic studies due to its relatively simple structure, which is suitable for cytological and developmental analysis in addition to the establishment of protocols for axenic culture with a synthetic medium (Ashton *et al.*, 1979; Cove *et al.*, 1997). An increasing number of plant biologists have been using this moss recently due to the establishment of molecular techniques for genetic transformation (Schaefer *et al.*, 1991). Its high rate of homologous recombination makes it possible to analyse a gene of interest by targeted gene disruption, which is not possible in other plant systems (Schaefer, 2002). In addition, large-scale sequencing of expressed sequence tag (EST) clones of *P. patens*, and future genome projects, are making this organism more attractive as a tool for functional genomics (Nishiyama *et al.*, 2003).

As is the case in many other mosses, the vegetative phase of *P. patens* is haploid gametophytes such as thread-like protonema, leaf-like gametophore or root-like rhizoid. Most studies on the stress resistance of *P. patens* have been done using cultured protonema tissues with relatively uniform chloroplast-containing cells. It has been shown that survival of *P. patens* protonema cells after cryopreservation was improved following treatment with 0.5 M mannitol (Grimsley and Withers, 1983) or ABA (Christianson, 1998). Knight *et al.* (1995) revealed that protonema cells of *P. patens* accumulate boiling-soluble polypeptides in response to ABA and osmotic stress. They also showed that the wheat *Em* promoter fused to the GUS reporter gene in protonemata of transgenic *P. patens* was activated by treatment with 100 μM ABA. These results indicated that ABA- and osmotic stress-induced physiological changes, which accompanied expression of stress-related genes, might lead to acquisition of freezing tolerance. ABA has been shown to trigger morphological changes in moss protonemata, leading to the formation of round shaped 'brood cells' with thick cell walls, indicating that ABA-induced morphological changes are also important for cells to exert stress resistance (Goode *et al.*, 1992; Schnepf and Reinhard, 1997). Preliminary observation indicated that *P. patens* protonemata also form brood cells by incubation with ABA. These results indicated that *P. patens* is potentially a suitable model to study physiological and morphological changes associated with freezing tolerance at the cellular level.

ABA-induced Freezing Tolerance in Protonema Cells of *P. patens*

We carried out experiments using *P. patens* protonema cells to clarify their behaviour at freezing temperatures and other physiological alterations associated with changes in freezing tolerance. First, we examined the effect of freezing on survival of *P. patens* protonema cells that had been cultured under continuous light at 25°C on a cellophane-overlaid agar medium. Unfrozen protonema cells have nearly round-shaped chloroplasts around a central vacuole. When the protonema cells were frozen at −6°C and thawed, the protoplasmic structure was apparently disrupted with aggregated cellular materials. This indicated that protonema cells cultured under normal growth temperatures are quite sensitive to freezing. Estimation of freezing tolerance by measurement of electrolyte leakage indicated that the cells have a lethal freezing temperature (LT_{50}) value of about −2°C (Nagao *et al.*, 2005). When the protonema cells were incubated in the presence of ABA, the freezing tolerance remarkably changed within a short period of time. Estimation by measurement of electrolyte leakage indicated that survival after freezing at −4°C, −6°C and −8°C of the protonema cells was dramatically improved by incubation with 1 μM ABA for 1 day (Fig. 10.1). Incubation with 10 μM ABA was more effective than incubation with 1 μM ABA, and 1-day treatment resulted in changes in the LT_{50} value from −2°C to −10°C (Nagao *et al.*, 2005). Threshold ABA levels that increase freezing tolerance of protonema cells have been examined and found to be in the range of 10 nM to 100 nM in the case of 1-day treatment at 25°C (Minami *et al.*, 2005).

Morphological and Metabolic Changes Associated with ABA-induced Freezing Tolerance in *P. patens*

Plant cells are known to undergo dramatic morphological changes, such as formation of small-fragmented vacuoles, small vesicles of endoplasmic reticulum and thickening of the cell wall, when they develop freezing tolerance (Siminovitch *et al.*, 1968; Ristic and Ashworth, 1993; Fujikawa and Takebe, 1996). It has been suggested that these morphological changes somehow mitigate irreversible damage of cellular membranes caused by freeze-induced dehydration. We examined morphological changes of protonema cells in the course of treatment with ABA by electron microscopic observation of both ultra-thin sections and freeze-fracture replicas of ABA-treated and non-treated protonema cells. The results of the observations indicated that ABA-treated cells had fragmented vacuoles and slender chloroplasts with reduced starch grains in comparison with the non-treated cells (Nagao *et al.*, 2005). However, we did not observe typical 'brood cells' by incubation with ABA for 1 day, indicating that formation of brood cells is not necessary for rapid increase in freezing tolerance induced by ABA. When cells were frozen to −4°C and observed by freeze-fracture electron microscopy, both control and ABA-treated cells were shrunken and deformed due to dehydration caused by growing extracellular ice crystals. Observation of plasma membrane structures indicated that membranes of control cells had ultrastructures known to be associated with freezing injury, such as aparticulate domains and fracture-jump lesions (Steponkus *et al.*, 1993). However, membranes of most cells treated with ABA for 1 day did not have such characteristic structures, suggesting that ABA-induced cellular changes lead to protection of the plasma membrane under freezing conditions (Nagao *et al.*, 2005).

In the process of cold acclimation, higher plant species accumulate soluble sugars (Sakai, 1961); these sugars have been suggested to play a role in protection of cellular membranes and proteins from freezing injury (Anchordoguy *et al.*, 1987; Carpenter and Crowe, 1988; Leslie *et al.*, 1995). We analysed accumulation of soluble sugars in *P. patens* protonema cells and found that ABA treatment resulted in a remarkable increase in the amount of soluble sugars. Accordingly, there was an increase in osmotic concentration from 0.3 osmoles to 0.5 osmoles after 1 day (Nagao *et al.*, 2005). These changes are in agreement with the results of previous studies showing accumulation of sugars during cold seasons in moss species as well as in

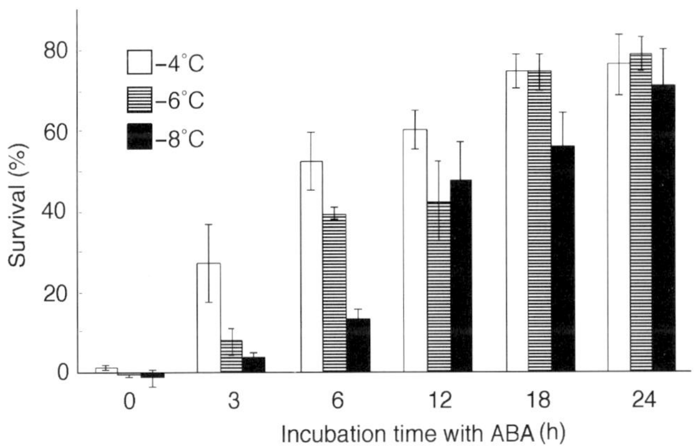

Fig. 10.1. Changes in freezing tolerance of *P. patens* protonema cells caused by abscisic acid (ABA) treatment. One-week-old protonema tissues were transferred onto a medium containing 1 μM ABA. After incubation for different time periods (h), the tissues were subjected to equilibrium freezing to the indicated temperatures at a cooling rate of −2.4°C/h. After thawing, survival of cells was determined by measurement of electrolyte leakage from dead cells, assuming values of unfrozen tissues to be 100%.

higher plants (Rütten and Santarius, 1992a). These results suggest that, as is the case in higher plants, soluble sugars play a role in freezing tolerance in mosses. However, it has also been shown that accumulation of soluble sugars is not necessarily correlated with seasonal variations in freezing tolerance or with differences between degrees of tolerance of older and younger tissues (Rütten and Santarius, 1992b, 1993). This result raises the possibility that other factors such as functional proteins may also contribute to freezing tolerance in moss cells.

ABA-responsive Genes in *P. patens* Encode Proteins with Similarity to Stress-associated Proteins of Higher Plants

To examine the importance of *de novo* synthesis of proteins in freezing tolerance of *P. patens* cells, ABA treatment was carried out in the presence or absence of protein-synthesis inhibitors. Cycloheximide (CHX), which inhibits synthesis of nuclear-encoded proteins, dramatically reduced the degree of ABA-induced freezing tolerance of *P. patens* (Fig. 10.2). On the other hand, chloramphenicol (CAP), which inhibits protein synthesis in plastids and mitochondria, did not have a significant effect on freezing tolerance. This indicated that nuclear-encoded genes are likely to play a critical role in the development of freezing tolerance (Minami *et al.*, 2003b).

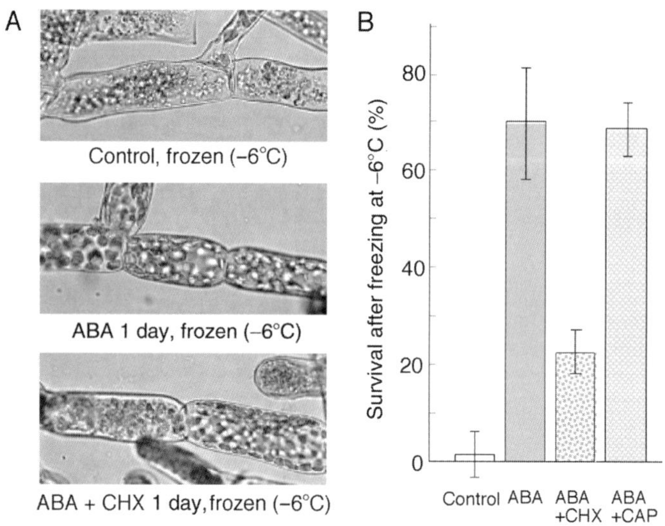

Fig. 10.2. Effects of protein-synthesis inhibitors on abscisic acid (ABA)-induced freezing tolerance of *P. patens* protonema cells. (A) Non-treated protonema cells (control) or cells treated with either 1 μM ABA or 1 μM ABA plus 5 μM cycloheximide (CHX) for 1 day were frozen to –6°C and thawed. (B) Survival of cells after freezing was determined by electrolyte-leakage measurement. Cycloheximide at 5 μM significantly decreased the ABA-induced freezing tolerance, whereas chloramphenicol (CAP) at 50 μM did not significantly change freezing tolerance. Incubation with protein inhibitors did not affect survival of unfrozen cells (data not shown).

It has been shown in higher plants that late-embryogenesis-abundant (LEA) and LEA-like hydrophilic proteins with heat-soluble traits accumulate when cells acquire freezing and desiccation tolerance (Lin *et al.*, 1990; Wise and Tunnacliffe, 2004). When heat-soluble proteins of ABA-treated and non-treated protonema cells were analysed by sodium dodecyl sulphate-polyacrylamide gel electrophoresis (SDS-PAGE), several bands of proteins that were significantly increased by ABA were detected (Fig. 10.3A). The amounts of heat-soluble proteins were remarkably reduced when cells were treated with ABA in the presence of CHX (data not shown). Antibodies raised against LEA-like 17B9 protein with moderate similarity to cold-induced CAP160 of spinach (Table 10.1) reacted with a specific protein induced by ABA treatment (Fig. 10.3B). The 17B9 protein was also detected in heat-soluble fractions, although the amount was significantly reduced (Fig. 10.3C).

In order to identify genes involved in the development of freezing tolerance, we analysed genes induced by ABA in protonema cells. Dozens of ABA-induced genes were isolated by differential display and differential screening using cDNA filter arrays. The results indicated that a large proportion of the ABA-induced genes encoded proteins with sequence similarity to higher plant proteins that had been shown to be induced during cold acclimation, desiccation and ABA treatment (Table 10.1). We identified several different genes encoding hydrophilic proteins, such as LEAIII (6A5, 22F6 and 28A6), dehydrin (19C6), spinach

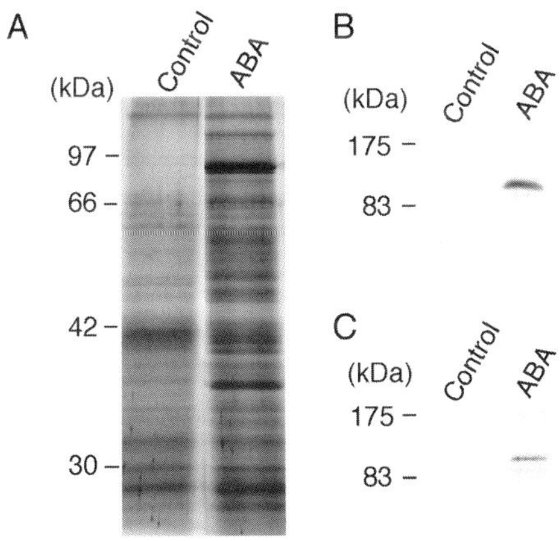

Fig. 10.3. Abscisic acid (ABA) induces heat-soluble proteins in the protonema cells. (A) Protonema cells were treated with 1 μM ABA for 1 day, and total soluble proteins were extracted. To analyse heat-soluble protein fractions, the total soluble proteins were heated at 100°C for 10 min and the fraction that was not precipitated after centrifugation was analysed by SDS-PAGE. (B) ABA-induced accumulation of a late-embryogenesis-abundant (LEA)-family protein 17B9 in *P. patens*. Total soluble fractions of control and ABA-treated protonema cells were separated by SDS-PAGE, blotted onto a PVDF membrane, and then reacted with antiserum raised against recombinant 17B9 protein. (C) Proteins that were not precipitated by heat treatment (70°C, 3 min) were analysed in a similar way to detect 17B9 protein in the heat-soluble fraction.

Table 10.1. Abscisic acid (ABA)-induced genes in *P. patens.*

Clone name	Homology	Organism
Hydrophilic proteins		
4G8	LEA-like protein	*Arabidopsis thaliana*
6A5	ABA-inducible protein	*Triticum aestivum*
13A9	Water stress-inducible protein (StDS2)	*Solanum tuberosum*
17B9	Cold-acclimation protein (CAP160)	*Spinacia oleracea*
19C6	85-kDa cold-acclimation protein	*Spinacia oleracea*
22F6	Group III LEA	*Triticum aestivum*
28A6	Group III LEA (pGmPM8)	*Glycine max*
Membrane proteins		
PPAR5	Low temperature-induced protein (blt101)	*Hordeum vulgare*
PPAR8	Desiccation-associated protein (pSD13)	*Sporobolus stapfianus*
4E4	Cold acclimation-associated protein (Wcor413)	*Triticum aestivum*
6A8	Chlorophyll *a/b* binding protein-like protein (LI818)*	*Chlamydomonas reinharditii*
16E9	Green spore chloroplast 22-kDa protein*	*Osmunda japonica*
17C6	Tonoplast-intrinsic protein	*Arabidopsis thaliana*
20E2	Putative transmembrane protein*	*Tortula ruralis*
21F2	UV-inducible gene 1-like protein*	*Leptosphaeria maculans*
26E1	ABA-inducible AWPM19-like protein*	*Triticum aestivum*
Other proteins and enzymes		
PPAR4	Ribulose 5-phosphate-3 epimerase	*Spinacia oleracea*
PPAR3	Protein serine/threonine kinase	*Arabidopsis thaliana*
PPAR11	NADHP-quinone oxidoreductase	*Arabidopsis thaliana*
PPAR13	Glutathione S-transferase	*Petunia* × *hybrida*
3E2	Enolase	*Spinacia oleracea*
12H2	Nodule development-associated MtN19-like protein	*Oryza sativa*
14B8	Carbonic anhydrase	*Zea mays*
15D12	Plastid phosphoglycerokinase	*Hordeum vulgare*
16A12	Chalcone synthase	*Lilium hybrid*

Table 10.1. *(continued)*

17A4	RNA-binding protein	*Beta vulgaris*
19G9	Nucleotide diphosphate kinase	*Pinus pinaster*
20A6	Heat-shock protein (HSP70-3)	*Lycopersicon esculentum*
22D1	Nuclear RNA-binding protein	*Oryza sativa*
27C9	Mitochondrial formate dehydrogenase	*Oryza sativa*

Asterisks (*) indicate proteins not conserved in higher plants.

CAP160 (17B9) and the *Solanum* acidic protein (13A9). Not only genes encoding hydrophilic proteins but also several genes of membrane proteins were found to be induced by ABA, most of which also have homologues in higher plants. These membrane proteins include Cor413 (Breton *et al.*, 2003), wheat plasma membrane protein AWPM19 (Koike *et al.*, 1997), barley blt101 (Goddard *et al.*, 1993) and the tonoplast-intrinsic protein (Maurel *et al.*, 1993). These membrane proteins of higher plants were also shown to be induced in the process of acquisition of freezing tolerance or desiccation tolerance, although their precise functions in stress tolerance have not been determined. We also identified genes encoding a heat-shock protein and different types of RNA-binding proteins, some of which have also been shown to be induced by cold, and have been suggested to play a role in stabilization of RNA or protein molecules under stress conditions (Anderson *et al.*, 1994; Carpenter *et al.*, 1994). These results suggest that *P. patens* and higher plant cells share common features with respect to freezing behaviour as well as genes required for development of freezing tolerance. We also identified several ABA-induced genes that are not conserved in higher plants but, rather, have similarity to genes of fungi, alga, mosses or ferns. These genes are possibly related to traits of stress tolerance specific to lower plants, which have been lost in higher plants in the evolution process.

Moss Senses Cold and Osmotic Stress to Increase Freezing Tolerance

By detailed molecular analyses of promoter and transcription factors for cold- and desiccation stress-induced genes of *Arabidopsis*, it has been shown that different stress signals, i.e. cold and dehydration/osmotic stress, activate common or distinct transcription factors, enabling these signals to differentially activate transcription of the same gene (Shinozaki and Yamaguchi-Shinozaki, 2000). This leads to questions like whether mosses could increase freezing tolerance in response to environmental stresses such as cold and osmotic stress, and whether these processes require activation of ABA-induced genes such as those shown in Table 10.1. Changes in freezing tolerance by incubation at non-freezing low temperatures in *P. patens* protonema cells were examined. Only 16% of the protonema cells grown at room temperature survived after freezing to −3°C; whereas cold treatment at 0°C increased the survival to 57% after 7 days and over 80% after 10 days (Fig. 10.4). Incubation at 4°C also increased freezing tolerance, but it was less effective than incubation at 0°C. Incubation at 10°C only transiently increased freezing tolerance, and incubation at 15°C had no effect on freezing tolerance (Minami *et al.*, 2005). These results indicate that *P. patens* cells are capable of acclimating to a low temperature below 10°C in order to enhance their freezing tolerance.

We analysed expression of ABA-induced genes in response to cold stresses and found that expression of several ABA-induced genes was found to be responsive to low temperatures (Minami *et al.*, 2003a). Further analysis indicated that *P. patens* cells distinguished different cold temperatures. For instance, a transcript of the *LEAIII*-like gene *6A5* was most strongly induced by a temperature of 0°C, at which freezing tolerance was most remarkably increased. On the other hand, a transcript of the *19C6* gene was induced more at 15°C than at 0°C (Minami *et al.*, 2005). Since freezing tolerance was not induced by incubation at 15°C, the *19C6* gene might not be directly involved in freezing tolerance but might have a function for resistance to chilling stress.

We also analysed changes in freezing tolerance and expression of ABA-induced genes in response to osmotic and salinity stress. Freezing tests indicated that treatment with hyperosmotic concentrations of mannitol (0.5 M) and NaCl (0.25 M) increased freezing tolerance. Expression analysis of ABA-inducible '*PPAR*' genes indicated that transcripts of several genes were induced in response to hyperosmotic concentrations of mannitol and NaCl. In contrast, treatment with hypo-osmotic concentrations of these solutes (0.25 M mannitol and 0.125 M NaCl) had much less effect on both freezing tolerance and expression of ABA-induced genes (Minami *et al.*, 2003a). Recently, stress-induced genes of *P. patens* have been analysed, and genes encoding AWPM19-like and blt101-like proteins have been found to be induced by desiccation, salinity and osmotic stress (Kroemer *et al.*,

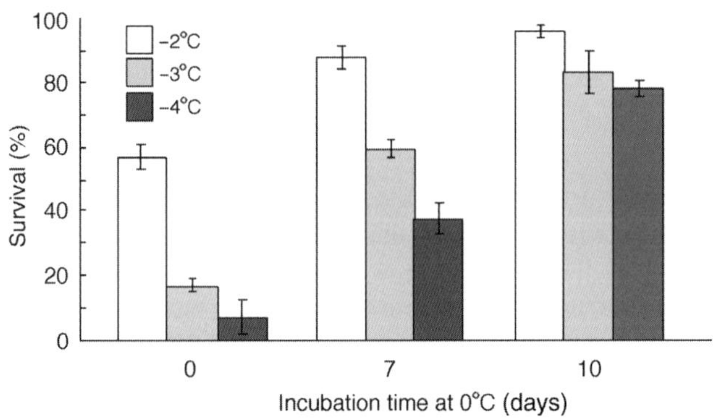

Fig. 10.4. Acclimation of *P. patens* protonema cells to low temperature. Protonema cells were acclimated at 0°C for the indicated days and frozen to −2°C, −3°C and −4°C. Survival of cells was determined by measurement of electrolyte leakage after thawing.

2004; Frank *et al.*, 2005). These results indicate that not only higher plants but also mosses have a molecular machinery controlling signal crosstalk among ABA, osmotic, salinity and cold stress, which leads to the development of tolerance to freezing and osmotic stress.

In order to clarify the role of ABA in cold- and osmotic stress-induced enhancement of freezing tolerance in *P. patens*, we determined endogenous ABA levels by gas chromatography–mass spectrometry (GC–MS). We analysed changes in ABA levels in protonema tissues as well as in culture medium in which the tissues were cultured. Results of our measurements indicated that low-temperature treatment (0°C) for up to 1 week, which significantly increased freezing tolerance, did not specifically increase levels of ABA in either tissues or the culture medium. This indicated that, as is the case in higher plants, there exists an ABA-independent cold signalling pathway leading to freezing tolerance in *P. patens*. On the other hand, osmotic stress treatment by 0.5 M mannitol increased ABA levels in tissues by approximately twofold, indicating that osmotic stress signalling in *P. patens* might, in part, be mediated by endogenously produced ABA (Minami *et al.*, 2005).

Alteration of Stress Resistance of Moss Cells by Transient Expression of an *ABI3* Homologue

Several transgenic studies on higher plants have revealed that stress resistance, including freezing tolerance, of plants can be changed by introduction of genes encoding transcription factors that drive expression of stress-induced genes (e.g. Jaglo-Ottosen *et al.*, 1998). To establish a system for evaluation of an exogenously introduced gene in protonema cells, we used a *P. patens PpABI3* gene, a homologue of *Arabidopsis ABI3* encoding a transcriptional regulator. *ABI3* is essential for the legitimate seed-maturation process leading to dormancy and plays a role in activation of *LEA*-family genes expressed during late stages of embryogenesis (Giraudat *et al.*, 1992). Although an *ABI3* transcript is exclusively expressed in developing seeds in wild-type plants, ectopic expression of *ABI3* in transgenic plants resulted in enhanced expression of *rab18* or seed-specific *LEA* genes in vegetative tissues (Parcy and Giraudat, 1997) and enhancement of freezing tolerance in response to ABA or cold (Tamminen *et al.*, 2001). It has been reported that *P. patens* has an *ABI3*-like gene (*PpABI3*) and that its transcript levels in protonema cells were increased by osmotic stress and ABA treatment (Sakata *et al.*, 2003; Frank *et al.*, 2005). *PpABI3* is likely to have a function similar to that of *Arabidopsis ABI3*, since its transient expression in barley aleurone cells resulted in enhanced expression of the GUS reporter gene fused to the wheat *Em* promoter (Sakata *et al.*, 2003). It has also been shown that transient overexpression of *PpABI3* in *P. patens* protonema protoplasts had an inhibitory effect on protoplast germination (Fujita *et al.*, 2003). In order to analyse the role of *PpABI3* in freezing tolerance of *P. patens*, we transiently expressed *PpABI3* in protonema cells and froze the cells at different temperatures. The results of the freezing tests indicated that freezing to –9°C had different effects on *PpABI3*- and vector-introduced protonema cells. Culture of transformed protonema cells after freezing to –9°C resulted in survival of only a small number of vector-introduced cells, whereas substantial numbers of *ABI3*-introduced cells formed protonema colonies (Fig. 10.5). These results suggest that ABA-activated genes identified (Table 10.1), possibly controlled by a specific transcriptional regulator(s), play a role in the development of freezing tolerance in *P. patens*. These results also suggest that a transient expression system of *P. patens* can be useful for evaluating functions of stress-induced genes in freezing tolerance.

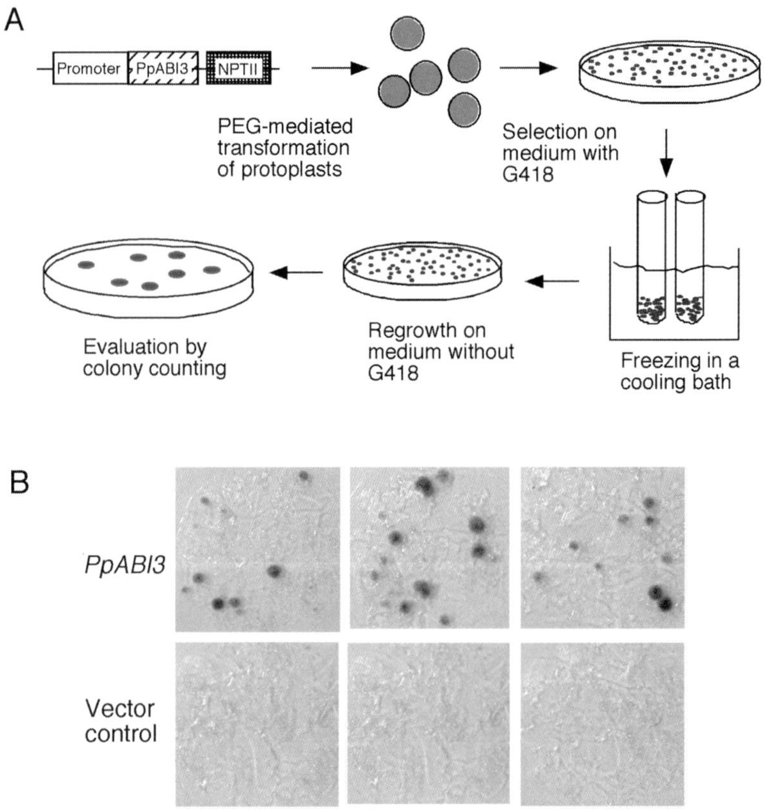

Fig. 10.5. Evaluation of an introduced gene by a transient expression system of *P. patens*. (A) *PpABI3* driven by the rice actin promoter was introduced into protoplasts of protonema cells with a neomycin phosphotransferase gene cassette (NPTII) for G418 resistance. The protoplasts were spread over a cellophane-overlaid agar plate of selection medium containing G418. When colonies of transformed cells had grown for 1 week, the cells were transferred into test tubes that were placed in a cooling bath. The samples in the test tubes were inoculated with ice at –1°C and frozen to desired temperatures at a cooling rate of –2.4°C/h. After thawing, the frozen cells were plated onto a medium without G418 to observe regrowth. Two to three weeks later, the effect of introduced genes was evaluated by counting the number of growing colonies. (B) Representative results showing that incubation for 3 weeks after freezing at –9°C resulted in regrowth of dozens of colonies in *PpABI3*-introduced cells but no colonies in vector-introduced cells. The experiment was carried out in triplicate.

Concluding Remarks

Our results indicate that a number of traits are common in higher plants and mosses in terms of physiological aspects associated with freezing tolerance at cellular levels, including ABA responsiveness, organelle morphology and accumulation of soluble sugars and hydrophilic proteins. Analysis of ABA-induced genes of *P. patens* revealed that several hydrophilic and membrane proteins are common in higher plants and mosses. Evolutionary conservation of these features with respect to freezing tolerance suggests that these traits might have been necessary for their common ancestor to adapt successfully to terrestrial regions. Further physiological studies and molecular analysis employing an expression system of protonema cells will help us clarify the common mechanism underlying freezing tolerance at the cellular level of lower and higher plants.

Acknowledgements

We thank Dr Tomomichi Fujita and Dr Mitsuyasu Hasebe for providing an expression construct of the *PpABI3* gene. We also thank Takako Matsushita and Shuri Kasai for their technical assistance in moss cultures.

References

Anchordoguy, T.J., Rudolph, A.S., Carpenter, J.F. and Crowe, J.H. (1987) Modes of interaction of cryoprotectants with membrane phospholipids during freezing. *Cryobiology* 24, 324–331.

Anderson, J.V., Li, Q.B., Haskell, D.W. and Guy, C.L. (1994) Structural organization of the spinach endoplasmic reticulum-luminal 70-kilodalton heat-shock cognate gene and expression of 70-kilodalton heat-shock genes during cold acclimation. *Plant Physiology* 104, 1359–1370.

Ashton, N.W., Grimsley, N.H. and Cove, D.J. (1979) Analysis of gametophytic development in the moss *Physcomitrella patens*, using auxin and cytokinin resistant mutants. *Planta* 144, 427–435.

Breton, G., Danyluk, J., Charron, J.-B.F. and Sarhan, F. (2003) Expression profiling and bioinformatic analysis of a novel stress-regulated multispanning transmembrane protein family from cereals and *Arabidopsis*. *Plant Physiology* 132, 64–74.

Burch, J. (2003) Some mosses survive cryopreservation without prior pretreatment. *Bryologist* 106, 270–277.

Carpenter, C.D., Krepes, J.A. and Simon, A.E. (1994) Genes encoding glycine-rich *Arabidopsis thaliana* proteins with RNA binding motifs are influenced by cold treatment and an endogenous circadian rhythm. *Plant Physiology* 104, 1015–1025.

Carpenter, J.F. and Crowe, J.H. (1988) The mechanism of cryoprotection of protein solutes. *Cryobiology* 25, 244–255.

Chen, H.H., Li, P.H. and Brenner, M.L. (1983) Involvement of abscisic acid in potato cold acclimation. *Plant Physiology* 71, 362–365.

Christianson, M.L. (1998) A simple protocol for cryopreservation of mosses. *Bryologist* 101, 32–35.

Cove, D.J., Knight, C.D. and Lamparter, T. (1997) Mosses as model systems. *Trends in Plant Science* 2, 99–105.

Dilks, T.J.K. and Proctor, M.C.F. (1975) Comparative experiments on temperature responses of bryophytes: assimilation, respiration and freezing damage. *Journal of Bryology* 8, 317–336.

Fowler, S. and Thomashow, M.F. (2002) *Arabidopsis* transcriptome profiling indicates that multiple regulatory pathways are activated during cold acclimation in addition to the CBF cold response pathway. *Plant Cell* 14, 1675–1690.

Frank, W., Ratnadewi, D. and Reski, R. (2005) *Physcomitrella patens* is highly tolerant against drought, salt, and osmotic stress. *Planta* 220, 384–394.

Fujikawa, S. and Takabe, K. (1996) Formation of multiplex lamellae by equilibrium slow freezing of cortical parenchyma cells of mulberry and its possible relationship to freezing tolerance. *Protoplasma* 190, 189–203.

Fujita, T., Hiwatashi, Y., Nishiyama, T., Murata, T. and Hasebe, M. (2003) Identification of genes involved in cell polarity, asymmetric cell division or cell differentiation by overexpression of full-length cDNAs in the moss *Physcomitrella patens*. *Plant and Cell Physiology* Suppl. 44, s128.

Giraudat, J., Hauge, B.M., Valon, C., Smalle, J., Parcy, F. and Goodman, H.M. (1992) Isolation of the ABI3 gene by positional cloning. *Plant Cell* 4, 1251–1261.

Goddard, N.J., Dunn, M.A., Zhang, L., White, A.J., Jack, P.L. and Hughes, M.A. (1993) Molecular analysis and spatial expression pattern of a low-temperature-specific barley gene, blt101. *Plant Molecular Biology* 23, 871–879.

Goode, J.A., Stead, A.D. and Duckett, J.G. (1992) Redifferentiation of moss protonemata: an experimental and immunofluorescence study of brood cell formation. *Canadian Journal of Botany* 71, 1510–1519.

Grimsley, N.H. and Withers, L.A. (1983) Cryopreservation of cultures of the moss *Physcomitrella patens*. *Cryo-Letters* 4, 251–258.

Heino, P., Sandman, G., Lång, V., Nordin, K. and Palva, E.T. (1990) Abscisic acid deficiency prevents development of freezing tolerance in *Arabidopsis thaliana* (L.) Heynh. *Theoretical and Applied Genetics* 79, 801–806.

Jaglo-Ottosen, K.R., Gilmour, S.J., Zarka, D.G., Schabenberger, O. and Thomashow, M.F. (1998) *Arabidopsis* CBF1 overexpression induces COR genes and enhances freezing tolerance. *Science* 280, 104–106.

Knight, C.D., Sehgal, A., Atwal, K., Wallace, J.C., Cove, D.J., Coates, D., Quatrano, R.S., Bahadur, S., Stockley, P.G. and Cuming, A.C. (1995) Molecular responses to abscisic acid and stress are conserved between moss and cereals. *Plant Cell* 7, 499–506.

Koike, M., Takezawa, D., Arakawa, K. and Yoshida, S. (1997) Accumulation of a 19 kDa plasma membrane polypeptide during induction of freezing tolerance in wheat suspension-cultured cells by abscisic acid. *Plant and Cell Physiology* 38, 707–716.

Kroemer, K., Reski, R. and Frank, R. (2004) Abiotic stress response in the moss *Physcomitrella patens*: evidence for an evolutionary alteration in signaling pathways in land plants. *Plant Cell Reports* 22, 864–870.

Lalk, I. and Dörfflung, K. (1985) Hardening abscisic acid, proline and freezing resistance in two winter wheat varieties. *Physiologia Plantarum* 63, 287–292.

Lång, V., Mäntylä, E., Welin, B., Sundberg, B. and Palva, E.T. (1994) Alterations in water status, endogenous abscisic acid content, and expression of *rab18* gene during the development of freezing tolerance in *Arabidopsis thaliana*. *Plant Physiology* 104, 1341–1349.

Leslie, S.B., Israeli, E., Lightart, B., Crowe, J.H. and Crowe, L.M. (1995) Trehalose and sucrose protect both membranes and proteins in intact bacteria during drying. *Applied Environmental Microbiology* 61, 3592–3597.

Lin, C., Guo, W.W., Everson, E. and Thomashow, M.F. (1990) Cold acclimation in *Arabidopsis* and wheat. A response associated with expression of related genes encoding 'boiling-stable' polypeptides. *Plant Physiology* 94, 1078–1083.

Maurel, C., Reizer, J., Schroeder, J.I. and Chrispeels, M.J. (1993) The vacuolar membrane protein gamma-TIP creates water specific channels in Xenopus oocytes. *EMBO Journal* 12, 2241–2247.

Minami, A., Nagao, M., Arakawa, K., Fujikawa, S. and Takezawa, D. (2003a)

Abscisic acid-induced freezing tolerance in the moss *Physcomitrella patens* is accompanied by increased expression of stress-related genes. *Journal of Plant Physiology* 160, 475–483.

Minami, A., Nagao, M., Arakawa, K., Fujikawa, S. and Takezawa, D. (2003b) Physiological changes associated with abscisic acid-induced freezing tolerance in *Physcomitrella patens*. *Cryobiology and Cryotechnology* 49, 179–183.

Minami, A., Nagao, M., Ikegami, K., Koshiba, K., Arakawa, K., Fujikawa, S. and Takezawa, D. (2005) Cold acclimation in bryophytes: low temperature-induced freezing tolerance in *Physcomitrella patens* is associated with increases in expression levels of stress-related genes but not with increase in levels of endogenous abscisic acid. *Planta* 220, 414–423.

Nagao, M., Minami, A., Arakawa, K., Fujikawa, S. and Takezawa, D. (2005) Rapid degradation of starch in chloroplasts and concomitant accumulation of soluble sugars associated with ABA-induced freezing tolerance in the moss *Physcomitrella patens*. *Journal of Plant Physiology* 162, 169–180.

Nishiyama, T., Fujita, T., Shin-I, T., Seki, M., Nishide, H., Uchiyama, I., Kamiya, A., Carninci, P., Hayashizaki, Y., Shinozaki, K., Kohara, Y. and Hasebe, M. (2003) Comparative genomics of *Physcomitrella patens* gametophytic transcriptome and *Arabidopsis thaliana*: implication for land plant evolution. *Proceedings of the National Academy of Sciences USA* 100, 8007–8012.

Parcy, F. and Giraudat, J. (1997) Interactions between the ABI1 and the ectopically expressed *ABI3* genes in controlling abscisic acid responses in *Arabidopsis* vegetative tissues. *Plant Journal* 11, 693–702.

Ristic, Z. and Ashworth, F.N. (1993) Changes in leaf ultrastructure and carbohydrates in *Arabidopsis thaliana* L. (Heyn) cv. Columbia during rapid cold acclimation. *Protoplasma* 172, 111–123.

Rütten, D. and Santarius, K.A. (1992a) Relationship between frost tolerance and sugar concentration of various bryophytes in summer and winter. *Oecologia* 91, 260–265.

Rütten, D. and Santarius, K.A. (1992b) Age-related differences in frost sensitivity of the photosynthetic apparatus of two *Plagiomnium* species. *Planta* 187, 224–229.

Rütten, D. and Santarius, K.A. (1993) Seasonal variation in frost tolerance and sugar content of two *Plagiomnium* species. *Bryologist* 96, 564–568.

Sakai, A. (1961) Studies on the frost-hardiness of woody plants. The causal relation between sugar content and frost-hardiness. *Low Temperature Science Series B, Biological Science* 21, 1–40.

Sakata, Y., Tanaka, S. and Quatrano, R. (2003) Comparative analysis of ABI3/VP1-mediated transcriptional regulation of an ABA-inducible promoter between moss and higher plants. *Plant and Cell Physiology* Suppl. 45, s52.

Schaefer, D.G. (2002) A new moss genetics: targeted mutagenesis in *Physcomitrella patens*. *Annual Review of Plant Biology* 53, 477–501.

Schaefer, D.G., Zryd, J., Knight, C.D. and Cove, D.J. (1991) Stable transformation of the moss *Physcomitrella patens*. *Molecular and General Genetics* 226, 418–424.

Schnepf, E. and Reinhard, C. (1997) Brachycytes in *Funaria* protonemata: introduction by abscisic acid and fine structure. *Journal of Plant Physiology* 151, 166–175.

Seki, M., Narusaka, M., Abe, H., Kasuga, M., Yamaguchi-Shinozaki, K., Carninci, P., Hayashizaki, Y. and Shinozaki, K. (2001) Monitoring the expression pattern of 1300 *Arabidopsis* genes under drought and cold stresses by using a full-length cDNA microarray. *Plant Cell* 13, 61–72.

Shinozaki, K. and Yamaguchi-Shinozaki, K. (2000) Molecular responses to dehydration and low temperature: differences and cross-talk between two stress signaling pathways. *Current Opinion in Plant Biology* 3, 217–223.

Siminovitch, D., Rheaume, K., Pomeroy, K. and Lepage, M. (1968) Phospholipid, protein, and nucleic acid increases in protoplasm and membrane structure associated with development of extreme freezing resistance in black locust tree cells. *Cryobiology* 5, 202–225.

Steponkus, P.L., Uemura, M. and Webb, M.S. (1993) A contrast of the cryostability of the plasma membrane of winter rye and spring oat. Two species that widely differ in their freezing tolerance and plasma membrane lipid composition. In: Steponkus, P.L. (ed.) *Advances in Low-Temperature Biology*, Vol. 2. JAI Press, London, 211–304.

Tamminen, I., Malela, P., Heino, P. and Palva, T.E. (2001) Ectopic expression of *ABI3* gene enhances freezing tolerance in response to abscisic acid and low temperature in *Arabidopsis thaliana. Plant Journal* 25, 1–8.

Thomashow, M.F. (1999) Plant cold acclimation: freezing tolerance genes and regulatory mechanisms. *Annual Review of Plant Physiology and Plant Molecular Biology* 50, 571–599.

Wise, M.J. and Tunnacliffe, A. (2004) POPP the question: what do LEA proteins do? *Trends in Plant Science* 9, 13–17.

Xiong, L., Schumaker, K.S. and Zhu, J.K. (2002) Cell signaling during cold, drought, and salt stress. *Plant Cell* 14, S165–S183.

11 Control of Growth and Cold Acclimation in Silver Birch

M.K. Aalto and E.T. Palva*

Viikki Biocenter, Department of Biological and Environmental Sciences, Division of Genetics, PO Box 56, FI-00014, Helsinki, Finland

Introduction

Plants, being sessile and poikilothermous, are unable to escape unfavourable conditions present in their growth environment. Furthermore, they are strictly dependent on the light environment for growth and development. Consequently, plants have developed a number of adaptive strategies to cope with the environmental stresses and to ensure optimal growth and survival in adverse conditions. The importance of light perception in plant life has led to the development of an elaborated machinery consisting of families of phytochromes and cryptochromes, among many other light-sensing proteins, in order to extract as much information as possible from the light environment. These cues must then be integrated with other environmental signals to ensure optimal plant response. This integration of environmental cues is particularly important for tree growth in boreal forests governed by the special combination of light and temperature. In northern Europe, for example, the high latitudes cause not only very long days together with relatively low temperatures at the beginning of the growth period, but also the drastic changes in the length of the photoperiod during active growth, set up a very special environment for perennial plants. The photoperiod may vary as much as 12 h from summer to autumn. In the northernmost part of Finland, the photoperiod varies from 24 h to 12 h and the variation is about 7 h even in the south of Finland (from 19 h to 12 h, http://almanakka.helsinki.fi). This is about 4 h more than in Ohio, USA (a 3-h change in day length during the same growth period; http://www.almanac.com).

Consequently, there is a very steep latitudinal cline in the critical photoperiod leading to growth cessation within Finland and this has led to the development of local populations showing differential responses to photoperiod (Li *et al.*, 2003a,b).

*Corresponding author (e-mail: Tapio.Palva@helsinki.fi; Fax: +358-9-191-59076)

Overwintering of perennial plants is dependent on their ability to develop high levels of cold hardiness. It is well established that winter survival requires proper timing of cold hardening executed by perception of two main environmental cues, photoperiod and temperature. Freezing stress occurs both during the active growth and as seasonal stress (winter). In both cases recognition or anticipation of stress triggers an adaptive response, cold hardening or cold acclimation, leading to enhanced tolerance to freezing stress (Guy, 1990; Thomashow, 1998). At the physiological level this includes alterations of gene expression, changes in hormone levels, accumulation of protective proteins, soluble sugars and osmolytes, and modifications of the membrane lipid composition (Thomashow, 1999; Heino and Palva, 2003).

Furthermore, in particular in perennial plants, these physiological alterations are accompanied by a number of developmental changes enhancing freezing survival. These include cessation of growth, leaf senescence and abscission, bud formation and dormancy development (Heino and Palva, 2003; Olsen *et al.*, 2004). In summary, cold acclimation and development of winter hardiness is an outcome of several complex signalling networks triggered by distinct environmental cues. Perception of these cues triggers separate but interacting signalling pathways, each controlled both by positive and negative factors, controlling in turn a number of distinct stress regulons. Activation of these regulons will result in synthesis of protective proteins and metabolites which promote freezing tolerance development (Thomashow, 1999; Chinnusamy *et al.*, 2004).

Stress Regulons

Low-temperature-responsive regulons are best characterized in annual herbaceous plants. Hormonal signalling is central to many stress responses and numerous studies have shown that an increase in the level of phytohormone abscisic acid (ABA) is not only one of the early responses to water stress (Ingram and Bartels, 1996), but it is also involved in low-temperature signalling (Daie and Campbell, 1981; Zeevaart and Creelman, 1988; Li *et al.*, 2002, 2003a; Chinnusamy *et al.*, 2004). ABA application increases freezing tolerance in non-acclimated plants and specifically induces expression of genes associated with cold tolerance.

However, many of the signal pathways central to plant cold acclimation are independent of ABA. Of these, the best characterized is the pathway controlled by the DRE/CRT-binding proteins CBF/DREB family of transcription factors (Thomashow, 2001; Maruyama *et al.*, 2004).

A recent estimate is that up to 25% of the transcriptome of *Arabidopsis thaliana* is affected by low non-freezing temperature (Kreps *et al.*, 2002). Most of these genes are expressed at ambient temperatures and are two to five times upregulated at low temperatures. Some of these have been earlier identified as *COR* (cold-regulated) or *LTI* (low-temperature-induced) genes (Thomashow, 1999).

Deletion analysis was used to identify the *cis*-acting elements required for the low-temperature control of promoters of some of the *COR* genes (reviewed in Thomashow, 1999). These elements designated low temperature-responsive elements (LTREs), drought-responsive elements (DREs) or C-repeat (CRT) contained the core sequence of CCGAC. These elements were subsequently used in yeast one-hybrid analysis as baits in successfully cloning members of a gene family, encoding related transcription factors designated DREB (for DRE-binding factors) or CBF (for CRT-binding factor) (Stockinger *et al.*, 1997; Liu *et al.*, 1998). Three of these, CBF1/DREB1b, CBF2/DREB1c and CBF3/DREB1a (Stockinger *et al.*, 1997; Gilmour *et al.*, 1998; Liu *et al.*, 1998; Shinwari *et al.*, 1998), are central to cold signalling and are estimated to directly control expression of up to 100 genes (Seki *et al.*, 2001; Fowler and Thomashow, 2002; Maruyama *et al.*, 2004; Vogel *et al.*, 2005), as well as control cellular metabolism. In a recent study, a total of 256 metabolites out of 434 examined were shown to be directly influenced by CBF overproduction (Cook *et al.*, 2004).

Birch as a Model for Frost-hardiness Studies

Silver birch (*Betula pendula* Roth) is economically the most important hardwood species in Finland. Due to the long tradition of birch research in Finland, existing breeding material and programmes, as well as elite clones, are readily available. Silver birch is a diploid organism, having a small genome about four times that of *Arabidopsis* (470 Mb). Birch is easy to grow, also *in vitro*, and it is readily transformable with *Agrobacterium*. It has a remarkably short generation time and can be made to flower in less than 1 year, greatly facilitating breeding programmes.

Birch is an extremely freezing-tolerant species and when fully hardened can tolerate temperatures down to −196°C. This is in contrast to cold hardy herbaceous plants that usually tolerate temperatures down to −10°C to −20°C. Such differences in tolerance between plant species may help to identify crucial regulons and specific protection mechanisms that explain extreme tolerance, and may provide targets for improving plant frost hardiness and overwintering. The moderate tolerance achieved by annuals is in sharp contrast to that in woody perennials, especially trees that may survive hundreds of winters. A central feature of such survival is the proper control of the annual growth cycle, and it includes proper timing of growth cessation and development of winter hardiness in the autumn, as well as timely bud burst in the spring.

The first cue for development of winter hardiness in birch and other boreal tree species is the shortening of the photoperiod. After the night (dark) period exceeds a critical time, birch stops height growth and starts to prepare for the winter. This includes initiation of dormancy, bud maturation, leaf senescence and abscission and priming for freezing tolerance development (Chen *et al.*, 2002; Li *et al.*, 2003a, 2004a; Puhakainen *et al.*, 2004a).

Northern populations of birch set bud and develop frost hardiness earlier than the southern ones, even when grown under identical environments. This adaptation to local environment, a generation of so-called photoperiodic ecotypes (Håbjørg, 1978), is genetically determined (Junttila, 1980; Howe *et al.*, 2003; Li *et al.*, 2004b).

The steep latitudinal cline observed in such photoperiodic ecotypes in Finland is exemplified by the following: for the northern ecotype 'Kittilä' (67°44′N) the critical day-length, or SD (short day)-signal, is 22 h (2 h of night), whereas it is 18 h for the 'Viitasaari' ecotype originating from the middle of Finland (63°14′N) and only 14 h for the southernmost, Estonian 'Viljandi' ecotype (58°10′N). The genetic basis for these differences is demonstrated by common garden experiments, where the trees are grown under identical environments (Junttila, 1980; Chen *et al.*, 2002; Viherä-Aarnio *et al.*, 2005). Such experiments have shown that the populations still differ from each other in timing of the growth cessation, onset of bud dormancy and even in the level of freezing tolerance achieved (Li *et al.*, 2002; Fig.11.1). This is also reflected in hormonal responses; for example, ABA levels increase and auxin levels decrease upon shortening of the day length in an ecotype-dependent fashion (Li *et al.*, 2003a; Fig. 11.2).

A prominent feature of the acclimation response in birch is the synergistic effect of SD and low temperature (LT) on freezing tolerance. Birch develops freezing tolerance much faster and the level of tolerance reached is higher if the plants have been exposed to SD signal prior to LT treatment. The SD treatment alone increased the freezing tolerance (as judged by ion-leakage measurements) about 4°C, whereas low temperature increased the freezing tolerance roughly by 9°C. One week in SD, followed by 1 week in low temperature, improved the freezing tolerance by 13°C (Li *et al.*, 2002; Fig. 11.3). The SD signal may thus help the birch to distinguish between acute and seasonal cold. However, the decision to initiate dormancy and winter hardiness-related changes is also affected by the maturity of the seedlings (Li *et al.*, 2002; Viherä-Aarnio *et al.*, 2005).

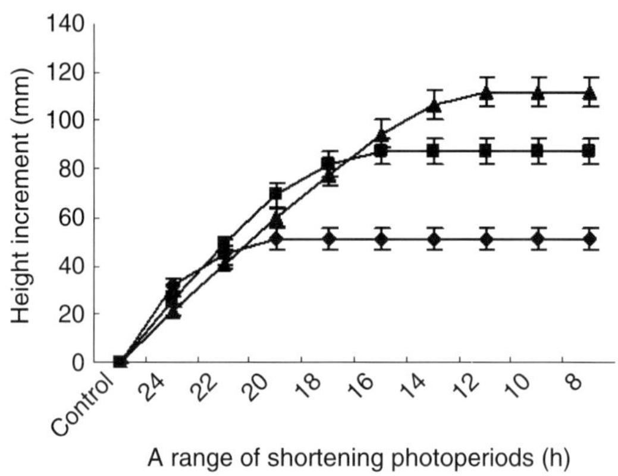

Fig. 11.1A. Apical shoot elongation in seedlings of three ecotypes of *Betula pendula* (▲, the southern ecotype; ■, the central ecotype; ◆, the northern ecotype) after transfer to a range of shortening photoperiods. Values are the means ± SE of 10 replicates. (Li *et al.*, 2003a. Reproduced with permission.)

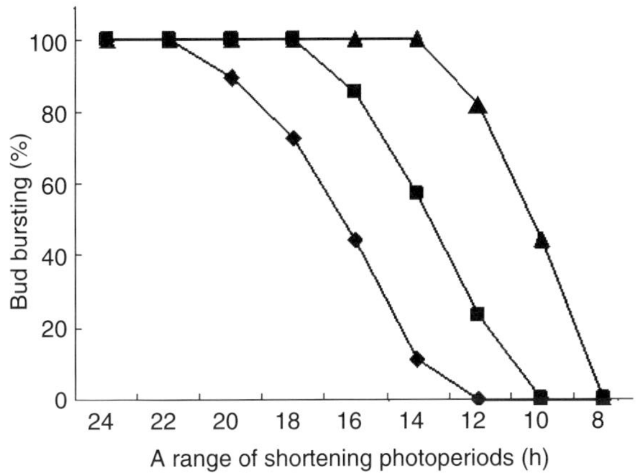

Fig. 11.1B. Bud bursting (%) in seedlings of three ecotypes of *Betula pendula* (▲, the southern ecotype; ■, the central ecotype; ◆, the northern ecotype) after transfer to a range of shortening photoperiods. (Li *et al.*, 2003a. Reproduced with permission.)

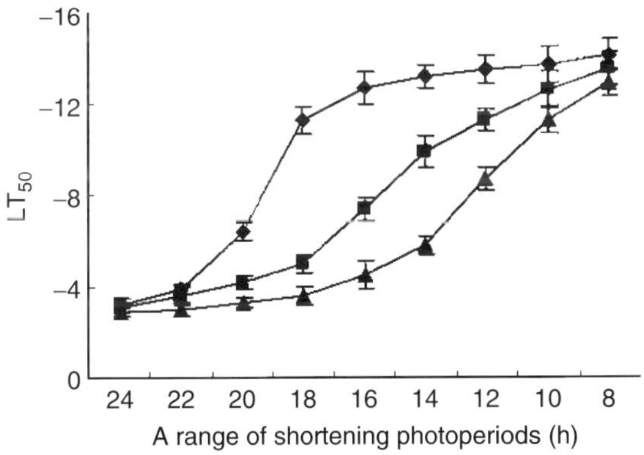

Fig. 11.1C. Freezing tolerance of the buds (LT_{50}, °C) in seedlings of three ecotypes of *Betula pendula* (▲, the southern ecotype; ■, the central ecotype; ◆, the northern ecotype) after transfer to a range of shortening photoperiods. Values are the means ± SE of 5 replicates. (Li *et al.*, 2003a. Reproduced with permission.)

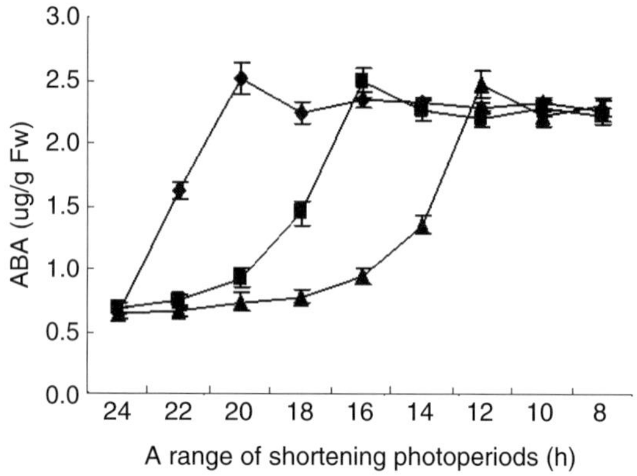

Fig. 11.2A. Changes in ABA content of the shoot apex in seedlings of three ecotypes of *Betula pendula* (▲, the southern ecotype; ■, the central ecotype; ◆, the northern ecotype) after transfer to a range of shortening photoperiods. Values are the means ± SE of three replicates. (Li *et al.*, 2003a. Reproduced with permission.)

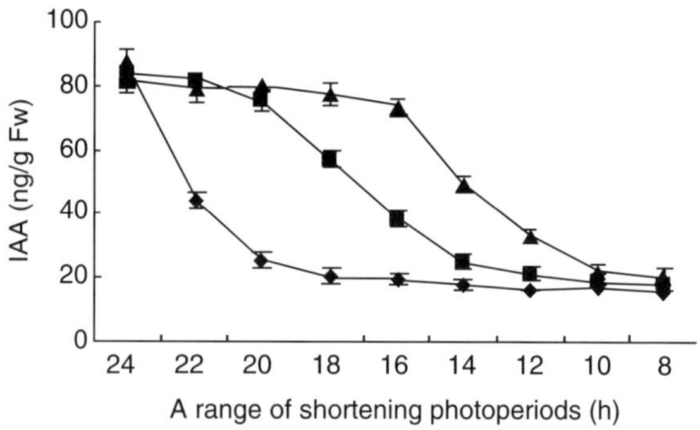

Fig. 11.2B. Changes in IAA content of the shoot apex in seedlings of three ecotypes of *Betula pendula* (▲, the southern ecotype; ■, the central ecotype; ◆, the northern ecotype) after transfer to a range of shortening photoperiods. Values are the means ± SE of three replicates. (Li *et al.*, 2003a. Reproduced with permission.)

Expressed Sequence Tags from Birch

To elucidate the molecular basis of birch responses to temperature and photoperiod as well as development of winter hardiness, the regulons that are responsive to these cues need to be identified. To facilitate such analysis, birch has been subjected to high throughput expressed sequence tag (EST) sequencing. This project is part of a coordinated joint effort of Finnish research groups working in plant molecular biology and forest biotechnology at the University of Helsinki (Academy of Finland Centre of Excellence programme 2000–2005).

Within this project, over 75,000 ESTs were generated from ten different cDNA libraries representing a wide variety of tissues and growth conditions.

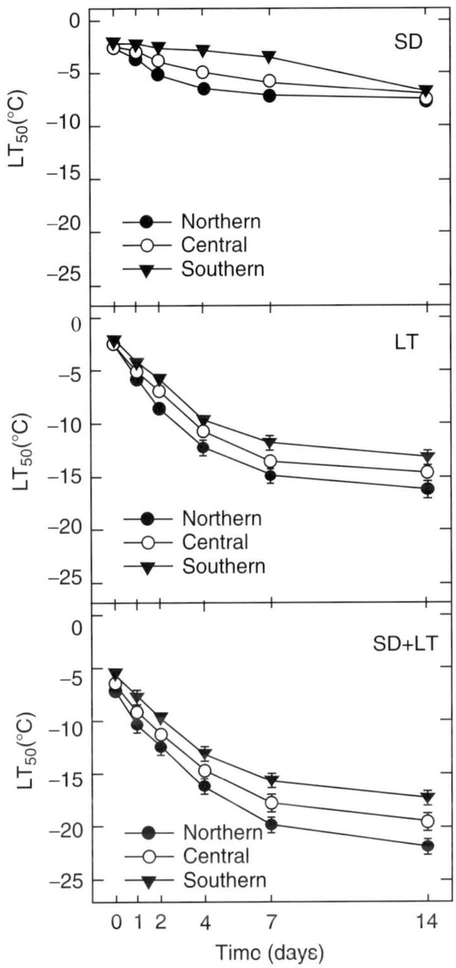

Fig. 11.3. Freezing tolerance (LT_{50}) of leaves in seedlings of three ecotypes of *Betula pendula* as affected by the different treatments. Values are the means ± SE of 5 replicates. (Li *et al.*, 2002. Reproduced with permission.)

Among them, leaf, bud and stem libraries were made both from low-temperature treated and short-day samples (manuscript in preparation). These libraries are thus expected to be enriched for genes involved in freezing tolerance, dormancy, overwintering, growth control, photoperiod, leaf abscission and senescence. Moreover, the genes identified from all of the libraries represent a good coverage of the majority of the actively expressed part of birch transcriptome, giving us a firm basis for accurate genomic scale studies.

Out of 74,000 accepted final sequences, a unigene set of about 19,000 clusters was predicted, out of which about 10,000 were singletons (e.g. represented only once). A microarray containing 1000 EST sequences was designed and is currently being used for initial characterization of stress-responsive genes in birch. An additional oligonucleotide microarray representing over 8000 ESTs is under construction. Annotation of the ESTs has brought up the remarkable fact that still, even after the genomic sequencing of *Arabidopsis* and Poplar, almost 30% of the genes in birch unigene set are catalogued as genes with unknown function, according to MIPS (http://mips.gsf.de/projects/funcat).

Comparison of the cold and short-day libraries with a reference library showed that the libraries exhibited 50–65% overlap with each other, with 35–50% ESTs being unique to each library. Statistical analysis of the expression differences of the ESTs between libraries (electronic northerns; Romualdi *et al.*, 2003) has effectively identified genes that are up- or downregulated in low temperature. Of the 655 clusters analysed, each containing at least 5 ESTs per cluster, 177 responded to cold and 82 clusters were upregulated and 95 downregulated. Many of the clusters show homology to known, cold-regulated genes in other plants, like *Arabidopsis* (manuscript in preparation). Thus, the EST project is providing tools for effective analysis of the birch low-temperature transcriptome.

Low-temperature-induced Genes in Birch

Unravelling of SD- and LT-responsive signalling pathways and regulons is of prime interest for elucidating of the birch acclimation response. Modulation of the expression of a particular gene may be under one or several distinct signalling pathways. Understanding how these signalling pathways lead to changes in metabolism, and result in growth cessation and development of dormancy and freezing tolerance, requires identification of target genes and signalling components. This can be achieved by genome-wide expression analysis combined with functional characterization of candidate genes by transgenic approaches.

Low-temperature-regulated gene expression is an intensively studied subject in plant molecular biology. Especially in *Arabidopsis thaliana*, the amount of data has rapidly increased our knowledge of the genetic regulation during cold acclimation (Fowler and Thomashow, 2002; Chinnusamy *et al.*, 2004; Maruyama *et al.*, 2004; Vogel *et al.*, 2005). Among the cold-inducible genes, those encoding dehydrin genes are a well-known group induced in response to several abiotic stresses (Nordin *et al.*, 1993; Close, 1996, 1997; Ingram and Bartels, 1996; Heino and Palva, 2003). Dehydrins are believed to confer low-temperature resistance when upregulated (Puhakainen *et al.*, 2004a,b).

Dehydrins were originally described as late embryogenesis-abundant (LEA)-related group II proteins. Typically, they accumulate late in embryogenesis or in vegetative tissue in response to low temperature or drought stress and they are thought to protect cellular structures against dehydration (Close, 1997; Nylander *et al.*, 2001). One of the cold-induced target genes identified from birch encodes a member of the dehydrin family, Bplti36 (Puhakainen *et al.*, 2004a).

Bplti36 is a typical YSK2 dehydrin (Close, 1996) and the corresponding gene was found to be represented in a number of copies in the EST library generated from plants exposed to low temperature. *Bplti36* was found to be strongly upregulated by LT, ABA and drought in northern analysis. The promoter region of *Bplti36* contains one ABA-response element (Marcotte *et al.*, 1989) and several CRT/DRE consensus sequences, suggesting that the gene is regulated by CBF/DREB1 transcription factors during LT (Puhakainen *et al.*, 2004a). Functional promoter analysis done in *Arabidopsis* showed that the LT-responsive expression of *Bplti36* was retained also in *Arabidopsis*, and deleting the CBF/DREB1-binding sites abolished the LT inducibility. Moreover, *BpLTI36* was shown to be constitutively expressed in *CBF3*-overproducing *Arabidopsis* (Puhakainen *et al.*, 2004a).

Three genes encoding CBF/DREB1-like transcription factors having over 50% amino acid identity with the corresponding *Arabidopsis* proteins were identified in birch (Welling *et al.*, manuscript in preparation). Northern analysis showed clear increase in the levels of CBF mRNAs following LT treatment. Also, one drought-specific DREB2-like EST was identified (Welling *et al.*, manuscript in preparation).

The rather limited analysis of birch ESTs characterized so far, and comparison with *Arabidopsis*, indicates that *Arabidopsis* and birch share common components in cold acclimation, including common target genes, e.g. genes encoding dehydrins and transcription factors of the CBF family (Puhakainen *et al.*, 2004a; Welling *et al.*, manuscript in preparation).

In addition to LT, the *Bplti36* gene is also controlled by photoperiod. When plants were exposed to SD photoperiod alone the gene was only slightly induced. However, the SD photoperiod appears to potentiate the moderate LT induction, since prior exposure to SD photoperiod resulted in enhancement of the subsequent LT induction, as compared to LT induction alone (Fig. 11.4). This SD-dependent potentiation of LT-induced gene expression appears to be distinct from the complex developmental regulation seen in overwintering cereals in which photoperiod sensitivity allows plants to maintain these genes in an upregulated level for longer periods of time under a short-day compared to a long-day situation (Fowler *et al.*, 2001; Puhakainen *et al.*, 2004a). The observed potentiation appeared specific to birch since there was no sign of enhancement of *Bplti36* induction following SD exposure of transgenic *Arabidopsis* harbouring a genomic copy of *Bplti36*, emphasizing the fact that there are basic differences in strategies between annual and perennial plant species. In birch, SD and LT treatment are both needed for maximal freezing tolerance, both in leaves as well as in overwintering tissues (Li *et al.*, 2002).

What could be the mechanism of SD potentiation of the low-temperature-induced gene expression in birch? Accumulation of transcripts of the birch *CBF* orthologues was not further enhanced in birch by SD treatment before LT, suggesting that potentiation might not involve *CBF* expression but might be through post-transcriptional processes or controlled via some other regulon (Puhakainen *et al.*, 2004a). It is intriguing, in this context, that phytohormones like ABA and auxin are clearly affected by the short-day treatment alone and at least ABA accumulation is further enhanced by low-temperature exposure (Li *et al.*, 2002, 2003a).

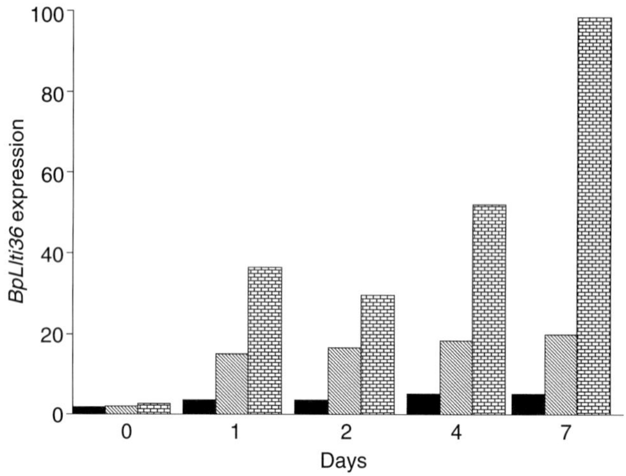

Fig. 11.4. Accumulation of *Bplti36* mRNA in birch leaves under SD (12 h light, 18°C), LT (24 h light, 4°C) and SD plus LT (12 h light, 4°C, LT preceded by 7 days in SD). (Puhakainen *et al.*, 2004a. Reproduced with permission.)

Photoperiod Mediated Control of Growth in Birch

Photoperiod perception is mediated by light-sensing proteins like phytochromes and cryptochromes. Studies with transgenic aspen overproducing Oat Phytochrome A (PhyA) show a strong correlation of the level of this photoreceptor with the developmental status of the plant (Olsen *et al.*, 1997). PhyA is involved in detection of changes in balance and fluence rate between red and far-red light. PhyA-overproducing aspen does not readily react to SD signal in growth chamber environment and thus is unable to stop growing, does not make buds and the freezing tolerance induced by SD treatment is abolished. No increase in the ABA levels due to SD treatment were found. However, the low-temperature-induced genetic and physiological responses were found to be normal (Welling *et al.*, 2002). The photoperiod phenotypes were partially restored by treating the plants with far-red light at the end of the light period, which is a sign of PhyA retaining the normal specificity and function despite overproduction (Olsen and Junttila, 2002).

In *Arabidopsis*, PhyA is involved in many light-directed processes such as seed germination, shade avoidance, seedling development and floral induction. A number of signalling components downstream of PhyA have been identified from *Arabidopsis* (Quail, 2002). One of these components is encoded by *CONSTANS* (*CO*) (Putterill *et al.*, 1995), which positively promotes flowering of *Arabidopsis* in long-day conditions. What makes CO especially interesting is that it has been postulated to converge the direct light signal with biological rhythms, especially the clock function (Hayama and Coupland, 2003).

In *Arabidopsis*, the *CO* gene expression is under circadian regulation. If the amount of CO protein peaks during daylight (long-day situation) then the protein positively enhances, as a transcriptional regulator, transition from vegetative growth to reproductive (flowering) phase. In a short-day situation, not enough protein is being produced during daylight to activate downstream processes. Thus, CO ensures that *Arabidopsis* flowers during the long-day situation (Hayama and Coupland, 2003).

The role of *CO* in integrating photoperiod with the circadian clock in *Arabidopsis* prompted us to search for genes homologous to *CO* and other genes known to be involved in photoperiod responses from the birch EST database. This led to identification of a family of genes that are related to the *Arabidopsis CO* gene. Currently, 12 genes having at least 1 B-box zinc finger have been identified. Furthermore, at least four of them also contain the CCT-domain, thus belonging to the CO family (Laine *et al.*, manuscript in preparation).

One of the birch *CO* homologues, *BpCOL2*, was present in leaf libraries, as may be expected when the gene is involved in light sensing and the expression of it is coupled with growth season. *Arabidopsis CO* transcript level is most strongly upregulated during dawn and dusk. A similar study conducted with *BpCOL2* in birch showed that, although transcript level responded to a circadian control, it peaked during the middle of the day, thus being the opposite to *CO* expression (Laine *et al.*, manuscript in preparation).

To elucidate the function of *BpCOL2*, transgenic *Arabidopsis* and birch lines either with sense or antisense constructs were generated (Laine *et al.*, manuscript in preparation). Unexpectedly, we found that both *BpCOL2*-transgenic birch and *Arabidopsis* plants generated showed an altered vegetative growth pattern. Overexpressing the gene shortens the length of the roots, and also inhibits shoot growth in both species, but otherwise the plants are healthy. Downregulation of the gene in birch resulted in longer roots. The likely explanation for the root phenotype is that the growth tip (apical meristem) of the root is affected. Microscopical analysis suggested that the elongation zone is shorter in the root.

Overexpression of *BpCOL2* in *Arabidopsis* suggested also a connection with flowering, as it delayed flowering of *Arabidopsis* both in short- and long-day situations, in contrast to *CO* function. Transgenic birches exhibited yet another phenotype, reduced responsiveness to the SD during growth cessation. These phenotypes strongly indicate that *BpCOL2* is involved in photoperiod regulation of physiological responses in birch. The exact function and target genes still remain to be determined.

Conclusions and Perspectives

Many physiological characteristics in plants are difficult, or impossible, to study in an herbaceous weed. Thus, alternative plant models are required for analysis of, e.g. seasonal responses in woody perennials. Birch provides an excellent model for such studies, made feasible by the rapid development of genomic techniques.

Studying trees at the genomic level may thus not only give answers to topics in wood formation, but also help to unravel mechanisms behind the seasonal control of growth in overwintering perennials in general. It promises a deeper understanding of the genetic regulation of plant responses at the physiological level, not only for the utilization of its genes for faster tree breeding, but also as both molecular markers and in transgenic approaches.

EST sequencing, clustering and annotation of genes provide the basic tools for such analysis aided by expression analysis through 'electronic northerns' and microarrays. This information, together with functional analysis of candidate genes, has helped us to elucidate basic processes in trees and will help to facilitate breeding of trees with improved growth characteristics.

Acknowledgements

The work in our laboratory is supported by the Academy of Finland, Biocentrum, Helsinki, and the National Technology Agency of Finland. We acknowledge the members of the laboratory for the experimental work involved and critical reading of the manuscript.

References

Chen, T.H.H., Howe, G.T. and Bradshaw, H.D. Jr (2002) Molecular genetic analysis of dormancy-related traits in poplars. *Weed Science* 50, 232–240.

Chinnusamy, V., Schumaker, K. and Zhu, J.K. (2004) Molecular genetic perspectives on cross-talk and specificity in abiotic stress signalling in plants. *Journal of Experimental Botany* 55, 225–236.

Close, T.J. (1996) Dehydrins: emergence of a biochemical role of a family of plant dehydration proteins. *Physiologia Plantarum* 97, 795–803.

Close, T.J. (1997) Dehydrins: a commonality in the response of plants to dehydration and low temperature. *Physiologia Plantarum* 100, 291–296.

Cook, D., Fowler, S., Fiehn, O. and Thomashow, M.F. (2004) A prominent role for the CBF cold response pathway in configuring the low-temperature metabolome of *Arabidopsis*. *Proceedings of the National Academy of Sciences USA* 101, 15243–15248.

Daie, J. and Campbell, W.F. (1981) Response of tomato plants to stressful temperatures. Increase in abscisic acid concentrations. *Plant Physiology* 67, 26–29.

Fowler, D.B., Breton, G., Limin, A.E., Mahfoozi, S. and Sarhan, F. (2001) Photoperiod and temperature interactions regulate low-temperature-induced gene expression in barley. *Plant Physiology* 127, 1676–1681.

Fowler, S. and Thomashow, M.F. (2002) *Arabidopsis* transcriptome profiling indicates that multiple regulatory pathways are activated during cold acclimation in addition to the CBF cold response pathway. *Plant Cell* 14, 1675–1690.

Gilmour, S.J., Zarka, D.G., Stockinger, E.J., Salazar, M.P., Houghton, J.M. and Thomashow, M.F. (1998) Low temperature regulation of the *Arabidopsis* CBF family of AP2 transcriptional activators as an early step in cold-induced COR gene expression. *Plant Journal* 16, 433–442.

Guy, C.L. (1990) Cold-acclimation and freezing stress tolerance – role of protein metabolism. *Annual Review of Plant Physiology and Plant Molecular Biology* 41, 187–223.

Håbjørg, A. (1978) Photoperiodic ecotypes in Scandinavian trees and shrubs. *Scientific Reports of the Agricultural University of Norway* Report No. 71.

Hayama, R. and Coupland, G. (2003) Shedding light on the circadian clock and the photoperiodic control of flowering. *Current Opinion in Plant Biology* 6, 13–19.

Heino, P. and Palva, E.T. (2003) *Signal transduction in plant cold acclimation.* In: Hirt, H. and Shinozaki, K. (eds) *Plant Responses to Abiotic Stress.* Springer-Verlag, Heidelberg, Berlin, pp. 151–186.

Howe, G.T., Aitken, S.N., Neale, D.B., Jermstad, K.D., Wheeler, N.C. and Chen, T.H.H. (2003) From genotype to phenotype: unraveling the complexities of cold adaptation in forest trees. *Canadian Journal of Botany* 81, 1247–1266.

Ingram, J. and Bartels, D. (1996) The molecular basis of dehydration tolerance in plants. *Annual Review of Plant Physiology and Plant Molecular Biology* 47, 377–403.

Junttila, O. (1980) Effect of photoperiod and temperature on apical growth cessation in two ecotypes of *Salix* and *Betula*. *Physiologia Plantarum* 48, 347–352.

Kreps, J.A., Wu, Y., Chang, H.S., Zhu, T., Wang, X. and Harper, J.F. (2002) Transcriptome changes for *Arabidopsis* in response to salt, osmotic, and cold stress. *Plant Physiology* 130, 2129–2141.

Li, C.Y., Puhakainen, T., Welling, A., Viherä-Aarnio, A., Ernstsen, A., Junttila, O., Heino, P. and Palva, E.T. (2002) Cold acclimation in silver birch (*Betula pendula*). Development of freezing tolerance in different tissues and climatic ecotypes. *Physiologia Plantarum* 116, 478–488.

Li, C., Junttila, O., Ernstsen, A., Heino, P. and Palva, E.T. (2003a) Photoperiodic control of growth, cold acclimation and dormancy development in silver birch (*Betula pendula*) ecotypes. *Physiologia Plantarum* 117, 206–212.

Li, C., Viherä-Aarnio, A., Puhakainen, T., Junttila, O., Heino, P. and Tapio Palva, E. (2003b) Ecotype-dependent control of growth, dormancy and freezing tolerance under seasonal changes in *Betula pendula* Roth. *Trees – Structure and Function* 17, 127–132.

Li, C., Junttila, O. and Palva, E.T. (2004a) Environmental regulation and physiological basis of freezing tolerance in woody plants. *Acta Physiologiae Plantarum* 2, 213– 222.

Li, C., Junttila, O., Heino, P. and Palva, E.T. (2004b) Low temperature sensing in silver birch (*Betula pendula* Roth) ecotypes. *Plant Science* 167, 165–171.

Liu, Q., Kasuga, M., Sakuma, Y., Abe, H., Miura, S., Yamaguchi-Shinozaki, K. and Shinozaki, K. (1998) Two transcription factors, DREB1 and DREB2, with an EREBP/AP2 DNA binding domain separate two cellular signal transduction pathways in drought- and low-temperature-responsive gene expression, respectively, in *Arabidopsis*. *Plant Cell* 10, 1391–1406.

Marcotte, W.R. Jr, Russell, S.H. and Quatrano, R.S. (1989) Abscisic acid responsive sequences from the Em gene of wheat. *Plant Cell* 1, 969–976.

Maruyama, K., Sakuma, Y., Kasuga, M., Ito, Y., Seki, M., Goda, H., Shimada, Y., Yoshida, S., Shinozaki, K. and Yamaguchi-Shinozaki, K. (2004) Identification of cold-inducible downstream genes of the *Arabidopsis* DREB1A/CBF3 transcriptional factor using two microarray systems. *Plant Journal* 38, 982–993.

Nordin, K., Vahala, T. and Palva, E.T. (1993) Differential expression of two related, low-temperature-induced genes in *Arabidopsis thaliana* (L.) Heynh. *Plant Molecular Biology* 21, 641–653.

Nylander, M., Svensson, J., Palva, E.T. and Welin, B.V. (2001) Stress-induced accumulation and tissue-specific localization of dehydrins in *Arabidopsis thaliana*. *Plant Molecular Biology* 45, 263–279.

Olsen, J.E. and Junttila, O. (2002) Far red end-of-day treatment restores wild type-like plant length in hybrid aspen overexpressing phytochrome A. *Physiologia Plantarum* 115, 448–457.

Olsen, J.E., Junttila, O., Nilsen, J., Eriksson, M.E., Martinussen, I., Olsson, O., Sandberg, G. and Moritz, T. (1997) Ectopic expression of oat phytochrome A in hybrid aspen changes critical daylength for growth and prevents cold acclimatization. *Plant Journal* 12, 1339–1350.

Olsen, J.E., Jensen, J.B., Molmann, J.A., Ernstsen, A. and Junttila, O. (2004) Photoperiodic regulation of apical growth cessation in northern tree species: the role of phytochrome and gibberellin. *Journal of Crop Improvement* 10, 77–112.

Puhakainen, T., Li, C., Boije-Malm, M., Kangasjärvi, J., Heino, P. and Palva, E.T. (2004a) Short-day potentiation of low temperature-induced gene expression of a C-repeat-binding factor-controlled gene during cold acclimation in silver birch. *Plant Physiology* 136, 4299–4307.

Puhakainen, T., Hess, M.W., Mäkelä, P., Svensson, J., Heino, P. and Palva, E.T. (2004b) Overexpression of multiple dehydrin genes enhances tolerance to freezing stress in *Arabidopsis*. *Plant Molecular Biology* 54, 743–753.

Putterill, J., Robson, F., Lee, K., Simon, R. and Coupland, G. (1995) The *CONSTANS* gene of *Arabidopsis* promotes flowering and encodes a protein showing similarities to zinc finger transcription factors. *Cell* 80, 847–857.

Quail, P.H. (2002) Phytochrome photosensory signalling networks. *Nature Reviews Molecular Cell Biology* 3, 85–93.

Romualdi, C., Bortoluzzi, S., D'Alessi, F. and Danieli, G.A. (2003) IDEG6: a web tool for detection of differentially expressed genes in multiple tag sampling experiments. *Physiological Genomics* 12, 159–162.

Seki, M., Narusaka, M., Abe, H., Kasuga, M., Yamaguchi-Shinozaki, K., Carninci, P., Hayashizaki, Y. and Shinozaki, K. (2001) Monitoring the expression pattern of 1300 *Arabidopsis* genes under drought and cold stresses by using a full-length cDNA microarray. *Plant Cell* 13, 61–72.

Shinwari, Z.K., Nakashima, K., Miura, S., Kasuga, M., Seki, M., Yamaguchi-Shinozaki, K. and Shinozaki, K. (1998) An *Arabidopsis* gene family encoding DRE/CRT binding proteins involved in low-temperature-responsive gene expression. *Biochemical and Biophysical Research Communication* 250, 161–170.

Stockinger, E.J., Gilmour, S.J. and Thomashow, M.F. (1997) *Arabidopsis thaliana* CBF1 encodes an AP2 domain-containing transcriptional activator that binds to the C-repeat/DRE, a cis-acting DNA regulatory element that stimulates transcription in response to low temperature and water deficit. *Proceedings of the National Academy of Sciences USA* 94, 1035–1040.

Thomashow, M.F. (1998) Role of cold-responsive genes in plant freezing tolerance. *Plant Physiology* 118, 1–8.

Thomashow, M.F. (1999) Plant cold acclimation: freezing tolerance genes and regulatory mechanisms. *Annual Review of Plant Physiology and Plant Molecular Biology* 50, 571–599.

Thomashow, M.F. (2001) So what's new in the field of plant cold acclimation? Lots! *Plant Physiology* 125, 89–93.

Viherä-Aarnio, A., Häkkinen, R., Partanen, J., Luomajoki, A. and Koski, V. (2005) Effects of seed origin and sowing time on timing of height growth cessation of *Betula pendula* seedlings. *Tree Physiology* 25, 101–108.

Vogel, J.T., Zarka, D.G., Van Buskirk, H.A., Fowler, S.G. and Thomashow, M.F. (2005) Roles of the CBF2 and ZAT12 transcription factors in configuring the low temperature transcriptome of *Arabidopsis*. *Plant Journal* 41, 195–211.

Welling, A., Moritz, T., Palva, E.T. and Junttila, O. (2002) Independent activation of cold acclimation by low temperature and short photoperiod in hybrid aspen. *Plant Physiology* 129, 1633–1641.

Zeevaart, J.A.D. and Creelman, R.A. (1988) Metabolism and physiology of abscisic acid. *Annual Review of Plant Physiology and Plant Molecular Biology* 39, 439–473.

12 The Role of the CBF-dependent Signalling Pathway in Woody Perennials

C. BENEDICT,[1] J.S. SKINNER,[2] R. MENG,[2] Y. CHANG,[2] R. BHALERAO,[3] C. FINN,[2,4] T.H.H. CHEN[2] AND V. HURRY[1,*]

[1]Umeå Plant Science Centre, Department of Plant Physiology, Umeå University, S-901 87 Umeå, Sweden; [2]Department of Horticulture, Oregon State University, Corvallis, OR 97331-7304, USA; [3]Umeå Plant Science Centre, Department of Forest Genetics, Swedish University of Agricultural Sciences, S-901 83 Umeå, Sweden; [4]US Department of Agriculture-Agricultural Research Service, Corvallis, OR 97330, USA

Introduction

In temperate climates, the meristematic tissues of woody perennials must be able to survive freezing temperatures in order to resume growth the following year, allowing the plant eventually to reach reproductive maturity. For many tree species of the temperate and boreal forests, this physiological challenge is amplified by the fact that stem and shoot tissues overwinter above the protective snow cover, and are therefore susceptible to extreme cold and winter drought. As the acquisition of deep freezing tolerance (FT) in woody perennials requires growth cessation (Frewen *et al.*, 2000; Li *et al.*, 2003), boreal tree species must synchronize their periods of growth with periods of favourable temperatures, and begin the lengthy processes of growth cessation and cold hardening early enough to ensure that maximal FT is obtained well in advance of winter temperature minima.

The tree's transition from active growth to winter dormancy is initiated by both shortening day length and low temperatures (Weiser, 1970; Welling *et al.*, 2002), and takes many weeks to complete. Boreal trees are most susceptible to freeze damage when freezing temperatures occur frequently while the plant is not fully dormant; and trees that survive freezing in liquid nitrogen (−196°C) while in deep endodormancy can be killed by temperatures as high as −2°C while actively

*Corresponding author (e-mail: Vaughan.Hurry@plantphys.umu.se; Fax: +46-90-786-6676)

 Cold Hardiness in Plants: Molecular Genetics, Cell Biology and Physiology (eds T.H.H. Chen *et al.*)

growing (Fuchigami *et al.*, 1971). Even when it is non-lethal, the accumulated lifetime costs of recurring freeze damage to tree fitness are significant; leaves prematurely lost to autumn frosts reduce the ability of the tree to cold harden (Howell and Stockhause, 1973), while spring frosts damage flushing buds and flowers, decreasing tree growth and seed production (Selas *et al.*, 2002). It is generally believed that tree FT increases in multiple stages and only partially overlaps with the development of dormancy, where this FT induction involves a low temperature-driven period of metabolic and transcriptomic adjustment known as cold acclimation.

In herbaceous annuals such as *Arabidopsis*, members of the CRT binding protein (CBF) family of transcriptional activators (also known as the DRE binding protein (DREB1) family) control the transcription of genes shown to play important roles in cold acclimation and the development of FT. The *Arabidopsis CBF/DREB1* family consists of six paralogues, but only three are cold inducible (Sakuma *et al.*, 2002). *CBF1* (*DREB1b*), *CBF2* (*DREB1c*) and *CBF3* (*DREB1a*) are low temperature-induced AP2 transcription factors that bind the *cis*-element known as the C-repeat/dehydration responsive element (CRT/DRE) (Stockinger *et al.*, 1997). These three low temperature-associated *CBF*s co-localize to an 8.7 kb region of chromosome 4. They share a conserved *CBF* AP2 DNA-binding region flanked by the characteristic amino acid motifs, PKKPAGRxKFxETRHP and DSAWR (Jaglo-Ottosen *et al.*, 2001). It is estimated that CBF regulons account for only 12% of the cold-responsive genes in *Arabidopsis* (Fowler and Thomashow, 2002; Vogel *et al.*, 2005). Whether or not the *CBF*s are induced by cold depends on the action of a constitutively expressed MYC-like bHLH transcriptional activator known as *ICE1* (for inducer of *CBF* expression 1) (Chinnusamy *et al.*, 2003). *ICE1* is expressed in all tissues in *Arabidopsis* and it is thought that low temperature induces a protein modification of either *ICE1* or an associated cofactor, allowing *ICE1* to bind MYC-recognition sites in the *CBF* promoters, activating *CBF* transcription.

Ectopic expression of *CBF1* and *CBF3* in *Arabidopsis* has been shown to be sufficient to improve FT (Jaglo-Ottosen *et al.*, 1998; Gilmour *et al.*, 2000), while a *cbf2* mutant is also more freezing tolerant due to higher sustained expression of *CBF1* and *CBF3* (Novillo *et al.*, 2004). *CBF1/3* upregulation in the *cbf2* mutant may occur because *CBF2* normally exerts direct negative feedback control on *CBF1/3*, as suggested by Novillo *et al.* (2004), or because a transcriptional repressor normally controlled by or associated with *CBF2* has its activity altered in the absence of *CBF2*. Of the two possibilities, the second seems more likely because neither *CBF1* nor *CBF3* contains the CRT/DRE in their promoter sequences and because a candidate *CBF2*-associated repressor, *ZAT12*, has now been shown to coordinately regulate members of the *CBF2* regulon and repress *CBF* expression (Vogel *et al.*, 2005). In either case, it is clear that feedback regulatory mechanisms exist to allow the plant to control the scale and duration of *CBF* response to the stresses it faces. This is important in the long term because the increase in FT-afforded plants through the constitutive expression of *CBF* factors in *Arabidopsis* (and their downstream *CBF*-regulated genes) appears to come at the expense of plant growth (Liu *et al.*, 1998; Kasuga *et al.*, 1999; Gilmour *et al.*, 2000; Haake *et al.*, 2002).

Following their original discovery in *Arabidopsis*, the list of annual plants with cold-inducible *CBF* orthologues has expanded to include both cold-sensitive and cold-tolerant annual species (Jaglo-Ottosen *et al.*, 2001), but detailed studies in tree species have been lacking. We have therefore characterized the *CBF* regulon in Poplar (*Populus*) to determine whether the *CBF*s are involved in the FT mechanism of temperate woody perennials. Our data demonstrate that: (i) *CBF*s are involved in the FT mechanism of temperate tree species such as poplar; (ii) stem and leaf tissue regulons in these deciduous trees are not the same; and (iii) the *Populus* genome contains multiple *CBF*s that are likely to be employed differentially in perennial (stem) and annual (leaf) tissues.

Ectopic Expression of *AtCBF1* in *Populus*

Three *Arabidopsis CBF* paralogues have been used to manipulate cold tolerance in a number of herbaceous plant species (Jaglo-Ottosen *et al.*, 2001; Hsieh *et al.*, 2002). To test whether the *CBF* regulon plays a role in the FT of overwintering woody perennials, transgenic *Populus tremula* × *Populus alba* (clone 717-1B4), overexpressing the well-characterized *AtCBF1* gene of *Arabidopsis*, were generated. Two independent lines ectopically expressing *AtCBF1* at rates equivalent to endogenous *AtCBF1* transcript levels in *Arabidopsis* after 3-h cold treatment were studied. The selected lines, with high *AtCBF1* transcript expression (*AtCBF1*-Poplar), had fewer roots (data not shown) and slower growth rates when cultivated *in vitro*, but fully recovered growth rates after 3–6 weeks in soil (Fig. 12.1).

Fig. 12.1. Average plant height during 6 weeks of growth on soil in wild-type (WT) and *AtCBF1*-Poplar line 1 and line 2.

Ectopic expression of *AtCBF1* in poplar leaves and stems significantly ($P \leq 0.01$) increased the FT of leaves from −3.9°C for non-acclimated wild-type (WT) to −6.9°C for *AtCBF1*-Poplar (Fig. 12.2). Ectopic expression of *AtCBF1* in stems of poplar trees also significantly ($P \leq 0.01$) improved stem FT in non-acclimated plants (−5.4°C for the *AtCBF1* lines versus −4.1°C for WT). Following cold acclimation for 4 weeks at 2°C, this trend of improved FT in the transgenic lines was maintained in both leaves and stems, but the differences were no longer statistically significant (Fig. 12.2). These data demonstrate that the cryoprotective benefit of ectopic expression of *AtCBF1* and *AtCBF3* shown previously in *Arabidopsis* leaves (Jaglo-Ottosen *et al.*, 1998; Gilmour *et al.*, 2000) is also conserved in both leaves and stems of *AtCBF1*-Poplar. These improvements in stem and leaf FT in *AtCBF1*-Poplar indicate that at least some part of the *CBF*-mediated signalling pathway and regulon are conserved in both annual and perennial tissues of temperate deciduous trees.

To determine the extent to which the CBF regulon was conserved, microarrays comparing the genes responding in poplar leaf and stem tissues ectopically expressing *AtCBF1* with genes responding to cold in WT poplar leaves and stems were analysed. The cDNA microarray experiments were repeated twice and gene lists were constructed comparing the average result of the two independent *AtCBF1*-Poplar lines from the two separate experiments. These experiments enabled us to determine: (i) which genes were susceptible to *AtCBF1* regulation; (ii) whether the genes that respond to *AtCBF1* control show differential expression in stem versus leaf tissue; and (iii) how much overlap there was between the *CBF* regulon and the endogenous (untransformed) response of poplar leaves and stems to

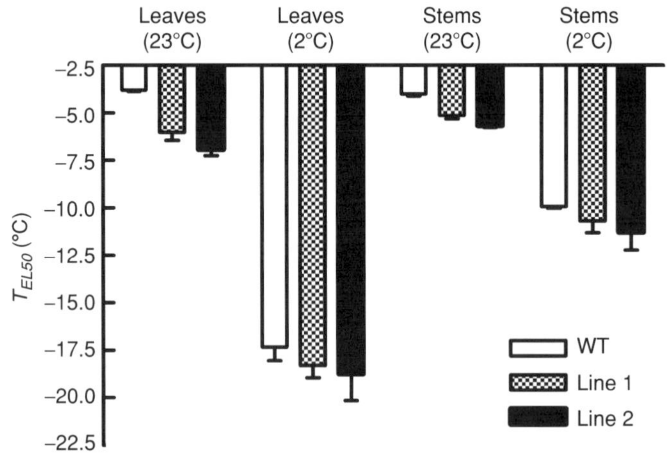

Fig. 12.2. Freezing tolerance, measured by electrolyte leakage (T_{EL50}), of 6-week-old wild-type (WT) and *AtCBF1*-Poplar grown under long days (16 h) at 23°C prior to cold acclimation, and then following 4 weeks of acclimation at 2°C under long days (16 h).

cold. From these experiments we identified 63 genes upregulated in leaves from warm-grown *AtCBF1*-Poplar plants where expression was significantly altered ($P \leq 0.05$) and that showed a greater than 1.8-fold change in expression level. Almost half of the *AtCBF1*-Poplar leaf regulon (29/63) was composed of novel 'expressed proteins' and other proteins with no known function. Transcripts encoding metabolic enzymes compose the majority (35%) of the 34 *AtCBF1*-Poplar leaf regulon genes with known functions (Fig. 12.3). Together, the classes 'cell rescue' (15%) and 'transcription' (21%) made up another third of the *AtCBF1*-Poplar leaf regulon, with the remaining classes represented being 'development' (9%), 'cellular communication/signal transduction' (6%), 'storage' (6%), 'cellular biogenesis' (6%) and 'protein fate' (3%).

In contrast to these results from annual leaves, the functional distribution of the *AtCBF1*-Poplar regulon genes in perennial stem tissue showed that 'transcription' comprised the dominant functional class (41%), followed by 'cellular communication/signal transduction' (18%), 'cell rescue' (12%) and 'metabolism' (12%) (Fig. 12.3). Thus, 59% of the *AtCBF1*-Poplar stem regulon was composed of genes involved in the processes of 'transcription' and 'cellular communication/signal

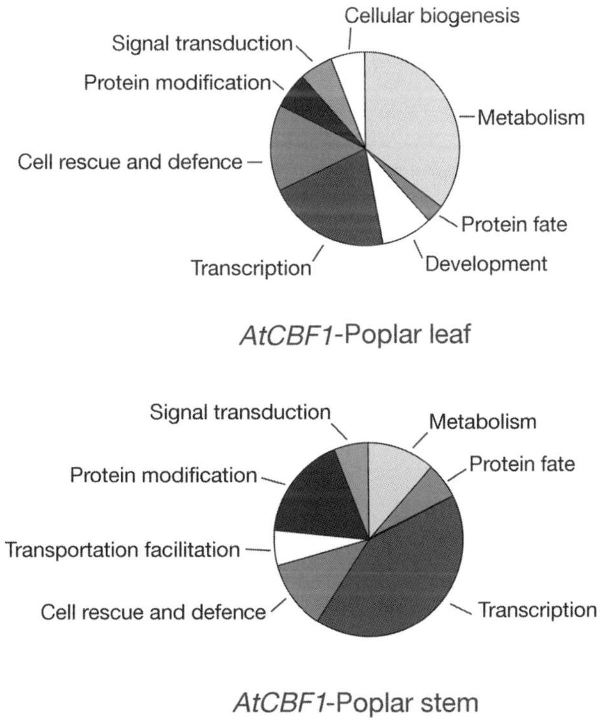

Fig. 12.3. Distribution of genes of known function that were significantly upregulated in warm-grown *AtCBF1*-Poplar leaves and stems.

transduction', while the same two processes comprised only 26% of the *AtCBF1*-Poplar leaf regulon. This difference could not be attributed to a general compositional difference in the low-temperature transcriptomes of the annual and perennial tissues, as the same comparison made between 7-day cold acclimated WT leaf and stem tissues demonstrated general agreement in the distribution of upregulated functional categories (data not shown). It could be argued that the *AtCBF1*-Poplar leaf/stem regulon functional differences were due to the small size of the stem regulon. However, when the null hypothesis of no functional difference between the leaf and stem *AtCBF1*-Poplar regulon is tested using the largest functional group, 'metabolism', only two metabolic genes were found, significantly (χ^2, $P \leq 0.05$) less than the expected six genes. Therefore, it seems that the initial perception of low temperature, which is biologically indicated by the accumulation of *CBF* transcripts in WT trees and artificially mimicked by ectopic expression of the *AtCBF1* transcript in *AtCBF1*-Poplar, results in the upregulation of a functionally different *CBF* regulon in perennial tissues.

As indicated by this functional analysis, the normalized expression data for all genes significantly upregulated by *AtCBF1* in at least one tissue show that both the leaf and stem *AtCBF1*-Poplar regulons contain clusters that are specific for that tissue. However, the leaf and stem regulons also share broad similarity, both with each other and with leaf and stem cold regulons (Table 12.1). This raises the question of just how similar are the *CBF* regulons of both annual and perennial tissues from poplar with those of *Arabidopsis*. Comprehensive transcriptome changes due to ectopic *AtCBF1* expression previously have been published only for *Arabidopsis* as a combined *CBF1/CBF2/CBF3* regulon (Fowler and Thomashow, 2002). However, paralogue-specific studies have identified gene members of the *AtCBF2* (Vogel *et al.*, 2005) and *AtCBF3* regulons in *Arabidopsis* (Maruyama *et al.*, 2004). We surveyed our *Populus* cDNA microarray database (Sterky *et al.*, 2004) to find all the closest protein homologues to previously reported *CBF*-responsive genes from the combined *AtCBF1–3* regulon and the *AtCBF2*- and *AtCBF3*-paralogue regulons. The *Arabidopsis CBF3* regulon shared the most similarity with the *CBF* regulon in *Populus*, suggesting that *AtCBF1* acts in a similar manner in *Populus* as *AtCBF3* does in *Arabidopsis*. Twelve of the 38 reported *AtCBF3*-upregulated genes had representative orthologues on our microarray and a gene comparison showed that most (7/12 having FC > 1.3) of the *AtCBF3*-upregulated gene orthologues present on the POP1 13K cDNA array were similarly upregulated in one or both *AtCBF1*-Poplar tissues and possessed DREs in their 1500 bp promoters. Examination of the induction of *AtCBF2*- and *AtCBF1/2/3*-regulon orthologues in *Populus* showed much greater disagreement. However, at this point it is unclear if these differences are due to changes in the composition of the species and/or paralogue-specific regulons, or whether they reflect the differing array technologies employed.

Our regulon lists from *Populus* contained 86 previously unreported *AtCBF1*-induced genes. Eleven of these had poor or no hits to orthologues in the *Arabidopsis* genome, suggesting possible unique roles for these genes in this perennial woody species. However, 21 of the 75 genes identified by our experiments had the DRE (RCCGAC) in the 500 bp promoters for the corresponding *Arabidopsis* gene. This is a significant enrichment (χ^2, $P \leq 0.001$) and it suggests that the stringent selection criteria in past experiments may have excluded these *CBF* regulon members.

Table 12.1. A selection of genes upregulated in both *AtCBF1-Poplar tremula* × *Populus alba* and 1-week cold acclimated wild-type (WT) *Populus tremula* × *Populus alba* (versus WT warm-grown *Populus tremula* × *Populus alba*). Normalized fold change (FC) values are shown in bold face for $P < 0.05$.

AtCBF1 leaf FC	At CBF1 stem FC	WT leaf FC	WT stem FC	Gene ID	Description	Category
2.1	1.1	**3.7**	**2.8**	At1g54410	SRC2	Cell rescue
2.0	1.0	**5.0**	**4.3**	At5g66780	Expressed protein	Unknown role
2.2	2.0	**3.5**	**3.4**	At2g21120	P0491F11.21 protein *Oryza sativa*	Metabolism
1.8	1.5	**3.1**	**2.2**	At2g19370	Expressed protein	Unknown role
2.0	1.4	**5.0**	**2.7**	At5g09390	Expressed protein	Unknown role
2.2	2.0	**12.3**	**5.2**	At3g06660	Expressed protein	Unknown role
2.0	1.9	**7.9**	**3.7**	At2g15970	WCOR413-like protein	Cell rescue
1.8	2.0	**7.8**	–	At4g39450	Expressed protein	Unknown role
1.9	1.9	**10.0**	**5.8**	At3g05880	LTI6A	Cell rescue
2.6	2.4	**14.5**	**4.1**	At5g38760	Pollen coat protein	Development
2.1	1.7	**9.4**	**3.2**	None	Arabinogalactan protein AGP21	Development
2.5	1.3	**2.5**	**2.8**	At5g61660	Glycine-rich protein	Unknown role
2.8	1.1	**3.5**	2.2	At1g71691	GDSL-motif lipase/ hydrolase	Metabolism
2.5	1.3	**9.4**	1.3	At4g29680	Nucleotide pyrophosphatase	Metabolism
2.6	2.5	**14.0**	1.3	At2g28680	Legumin-like protein	Storage
1.8	1.6	**8.5**	1.6	At5g67080	Expressed protein	Unknown role
2.2	1.8	**7.4**	1.5	At4g24220	Dehydrin	Cell rescue
2.1	1.5	**4.7**	2.5	At5g54470	CONSTANS B-box zinc finger protein	Transcription
1.9	1.1	**2.3**	–1.3	At5g55430	Expressed protein	Unknown role
1.8	–1.7	**2.3**	2.0	None	Expressed protein	Unknown role
1.0	**2.1**	1.3	**4.0**	At2g41380	Embryo-abundant protein	Unknown role

(Continued)

Table 12.1. *(continued)*

AtCBF1 leaf FC	At CBF1 stem FC	WT leaf FC	WT stem FC	Gene ID	Description	Category
1.0	**1.8**	1.2	**1.8**	At5g39670	Calcium-binding protein	Cellular communication and signal transduction
1.0	**1.8**	**1.7**	1.6	At5g62165	MADS-box transcription factor FBP22	Transcription
1.1	**2.0**	**4.6**	**4.8**	At1g06330	Copper-binding protein	Unknown role

AtCBF1 Regulon Promoter Analysis in *Populus* Reveals Conserved Linkage of the DRE and ABRE

Given this broad similarity between the *Populus* and *Arabidopsis CBF* regulons, we determined whether there is also a conserved enrichment of the DRE in the promoters of the *AtCBF1*-Poplar regulon genes. Sixteen of the 63 (25%) *AtCBF1*-Poplar leaf regulon gene promoters (defined as the 1500 bp of genomic sequence found upstream of the ATG site) contained the basic DRE nucleic acid sequence 'RCCGAC'. Eight of the 26 (31%) *AtCBF1*-Poplar stem regulon promoters contained the DRE. Positional analysis of the RCCGAC revealed no statistically significant enrichment along the 1500 bp promoter. However, it is worth noting that 7/33 DREs appeared between 200–400 bp upstream of ATG, leaving the possibility that a larger sample set, using currently unavailable transcriptional start-site positions instead of the ATG, would verify the −450/−51 positional enrichment of the DRE in *Populus* similar to that reported for *Arabidopsis* (Maruyama *et al.*, 2004). Furthermore, given the known role of *AtCBF1* as an interacting component of the chromatin-modifying ada2/gcn5 histone-acetylating complex (Stockinger *et al.*, 2001; Viachonasios *et al.*, 2003), the positional enrichment and close proximity of DREs to the transcriptional start-site in *Arabidopsis* may be a reflection of its compact genome size, rather than a distance requirement *per se*. Thus, the larger genome and intergenic size of *Populus*, relative to *Arabidopsis*, or an altered *AtCBF1*-binding consensus sequence, may have contributed to this apparent lack of enrichment for the DRE in the *AtCBF1*-Poplar regulon gene promoters; further experiments are required to clarify this.

In contrast, the ABA-responsive element (ABRE) 'ACGTGTC', previously shown to be enriched in *CBF3* regulon promoters and to impart ABA-responsiveness (Hattori *et al.*, 2002; Maruyama *et al.*, 2004), showed positional enrichment (between 100 bp and 400 bp upstream of the ATG) in 12/87 *AtCBF1*-Poplar regulon 1500 bp promoters, representing a statistically significant enrichment (χ^2, $P \leq 0.05$). There was also statistically significant overlap between ABRE- and DRE-containing gene promoters in *Populus* (4/12 ABRE-containing gene promoters also contained the RCCGAC consensus: χ^2, $P \leq 0.001$) and positional enrichment of the ABRE at positions 100–199 bp and 300–399 bp upstream of the start codon (χ^2, $P \leq 0.001$), indicating that this *cis*-element relationship is important. Interestingly, the ACGTGTC element has also been identified as one that responds to GA treatment in *Arabidopsis* seeds (Ogawa *et al.*, 2003), but the functional significance of this remains unclear.

Neither *AtCBF1*-repressed regulon was significantly enriched for the DRE (8/30 leaf gene promoters had RCCGAC while 2/24 stem gene promoters had RCCGAC). The ABRE (ACGTGTC) appeared in 0/30 *AtCBF1*-repressed leaf regulon promoters and in 2/24 *AtCBF1*-repressed stem regulon promoters – approximating the frequencies expected by random chance. There was no gene overlap between DRE- and ABRE-containing *AtCBF1*-repressed genes.

The *Populus trichocarpa* Genome Encodes Six CBF-like Transcription Factors

The *Populus* CBF regulon contained many genes previously linked to the process of cold acclimation in herbaceous annuals such as *Arabidopsis*, and the *AtCBF1*-Poplar transgenic data showed that ectopic expression of *AtCBF1* altered gene expression in poplar to mirror patterns seen in 1-week cold-acclimated trees. However, the question remains as to whether there are cold-inducible *CBF* transcription factors encoded by the *Populus* genome. Iterative tBLASTn searches of the *Populus trichocarpa* genome (http://genome.jgi-psf.org/Poptr1/Poptr1.home.html) using the conserved AP2 and flanking *CBF* ID sequences from *Arabidopsis* (Jaglo-Ottosen *et al.*, 2001) identified six potential CBF-encoding genes. An amino acid sequence alignment of these candidate genes showed that two of the six (*PtCBF1* and *PtCBF2*) showed 100% conservation of the previously reported *Arabidopsis* CBF consensus sequences (PKKR/PRAGRxKFxETRHP and DSAWR), two others (*PtCBF3* and *PtCBF4*) possessed a single amino acid substitution in the N-terminal consensus sequence (I instead of K in position 10), with the remaining two sequences (*PtCBFL1* and *PtCBFL2*) deviating from the consensus sequences by four and five amino acids, respectively. A phylogenetic analysis (data not shown) of known full-length CBF amino acid sequences from dicotyledonous plants using the *TINY* AP2 transcription factors as the outgroup comparison confirmed that *PtCBFL1* and *PtCBFL2* shared the least amino acid similarity to the characterized *CBF*s from *Arabidopsis*. Also of some note, a *Populus* gene previously reported to be 'CBF1-like' and upregulated in dormant cambium (*PtDRTY*) (Schrader *et al.*, 2004) grouped with *TINY*, not the *CBFs*, although, unlike *TINY*, it does share a C-terminal consensus with the CBFs. Investigation of the 1500 bp promoters of the *PtCBFs* also demonstrated that they possessed numerous potential *ICE*-binding sites (CANNTG) (Table 12.2). A tBLASTn search of the *Populus trichocarpa* genome also identified *Populus*-encoded ICE proteins with high similarity to *Arabidopsis ICE1*, indicating that *ICE*-mediation of the *CBF* signalling pathway may also have been conserved in this woody perennial.

A real-time polymerase chain reaction (RT-PCR) study of expression of *PtCBF1-4* in hybrid aspen (*Populus tremula* × *Populus tremuloides*, clone T89) demonstrated that all were cold inducible, though the kinetics of induction and tissue specificity differed between the orthologues (Fig. 12.4). *PtCBF1*, sharing 52% amino acid identity with *AtCBF1* (91% within the AP2 region), was cold inducible in both leaf and stem tissue. *PtCBF1* transcript levels in both tissue types peaked 6 h after transfer to cold. *PtCBF2*, sharing 77% amino acid identity with *PtCBF1* (99% within the AP2 region), also proved cold inducible in leaf tissue but was only weakly cold induced in the stem. Leaf *PtCBF2* expression peaked 9 h after transfer, with stem transcripts peaking between 3 h and 6 h. *PtCBF3*, with 58.5% identity to *PtCBF4* (94% within the AP2 region), was expressed in both leaf and stem tissue in exactly the same manner. *PtCBF3* transcript accumulated rapidly after cold shifting, peaking at 3 h, and returned to near starting levels within 6 h. *PtCBF4* expression also peaked at 3 h in leaves but was only marginally inducible in stem tissue (Fig. 12.4). Tissue specificity for *CBF* factors has not been reported previously, suggesting that possible differential roles in annual (leaf) and perennial (stem) tissues may have developed during the evolution of winter dormancy.

Our study of the native *Populus CBFs* showed that while they are as dissimilar at the amino acid level to the *AtCBFs* as are the DDF1 and DDF2 proteins, they are cold inducible. There is at least circumstantial evidence that the DDF proteins bind the DRE because DRE-containing genes are upregulated in DDF overexpressors without a concurrent overexpression of other *CBFs* (Magome *et al.*,

Table 12.2. *PtCBF* 1500 bp promoter *cis*-element frequencies.

	ICE boxes (CANNTG)	DREs (RCCGAC)	ABREs (ACGTGTC)
PtCBF1	12	0	2
PtCBF2	12	1	1
PtCBF3	7	0	1
PtCBF4	6	0	0
PtCBFL1	6	0	1
PtCBFL2	5	0	0

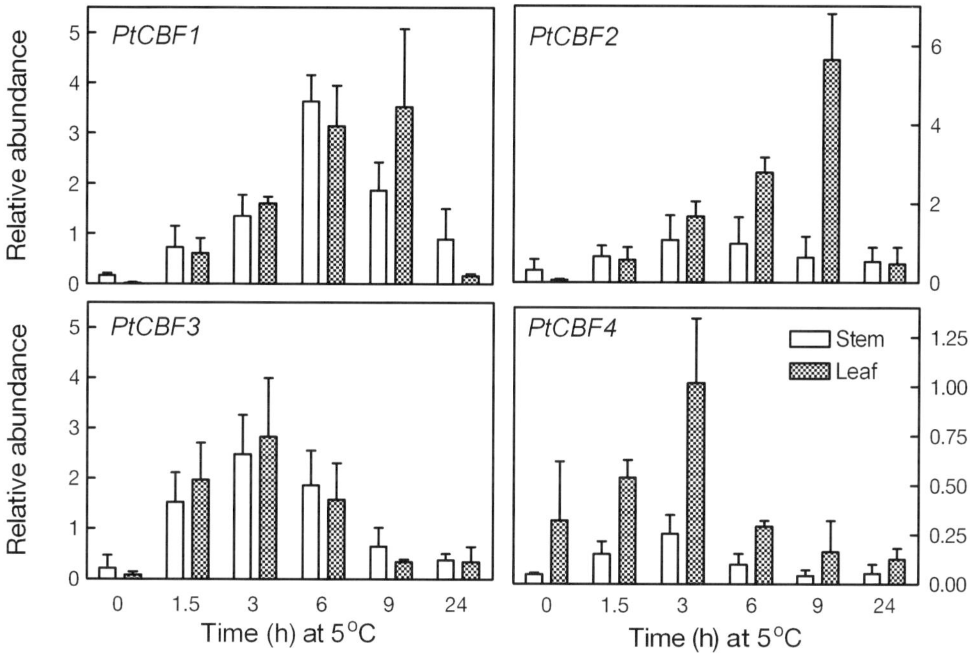

Fig. 12.4. Expression of *Populus* CBF paralogues in response to low temperature in wild-type trees determined by real-time polymerase chain reaction (RT-PCR). Each reaction used 30 ng first-strand cDNA generated from total RNA extracted from wild-type (WT) trees and normalized to 18S rRNA quantity.

2004). There is also a high (81–85%) conservation of sequence identity between the DDF and PtCBF DNA-binding AP2 region, including the critical V (and E) DRE-specific binding residues (Sakuma *et al.*, 2002; Qin *et al.*, 2004). It is therefore reasonable to assume that the *PtCBF*s also bind the DRE. The similarities between the DDFs and PtCBFs (protein sequence, overexpressor phenotype) raise the question of what may have caused the original *AtCBF1*-Poplar stunting observed for plantlets grown on agar. Plants ectopically expressing DDF have, like plants ectopically expressing *CBF*s, been observed to display dwarfism – a phenotype that has been linked to a gibberellin (GA) deficiency. Like the GA-deficient *lh* mutants in pea (Yaxley *et al.*, 2001), *AtCBF1*-Poplar trees have reduced internode elongation and root growth when grown on agar. Somewhat surprisingly, transfer to soil rescues the phenotype. We suggest that the observed upregulation of a *PHANTASTICA* orthologue in *AtCBF1*-Poplar may help explain why. Studies of leaf initiation in tobacco show that GA biosynthesis can be post-transcriptionally repressed (via repression of GA20 oxidase) by the class I KNOX gene *NTH15*. In turn, downregulation of *NTH15* at leaf initiation sites is associated with induction of GA synthesis, and *PHANTASTICA* orthologues act immediately downstream of this event to maintain the KNOX1-off/GA-on status in the developing leaf primordia. Overexpression of KNOX1/PHANTASICA-repression results in a disorganization of the strict patterns of anticlinal cell division in leaf mesophyll layers, leading to leaf phenotypes such as palisade proliferation (McHale and Koning, 2004). Interestingly, previously created *CBF* transgenics in *Arabidopsis* (Gilmour *et al.*, 2004) have also displayed altered palisade layer numbers, strengthening the hypothesis that the *PHANTASTICA*/KNOX1/GA status alteration is a common phenomenon associated with constitutive *CBF* overexpression. The closest *NTH15* orthologues in both *Populus* and *Arabidopsis* contain a CCGAC element in their 1500/1000 bp promoter suggesting that constitutive *AtCBF1* expression may cause overexpression of *NTH15*, leading to repression of GA synthesis. *Populus* trees grown on soil appear to be able to counteract this disequilibrium, by increasing the expression of *PHANTASTICA*, and grow normally.

Conclusions

We investigated the changes in phenotype and transcript profile of transgenic *Populus* constitutively expressing *CBF1* from *Arabidopsis* (*AtCBF1*). Ectopic expression of *AtCBF1* was sufficient to significantly increase the FT of non-acclimated leaves (−3.9°C for WT versus −6.7°C and −7.1°C for transgenic line 1 and 2, respectively) and stems (−4.1°C for WT versus −5.2°C and −5.7°C for transgenic line 1 and 2, respectively), comparable to WT plants after 1 week of cold acclimation. cDNA microarray experiments identified genes upregulated by ectopic *AtCBF1* expression in *Populus* and demonstrated the strong conservation of the *CBF* regulon between *Populus* and *Arabidopsis*. We also studied the induction kinetics and tissue specificity for four *CBF* paralogues identified from the *Populus trichocarpa* genome sequence (*PtCBFs*). All four *PtCBFs* are cold inducible in leaves, but only *PtCBF1* and *PtCBF3* show significant induction in stems. Our results suggest that the central role played by the *CBF* family of transcriptional activators in cold acclimation of *Arabidopsis* has been maintained in *Populus* in the 50 million years since their divergence. However, the differential expression of the *PtCBFs* in annual (leaf) and perennial (stem) tissues, and the stem- and leaf-specific clusters of CBF-responsive genes, suggests that the perennial-driven evolution of winter dormancy may have given rise to specific roles for these 'master-switches' in the different annual and perennial tissues of woody species. Studies of the *CBF* regulon in a perennially dormant woody plant species like *Populus* strengthen the conclusion that the *CBF* low-temperature signalling pathway is not only conserved, but similarly seems to work through ABA and GA signalling elements in woody dicots.

Acknowledgements

We thank Joel Davis and Zoran Jeknic for excellent technical assistance. We also thank Dr Mike Thomashow for providing the *AtCBF1* gene for this study. This study was supported in part by grants from the Swedish Forestry and Agricultural Research Council to Vaughan Hurry, from STEM to Rishikesh Bhalerao, and from the NSF Plant Genome Project (DBI 0110124) to Tony H.H. Chen.

References

Chinnusamy, V., Ohta, M., Kanrar, S., Lee, B.H., Hong, X.H., Agarwal, M. and Zhu, J.K. (2003) ICE1: a regulator of cold-induced transcriptome and freezing tolerance in *Arabidopsis*. *Genes and Development* 17, 1043–1054.

Fowler, S. and Thomashow, M.F. (2002) *Arabidopsis* transcriptome profiling indicates that multiple regulatory pathways are activated during cold acclimation in addition to the CBF cold response pathway. *Plant Cell* 14, 1675–1690.

Frewen, B.E., Chen, T.H.H., Howe, G.T., Davis, J., Rohde, A., Boerjan, W. and Bradshaw, H.D. (2000) Quantitative trait loci and candidate gene mapping of bud set and bud flush in *Populus*. *Genetics* 154, 837–845.

Fuchigami, L.H., Weiser, C.J. and Evert, D.R. (1971) Induction of cold acclimation in *Cornus stolonifera* Michx. *Plant Physiology* 47, 98–103.

Gilmour, S.J., Sebolt, A.M., Salazar, M.P., Everard, J.D. and Thomashow, M.F. (2000) Overexpression of the *Arabidopsis* CBF3 transcriptional activator mimics multiple biochemical changes associated with cold acclimation. *Plant Physiology* 124, 1854–1865.

Gilmour, S.J., Fowler, S.G. and Thomashow, M.F. (2004) *Arabidopsis* transcriptional activators CBF1, CBF2, and CBF3 have matching functional activities. *Plant Molecular Biology* 54, 767–781.

Haake, V., Cook, D., Riechmann, J.L., Pineda, O., Thomashow, M.F. and Zhang, J.Z. (2002) Transcription factor CBF4 is a regulator of drought adaptation in *Arabidopsis*. *Plant Physiology* 130, 639–648.

Hattori, T., Totsuka, M., Hobo, T., Kagaya, Y. and Yamamoto-Toyoda, A. (2002) Experimentally determined sequence requirement of ACGT-containing abscisic acid response element. *Plant Cell Physiology* 43, 136–140.

Howell, G.S. and Stockhause, S.S. (1973) The effect of defoliation time on acclimation and dehardening in tart cherry (*Prunus cerasus* L.). *Journal of the American Society for Horticultural Science* 98, 132–136.

Hsieh, T.H., Lee, J.T., Charng, Y.Y. and Chan, M.T. (2002) Tomato plants ectopically expressing *Arabidopsis* CBF1 show enhanced resistance to water deficit stress. *Plant Physiology* 130, 618–626.

Jaglo-Ottosen, K.R., Gilmour, S.J., Zarka, D.G., Schabenberger, O. and Thomashow, M.F. (1998) *Arabidopsis* CBF1 overexpression induces COR genes and enhances freezing tolerance. *Science* 280, 104–106.

Jaglo-Ottosen, K.R., Kleff, S., Amundsen, K.L., Zhang, X., Haake, V., Zhang, J.Z., Deits, T. and Thomashow, M.F. (2001) Components of the *Arabidopsis* C-repeat/dehydration-responsive element binding factor cold-response pathway are conserved in *Brassica napus* and other plant species. *Plant Physiology* 127, 910–917.

Kasuga, M., Liu, Q., Miura, S., Yamaguchi-Shinozaki, K. and Shinozaki, K. (1999) Improving plant drought, salt, and freezing tolerance by gene transfer of a single stress-inducible transcription factor. *Nature Biotechnology* 17, 287–291.

Li, C.Y., Junttila, O., Ernstsen, A., Heino, P. and Palva, E.T. (2003) Photoperiodic control of growth, cold acclimation and dormancy development in silver birch (*Betula pendula*) ecotypes. *Physiologia Plantarum* 117, 206–212.

Liu, Q., Kasuga, M., Sakuma, Y., Abe, H., Miura, S., Yamaguchi-Shinozaki, K. and Shinozaki, K. (1998) Two transcription factors, DREB1 and DREB2, with an EREBP/AP2 DNA binding domain separate two cellular signal transduction pathways in drought- and low-temperature-responsive gene expression, respectively, in *Arabidopsis*. *Plant Cell* 10, 1391–1406.

Magome, H., Yamaguchi, S., Hanada, A., Kamiya, Y. and Oda, K. (2004) Dwarf and delayed-flowering 1, a novel *Arabidopsis* mutant deficient in gibberellin biosynthesis because of overexpression of a putative AP2 transcription factor. *Plant Journal* 37, 720–729.

Maruyama, K., Sakuma, Y., Kasuga, M., Ito, Y., Seki, M., Goda, H., Shimada, Y., Yoshida, S., Shinozaki, K. and Yamaguchi-Shinozaki,

K. (2004) Identification of cold-inducible downstream genes of the *Arabidopsis* DREB1A/CBF3 transcriptional factor using two microarray systems. *Plant Journal* 38, 982–993.

McHale, N.A. and Koning, R.E. (2004) PHANTASTICA regulates development of the adaxial mesophyll in *Nicotiana* leaves. *Plant Cell* 16, 1251–1262.

Novillo, F., Alonso, J.M., Ecker, J.R. and Salinas, J. (2004) CBF2/DREB1C is a negative regulator of CBF1/DREB1B and CBF3/DREB1A expression and plays a central role in, stress tolerance in *Arabidopsis. Proceedings of the National Academy of Sciences USA* 101, 3985–3990.

Ogawa, M., Hanada, A., Yamauchi, Y., Kuwalhara, A., Kamiya, Y. and Yamaguchi, S. (2003) Gibberellin biosynthesis and response during *Arabidopsis* seed germination. *Plant Cell* 15, 1591–1604.

Qin, F., Sakuma, Y., Li, J., Liu, Q., Li, Y.Q., Shinozaki, K. and Yamagushi-Shinozaki, K.Y. (2004) Cloning and functional analysis of a novel DREB1/CBF transcription factor involved in cold-responsive gene expression in *Zea mays* L. *Plant Cell Physiology* 45, 1042–1052.

Sakuma, Y., Liu, Q., Dubouzet, J.G., Abe, H., Shinozaki, K. and Yamaguchi-Shinozaki, K. (2002) DNA-binding specificity of the ERF/AP2 domain of *Arabidopsis* DREBs, transcription factors involved in dehydration- and cold-inducible gene expression. *Biochemical and Biophysical Research Communications* 290, 998–1009.

Schrader, J., Moyle, R., Bhalerao, R., Hertzberg, M., Lundeberg, J., Nilsson, P. and Bhalerao, R.P. (2004) Cambial meristem dormancy in trees involves extensive remodelling of the transcriptome. *Plant Journal* 40, 173–187.

Selas, V., Piovesan, G., Adams, J.M. and Bernabei, M. (2002) Climatic factors controlling reproduction and growth of Norway spruce in southern Norway. *Canadian Journal of Forest Research* 32, 217–225.

Sterky, F., Bhalerao, R.R., Unneberg, P., Segerman, B., Nilsson, P., Brunner, A.M., Charbonnel-Campaa, L., Lindvall, J.J., Tandre, K., Strauss, S.H., Sundberg, B., Gustafsson, P., Uhlen, M., Bhalerao, R.P., Nilsson, O., Sandberg, G., Karlsson, J., Lundeberg, J. and Jansson, S. (2004) A Populus EST resource for plant functional genomics. *Proceedings of the National Academy of Sciences USA* 101, 13951–13956.

Stockinger, E.J., Gilmour, S.J. and Thomashow, M.F. (1997) *Arabidopsis thaliana* CBF1 encodes an AP2 domain-containing transcriptional activator that binds to the C-repeat/DRE, a cis-acting DNA regulatory element that stimulates transcription in response to low temperature and water deficit. *Proceedings of the National Academy of Sciences USA* 94, 1035–1040.

Stockinger, E.J., Mao, Y.P., Regier, M.K., Triezenberg, S.J. and Thomashow, M.F. (2001) Transcriptional adaptor and histone acetyltransferase proteins in *Arabidopsis* and their interactions with CBF1, a transcriptional activator involved in cold-regulated gene expression. *Nucleic Acids Research* 29, 1524–1533.

Viachonasios, K.E., Thomashow, M.F. and Triezenberg, S.J. (2003) Disruption mutations of ADA2b and GCN5 transcriptional adaptor genes dramatically affect *Arabidopsis* growth, development, and gene expression. *Plant Cell* 15, 626–638.

Vogel, J.T., Zarka, D.G., Van Buskirk, H.A., Fowler, S.G. and Thomashow, M.F. (2005) Roles of the CBF2 and ZAT12 transcription factors in configuring the low temperature transcriptome of *Arabidopsis. Plant Journal* 41, 195–211.

Weiser, C.J. (1970) Cold resistance and injury in woody plants. *Science* 169, 1269–1278.

Welling, A., Moritz, T., Palva, E.T. and Junttila, O. (2002) Independent activation of cold acclimation by low temperature and short photoperiod in hybrid aspen. *Plant Physiology* 129, 1633–1641.

Yaxley, J.R., Ross, J.J., Sherriff, L.J. and Reid, J.B. (2001) Gibberellin biosynthesis mutations and root development in pea. *Plant Physiology* 125, 627–633.

13 Functional Role of Winter-accumulating Proteins from Mulberry Tree in Adaptation to Winter-induced Stresses

S. Fujikawa,[1,*] N. Ukaji,[2] M. Nagao,[3] K. Yamane,[1] D. Takezawa[4] and K. Arakawa[1]

[1]Department of Environmental Resources, Graduate School of Agriculture, Hokkaido University, Sapporo 060-8589, Japan; [2]Institute of Low Temperature Science, Hokkaido University, Sapporo 060-0819, Japan; [3]Cryobiosystem Research Centre, Faculty of Agriculture, Iwate University, Morioka 020-8550, Japan; [4]Department of Regulation Biology, Faculty of Science, Saitama University, Saitama 338-8570, Japan

Introduction

Plants exposed to low, non-freezing temperatures increase freezing tolerance by a process known as cold acclimation. In the case of perennial woody plants, the freezing tolerance changes seasonally, depending mainly upon change of seasonal air temperature, known as a process of seasonal cold acclimation and deacclimation (Sakai and Larcher, 1987). For example, the cortical parenchyma cells of mulberry (*Morus bombycis* Koidz) grown in Sapporo, Japan, can barely survive a temperature of −5°C in summer, can survive to below −80°C during winter and the freezing tolerance acquired in winter decreases gradually towards spring (Niki and Sakai, 1981; Fujikawa, 1994; Ukaji *et al.*, 2004).

Cold acclimation of plants is associated with many physiological and biochemical changes as well as alterations in gene expression (Thomashow, 1999). During cold acclimation, diverse changes at cellular and molecular levels occur in a wide variety of plants including woody plants. These cold acclimation-induced changes include: ultrastructural changes in the protoplasts (Niki and Sakai, 1981; Fujikawa and Takabe, 1996); changes in the property of cell walls (Rajashekar and Lafta, 1996; Fujikawa and Kuroda, 2000; Yamada *et al.*, 2002); compositional changes in the plasma membranes including both lipids and proteins (Steponkus, 1984;

*Corresponding author (e-mail: sfuji@for.agr.hokudai.ac.jp; Fax: +81-11-736-1791)

Yoshida, 1984; Zhou *et al.*, 1994; Uemura *et al.*, 1995; Kawamura and Uemura, 2002); intracellular accumulation of compatible osmolytes such as soluble sugars, proline and betaine (Hare *et al.*, 1998); and accumulation of polypeptides and proteins (Thomashow, 1994, 1999). Among these cold acclimation-induced proteins, enzymes for synthesis of compatible solutes or for the modification of membrane lipids are included (Thomashow, 1999). Furthermore, a variety of proteins with structural similarity to late embryogenesis abundant (LEA) proteins (Thomashow, 1994, 1999), to heat-shock proteins (HSP) (Neven *et al.*, 1992; Ukaji *et al.*, 1999), and to pathogenesis-related (PR) proteins (Hon *et al.*, 1995) are included as an end product of cold-induced genes. Although almost all of these changes introduced by cold acclimation are associated with increased freezing tolerance (Guy, 1990; Fujikawa *et al.*, 1999; Thomashow, 1999), the precise functional roles in the acquisition of freezing tolerance of these cold acclimation-induced products, particularly with regard to cold-induced proteins as end products, are unclear.

In the present study, we analysed the functional roles of some winter-accumulating proteins in the acquisition of freezing tolerance in cortical parenchyma cells of mulberry. In the cortical parenchyma cells of mulberry grown in Sapporo, many proteins are specifically accumulated during winter; these are closely related to the acquisition of an extremely high level of freezing tolerance, as a result of seasonal cold acclimation. Among these winter-accumulating proteins in cortical parenchyma cells of mulberry, in this study, four distinct winter-accumulating proteins, including endoplasmic reticulum (ER)-localized 27-kDa proteins, ER-localized 20-kDa proteins, 18-kDa proteins localized in soluble fractions and a dehydrin (DHN), were selected for further characterization. Specifically, we analysed the functional roles of these proteins against winter-induced stresses *in vitro* and in transgenic *Arabidopsis* plants *in planta*.

Characteristic Properties of Winter-accumulating Mulberry Proteins

Winter-accumulating 27-kDa proteins (WAP27)

In cortical parenchyma cells of mulberry, the ERs show characteristic morphological and distributional changes induced both by seasonal cold acclimation as well as – although it is restricted only to winter cells – by freezing-induced dehydration (Fujikawa and Takabe, 1996). Sodium dodecyl sulphate-polyacrylamide gel electrophoresis (SDS-PAGE) and immunoblot analysis in ER-enriched fractions (Ukaji *et al.*, 2000) showed that 27-kDa proteins were most abundantly accumulated in ER during winter (Ukaji *et al.*, 1999). These proteins were named as winter-accumulating 27-kDa proteins (WAP27). Immunocytochemical electron microscopy using purified anti-WAP27 antibodies confirmed that WAP27 proteins were specifically distributed in ERs in cortical parenchyma cells of mulberry during winter (Ukaji *et al.*, 2001). Biochemical analysis also suggested that WAP27 proteins were located in the lumen of ER, since they were insensitive to trypsin treatment of ER-enriched fractions in the absence of Triton X-100 and urea (Ukaji *et al.*, 1999). Furthermore, WAP27 proteins appear to have a putative ER-retention signal (see below).

Immunoscreening using purified anti-WAP27 antibodies indicated the presence of two isoforms of WAP27 proteins, WAP27A and WAP27B (Ukaji *et al.*, 1999, 2001). Deduced molecular masses of mature forms of WAP27A and WAP27B proteins consisted of 194- and 190-amino acid residues, respectively (calculated molecular masses, 20,942 and 20,556, respectively), while the pI value of WAP27A (pI 5.00) was slightly more acidic than that of WAP27B (pI 5.12) (Ukaji *et al.*, 2001). These isoforms with different pI values were separately detected in two-dimensional gel electrophoresis (Ukaji *et al.*, 1999). WAP27 isoforms had twelve-time tandem repeats of 11-mer amino acid motif that form amphipathic alpha-helical secondary structures (Ukaji *et al.*, 2001), which are a common feature of group 3 LEA proteins (Dure, 1993). These results suggested that WAP27 proteins are members of group 3 LEA proteins. They had the transit peptide in the N-terminal region and a specific tetrapeptide sequence, DEEL, in the C-terminal region of the primary structure, which was supposed to be similar in function to the ER-retention signal, KDEL (Ukaji *et al.*, 2001). WAP27 was detected to be an oligomer with approximately 100-kDa in molecular mass (Ukaji *et al.*, 1999). When a mature form of recombinant WAP27A (rWAP27A) was prepared in a protein-expression system by *Escherichia coli* bacteria, the rWAP27A also formed the oligomeric structure with approximately 100-kDa in molecular mass as estimated by native-PAGE (results not shown).

The accumulation of WAP27 proteins as observed by both SDS-PAGE and immunoblotting was faint in summer (June to August), increased gradually in autumn (September to October), reached a maximum in winter (November to March) and decreased again in spring (April to May) (Ukaji *et al.*, 1999). Northern blot analysis showed that expression of *WAP27* genes exhibited a similar pattern with accumulation of proteins; although the maximum level of the gene expression was seen between November and December, it gradually reduced and then disappeared in summer (Ukaji *et al.*, 2001). The *WAP27* gene expression was rapidly induced to a high level by cold treatment and rapidly reduced by warming in the cortical tissues of twigs harvested in mid-October (Ukaji *et al.*, 2001). Moreover, in the cortical tissues of twigs harvested in mid-August, the level of *WAP27* gene expression was low by cold treatment even for a long duration of time (Ukaji *et al.*, 2001). As freezing tolerance of cortical tissue cells by cold treatment was quickly inducible in late autumn but not in summer in mulberry twigs (Sakai and Yoshida, 1968), the *WAP27* gene expression seems to be well controlled in response to the seasonal cold-acclimation process in parallel with the development of freezing tolerance. In cortical parenchyma cells of the mulberry twigs harvested in winter and then deacclimated for 3 weeks at 23°C, the *WAP27* gene expression was induced by treatments with abscisic acid (ABA) and dehydration (Ukaji *et al.*, 2001).

Winter-accumulating 20-kDa proteins (WAP20)

Similar to WAP27 (described above), 20-kDa proteins (WAP20) were also one of the most characteristically accumulated proteins during winter in ER-enriched fractions in cortical parenchyma cells of mulberry as analysed by SDS-PAGE and immunoblotting (Ukaji *et al.*, 1999). The specific localization of WAP20 in the lumen of ER was also confirmed by immunocytochemical electron microscopy using purified anti-WAP20 antibodies (results not shown), as well as by biochemical examinations (Ukaji *et al.*, 1999). WAP20 proteins also possessed putative ER-retention signals (see below).

Deduced amino acid sequences of the mature form of WAP20 consisted of 145-amino acid residues (calculated molecular mass, 18,635) and the pI value was 6.5, which corresponded to a polypeptide in the analysis by two-dimensional gel electrophoresis of ER-enriched fraction from cortical tissues of mulberry harvested in winter (Ukaji *et al.*, 1999). The primary structure of WAP20 had two consensus sequences of small heat-shock protein (smHSP), a transit peptide in the N-terminal region and a putative ER-retention signal, KQEL, in the C-terminal region. The primary sequence of WAP20 proteins had high homology to other ER-localized smHSP, including *PsHSP22.7* in pea (DNA accession number M33898; Lauzon *et al.*, 1990), *GmHSP22.0* in soybean (X63198; Helm *et al.*, 1993) and *AtHSP22.0* in *Arabidopsis* (U11501; Helm *et al.*, 1995). WAP20 proteins were detected to be oligomeric with approximately 260 kDa or more in the molecular mass as examined by native-PAGE analysis (Ukaji *et al.*, 1999). It has been reported that several smHSPs formed dodecamers (Waters *et al.*, 1996); WAP20 was also dodecameric in the native form (Ukaji *et al.*, 1999), suggesting that WAP20 is an ER-localized smHSP.

The accumulation of WAP20 as observed by both SDS-PAGE and immunoblotting was barely detected in summer (especially in August), gradually became apparent in autumn (September to October), reached the maximum level during winter to early spring (November to April) and gradually decreased from spring to summer (May to June) (Ukaji *et al.*, 1999). In addition, anti-WAP20 antibodies also reacted with 21-kDa proteins in ER-enriched fractions from cortical tissues of mulberry in a manner similar to the seasonal appearance of WAP20 (Ukaji *et al.*, 1999). *WAP20* genes were significantly induced in October to November, but the expression level was gradually reduced from December to February and remained undetectable in spring through summer. In the cortical tissue cells of mulberry twigs that were harvested in winter and subsequently deacclimated, the *WAP20* gene expression was induced by cold treatment for 14 days and by ABA treatment for 24 h, but not induced by dehydration, even for a prolonged period. Also, the gene expression was induced rapidly after 2 h by heat treatment of twigs at 37°C (results not shown).

Winter-accumulating 18-kDa proteins (WAP18)

SDS-PAGE and immunoblot analysis showed that 18-kDa proteins (WAP18) were most abundantly accumulated during winter in the soluble fractions of cortical tissues of mulberry (Ukaji *et al.*, 2004). Immunocytochemical electron microscopy using purified anti-WAP18 antibodies confirmed that WAP18 proteins were specifically distributed in the cytosol and nucleus in the cortical parenchyma cells of mulberry during winter (results not shown).

As a result of immunoscreening using purified anti- WAP18 antibodies, many clones were isolated from cortical parenchyma cells of mulberry, and they were classified into three distinct isoforms, being designated WAP18A, WAP18B and WAP18C proteins. All these isoforms had high sequence homology with each other and consisted of 156 amino acids in the mature forms (calculated molecular masses 16,621, 16,606 and 16,498, respectively). The calculated pI values of deduced amino acid sequences of WAP18A, WAP18B and WAP18C were 5.18, 5.35 and 5.28, respectively. These calculated pI values corresponded with results of two-dimensional gel electrophoresis analysis (Ukaji *et al.*, 2004). A search in the database confirmed high sequence similarity of these WAP18 proteins with the PR-10 protein family. WAP18 cDNAs showed high sequence homology with a member of PR-10, including Mal d 1 in apple (Vanek-Krebitz *et al.*, 1995), Bet v 1 in birch (Breiteneder *et al.*, 1989), Pru a 1 in cherry (Scheurer *et al.*, 1997), PR-1 in parsley (Somssich *et al.*, 1986), PR-10a in common bean (Walter *et al.*, 1996), ABR17 in pea (Iturriaga *et al.*, 1994) and SAM22 in soybean (Crowell *et al.*, 1992). Among these, WAP18 cDNA showed highest sequence homology to the PR-10/Bet v 1 protein family (Ukaji *et al.*, 2004).

The accumulation of WAP18 proteins increased gradually from August to September, reached maximum from October to March, reduced gradually from April to May and almost disappeared from June to July (Ukaji *et al.*, 2004). Northern blot analysis revealed that the transcript level of *WAP18* was detected at low levels in summer, markedly increased from September and reached maximum level in November, and maintained the maximum level until March, then rapidly decreased to an undetectable level in April (results not shown). The *WAP18* gene expression increased by cold treatment even in twigs harvested in August, although the expression by cold treatment became gradually more distinct in twigs harvested in September towards October. The *WAP18* transcript level decreased by warming of these twigs (results not shown).

Winter-accumulating mulberry dehydrin (MbDHN)

DHN, a group 2 LEA protein, is known as a common stress-induced, and especially dehydration-induced, protein (Close, 1997; Rinne *et al.*, 1999). In cortical tissues of mulberry, accumulation of MbDHN in winter was detected by immunoblotting by using anti-DHN antibodies against the consensus sequence (Stressgen Biotechnologies Corp., Canada). For characterization of MbDHN, the *MbDHN* cDNA clone was isolated using the anti-DHN antibodies by immnoscreening. The primary sequence of MbDHN had typical DHN motifs, two Y-segments, one shorter than typical S-segment and two K-segments, showing high sequence homology with gene products of *Lea3-D147* in upland cotton (N816155), *NeERD10B* in tobacco (AB049336) (Hara *et al.*, 2003) and *Dhn1* in potato (Y15813). Deduced molecular mass of MbDHN was calculated to be 15,491 and the pI value was 9.27 (results not shown).

MbDHN proteins appeared in September and gradually increased to December, remaining at the maximum level from December till March, reduced in April and disappeared from May to September (results not shown). Transcript levels of *MbDHN* genes were markedly induced in winter (October to February) and disappeared in other seasons. The gene expression was induced by cold, dehydration and ABA treatments in twigs harvested in autumn. In contrast, *MbDHN* genes were rapidly reduced by warming in twigs harvested in winter (results not shown).

In Planta Functional Analysis of Mulberry Winter-accumulating Proteins

Generation of transgenic *Arabidopsis* expressing winter-accumulating mulberry genes

The binary Ti vector WPZ4 derived from pBI121 was used for transformation of *Arabidopsis thaliana* Columbia ecotype. The cDNAs of mulberry winter-accumulating proteins were inserted between restriction sites downstream to the cauliflower mosaic virus (CaMV) 35S promoter of the WPZ4 vector. These constructs were transferred to *Agrobacterium tumefaciens* strain C58C1 by the freeze–thaw method. *Agrobacterium*-mediated transformation was carried out by the floral-dipping method (Clough and Bent, 1998), and seeds of T1 transformants were selected on a germination medium containing 50 μg/ml kanamycin. Five to ten T2 plants with single insertion of transgenes were obtained from independent T1 lines of each construct. Homozygous T3 plants were obtained for each construct and those expressing high levels of transcripts of mulberry genes were used for subsequent phenotypic analysis.

Freezing tolerance of transgenic *Arabidopsis* leaves expressing winter-accumulating mulberry genes

Freezing tolerance of leaves from transgenic *Arabidopsis* expressing each winter-induced mulberry gene, *WAP27*, *WAP20*, *WAP18* or *MbDHN*, was analysed by an electrolyte-leakage method. In all experiments, plant-introduced vectors without specific genes were used as control. The control showed LT_{50} (freezing temperature for 50% injury) at −3.2°C before cold acclimation. After cold acclimation, the freezing tolerance in control plants increased to −5.3°C (Fig. 13.1). These LT_{50} values in control, before and after cold acclimation, were the same as those of wild-type (WT) plants.

Before cold acclimation, the freezing tolerance of all of the transgenic *Arabidopsis*, overexpressing *WAP27*, *WAP20*, *WAP18* or *MbDHN* cDNA, was not different from that of the control plants. After cold acclimation, however, only transgenic *Arabidopsis*-introduced *WAP27* showed clear increase of freezing tolerance (Fig. 13.1). The LT_{50} of WAP27 transgenic *Arabidopsis* after cold acclimation was −6.3°C, about 1°C increase in freezing tolerance from that of the control. In other transgenic *Arabidopsis* expressing *WAP20*, *WAP18* and *MbDHN* transgenes, a clear improvement of freezing tolerance was not detected (results not shown).

In all transgenic *Arabidopsis,* the freezing tolerance was also estimated by visible symptoms of freezing injury of seedlings after freeze–thaw. Seeds of the untransformed control and seeds of one transgenic line were cultivated in a single Petri dish with agar containing Murashige–Skoog (MS) medium for 20 days at 23°C

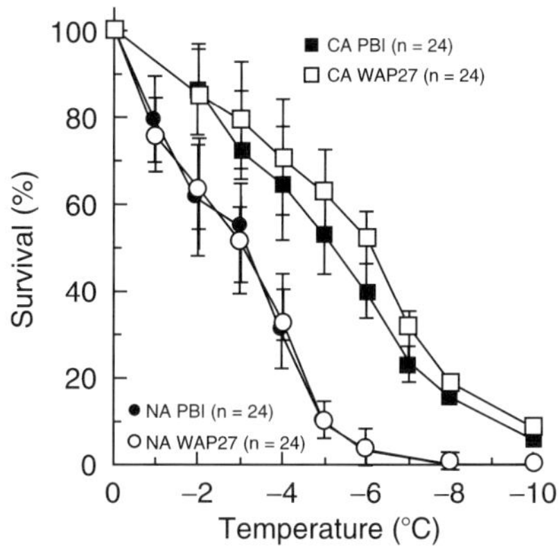

Fig. 13.1. *In planta* functional analysis of WAP27 from cortical parenchyma cells of mulberry to the freezing tolerance of transgenic *Arabidopsis*. *WAP27* transgenic (WAP27) and control (PBI) seeds of *Arabidopsis* were grown on 0.8% agar plate containing MS medium for 20 days at 23°C under continuous 24-h light. For cold acclimation, 20-day-grown plants (NA) were further grown for 2 days at 2°C under continuous 24-h light (CA). The leaves were cut from these plants. At −2°C, cut leaves were equilibrated for 5 min, frozen by seeding of ice, and left for 1 h. Then, they were cooled at a rate of 0.04°C/min to given temperatures, and thawed at 4°C overnight. The electrolyte leakages were measured and survival rate was expressed by per cent in comparison with those in unfrozen samples (0% injury) and with those in boiled samples (100% injury).

under continuous 24-h light. For cold acclimation, 20-day-old plants in Petri dishes were further grown at 2°C for 2 days under 24-h light. At −2°C, the Petri dishes with seedlings were equilibrated, frozen, and again equilibrated for 1 h. They were then cooled to each given temperature at a rate of 0.04°C/min, thawed at 4°C overnight, and subsequently regrown under normal growth conditions. These treatments provided the same growth and freeze–thaw conditions in order to compare freezing tolerance between control and transgenic plants. Although at whole-plant levels the absolute temperature for LT_{50} was different to that with electrolyte-leakage analysis, the visible symptoms of seedlings also showed improvement of freezing tolerance only in cold-acclimated *WAP27* transgenic *Arabidopsis*. Again, in this analysis, clear improvement in freezing tolerance was not detected in other transgenic *Arabidopsis*-introduced *WAP20*, *WAP18* or *MbDHN* cDNA (results not shown).

Recently, many studies have been performed to test *in planta* functional effects on freezing tolerance of low temperature-induced proteins by developing transgenic plants. However, significant improvement of freezing tolerance in transgenic plants was not obtained in these previous studies in which a single gene that produced a single end product was introduced. Transgenic tobacco overexpressing *CAP85* did not show any significant improvement in freezing tolerance (Kaye *et al.*, 1998). Overexpression or antisense-inhibition of *RAB18* had no effect on freezing and drought tolerance of *Arabidopsis* (Puhakainen *et al.*, 2004). Constitutive expression of *COR15* in transgenic *Arabidopsis* improved the freezing tolerance of chloroplasts frozen *in situ* and of protoplasts frozen *in vitro* (Artus *et al.*, 1996), but COR15a expression had little discernible effect on the survival of frozen whole plants (Jaglo-Ottosen *et al.*, 1998). Overproduction of a citrus DHN (CuCOR19) in tobacco leads to a slight decrease in ion leakage during chilling and freezing stresses (Hara *et al.*, 2003). Thus, in many *in planta* experiments to examine freezing tolerance in transgenic plants involving a single low-temperature-induced gene, whose end product is directly related to development of freezing tolerance, a clear increase of *in planta* freezing tolerance was rarely obtained. Similarly, temperature difference of freezing tolerance, as revealed by LT_{50} between control and transgenic plants, is not revealed in most of these studies. Among them, the improvement of *in planta* freezing tolerance in *Arabidopsis* by introduction of mulberry winter-accumulating *WAP27*, which showed clear increase of LT_{50}, is one of the most distinct results.

It is also interesting to note that, in contrast to the expression of a single gene, expression of two genes clearly increased freezing tolerance in transgenic plants. Simultaneous overexpression of two different types of DHNs from *Arabidopsis*, either *RAB18* and *COR47* or *LTI29* and *LTI30*, improved freezing tolerance of transgenic *Arabidopsis* (Puhakainen *et al.*, 2004). Constitutive expression of transcriptional activators that can express an entire battery of *COR* genes conferred high freezing tolerance to *Arabidopsis* (Jaglo-Ottosen *et al.*, 1998; Liu *et al.*, 1998; Kasuga *et al.*, 1999; Gilmour *et al.*, 2000) and to many other plants including hybrid aspen (S. Fujikawa *et al.*, unpublished results). However, for improvement of freezing tolerance by expression of transcriptional activators, multiple mechanisms appear to contribute to the enhancement of freezing tolerance, including the synthesis of many cryoprotective polypeptides and the accumulation of compatible solutes that have cryoprotective properties such as sucrose, raffinose and proline (Gilmour *et al.*, 2000). Thus, for the assessment of the functional role of individual low temperature-induced proteins in the acquisition of freezing tolerance, the results from the studies using transcriptional activator genes are less clear.

The relationship between increased freezing tolerance of WAP27 transgenic *Arabidopsis* and the morphology of ER in the leaf cells

In order to find out why *WAP27* transgenic *Arabidopsis* showed improved freezing tolerance only after cold acclimation, we have examined the ultrastructure of *Arabidopsis* leaf cells with special attention to the ER morphology where WAP27 proteins are specifically localized. This is because in cortical parenchyma cells of mulberry, a close relationship between the ER morphology and the freezing tolerance has been reported, as shown below (Fujikawa and Takabe, 1996). In cortical parenchyma cells of mulberry during summer (before cold acclimation), ERs in cisternal form were distributed randomly throughout the entire cytoplasm. The morphology and distribution of the ERs did not change when summer cells were subjected to freezing-induced dehydration. In winter (after cold acclimation), however, the ERs changed to vesicular form and were distributed randomly throughout the entire cytoplasm. Upon freezing-induced dehydration, these ERs in winter cells moved and fused to each other to produce multiplex lamellae just beneath the plasma membranes, with specific localization between plasma membrane and cytoplasmic organelles such as chloroplasts, vacuoles and mitochondria. Such specific localization of multiplex lamellar-form ER seemed to protect irreversible ultrastructural changes in the plasma membranes due to the close juxtaposition of plasma membrane to cytoplasmic organelle membrane, that was produced by shrinkage of cells during freezing-induced dehydration. It has been indicated that the inter-bilayer events due to this juxtaposition during cell shrinkage produce membrane fusion or lamellar-to-hexagonal II phase transitions of plasma membranes, and lead to severe injury of cells (Gordon-Kamm and Steponkus, 1984; Fujikawa and Miura, 1986; Steponkus and Lynch, 1989; Steponkus *et al*, 1993; Fujikawa, 1994; Uemura *et al.*, 1995; Fujikawa *et al.*, 1999). Quantitatively close relationships were observed between formation of ER multiplex lamellar by freezing-induced dehydration and acquisition of high freezing tolerance in the cortical parenchyma cells of mulberry throughout the seasons (Fujikawa and Takabe, 1996). Based upon these close relationships between change of ER morphology and acquisition of freezing tolerance, we have examined morphological changes of ER in both WT and *WAP27* transgenic *Arabidopsis*.

The ultrastructure of *Arabidopsis* leaf parenchyma cells (including the distribution of ER) was identical between WT and *WAP27* transgenic plants. The ER-specific distribution of WAP27 in transgenic *Arabidopsis* was confirmed by immunocytochemical electron microscopy (results not shown).

Before cold acclimation, the morphology of ER in *Arabidopsis* leaf parenchyma cells from both WT and the *WAP27* transgenic plants showed a cisternae form; a small number of these ERs were distributed randomly throughout the entire cytoplasm (Fig. 13.2A). After cold acclimation, small vesicles were produced from cisternae-form ER, but the basic morphology and distribution of cisternae-form ERs were similar to those in non-acclimated cells (Fig. 13.2B). The occurrence of fragments of cisternae-form ER in *Arabidopsis* leaf cells by cold acclimation has been previously reported (Ristic and Ashworth, 1993). By dehydration treatment with concentrated sorbitol solution that mimics freezing-induced dehydration, leaf parenchyma cells of *Arabidopsis* were shrunk and plasmolyzed (Fig. 13.2C,D). In non-acclimated cells (Fig. 13.2C), however, the fundamental morphology and the distribution of ERs in the shrunken cells were similar to those before dehydration (compare Fig. 13.2C with 13.2A). The ERs were still distributed randomly throughout the entire shrunken cytoplasm. On the other hand, ERs in cold-acclimated leaf cells showed drastic changes in the distribution after dehydration treatment (Fig. 13.2D). The majority of cisternae-form ERs were then found closely beneath the plasma membrane, distributed specifically between plasma membranes and cytoplasmic organelles, and formed single, plural or multiplex piles of ERs. The phenomenon was similar to the change of ERs in cortical parenchyma cells of mulberry in winter subjected to freezing-induced dehydration (Fujikawa and Takabe, 1996).

These ultrastructural and survival assays, thus, indicated that while *WAP27* transgenic *Arabidopsis* resulted in improvement of the freezing tolerance in which ERs with WAP27 were localized just beneath the plasma membrane (in cold-acclimated cells subjected to dehydration), the transgenic plants did not show improvement in freezing tolerance in which ERs with WAP27 were

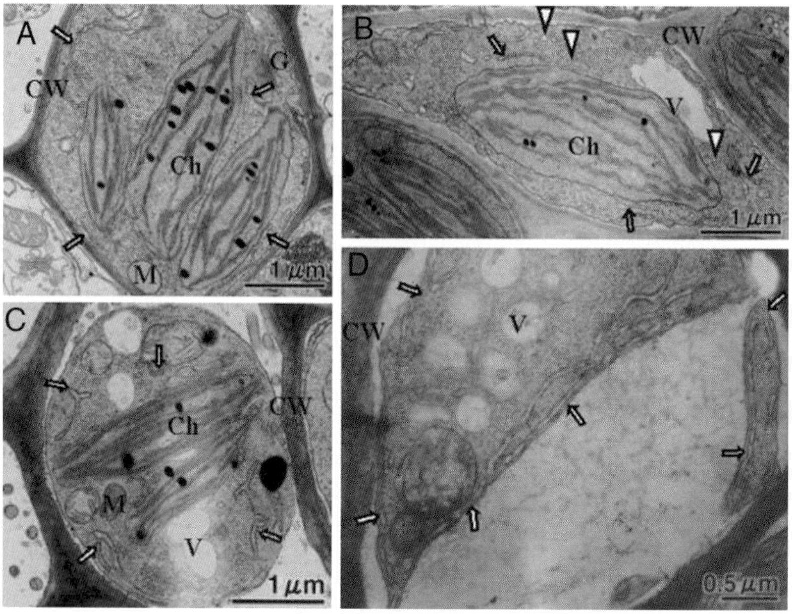

Fig. 13.2. Ultrastructure of leaf parenchyma cells in *WAP27* transgenic *Arabidopsis*. Plant growth and cold-acclimation conditions are as in Fig. 13.1. (A) Leaf cell before cold acclimation. Endoplasmic reticula (Ers) are indicated by arrows. CW, cell wall; G, Golji apparatus; M, mitochondrion; Ch, chloroplast. (B) Leaf cell after cold acclimation. Arrowheads show formation of small vesicles from cisternae-form ER. V, vacuole. (C) Dehydrated leaf cell before cold acclimation. Dehydration was performed by immersing cut leaves in 2.7 osmolal sorbitol solutions at 4°C for 2 h. Due to dehydration by concentrated solutions, plasmolysis occurs, instead of cytorrhysis by freezing-induced dehydration. (D) Dehydrated leaf cell after cold acclimation. Due to dehydration, plasmolysis occurs. ERs are distributed just beneath the plasma membranes forming a single, plural or multiplex pile.

localized in cytoplasm some distance from the plasma membranes (in non-acclimated cells subjected to dehydration). It has been reported that freezing tolerance increases in chloroplasts and/or protoplasts isolated from transgenic *Arabidopsis* in which COR15am was constitutively expressed in the stroma of chroloplasts (Artus *et al.*, 1996). It is hypothesized that the presence of COR15am in the stroma alters the intrinsic curvature of lipids in the inner envelope membranes of the chloroplasts and consequently reduces membrane destabilization in plasma membranes that are in close proximity to chloroplast envelope membranes (Steponkus *et al.*, 1998). The nature of WAP27 proteins resembles that of COR15am, which exhibits a four-times repeat of the 13mer-amino acid sequence motif. WAP27 is rich in Ala, Lys, Glu, Thr and Asp residues, which make up approximately 55% of the total amino acids. COR15am is also rich in Ala, Lys, Glu, Thr and Asp residues, which make up approximately 64% of the protein (Thomashow, 1994, 1999). Thus, we temporarily speculated that WAP27 in ER might also have a similar effect with COR15am in order to reduce plasma membrane destabilization from inter-bilayer events. Such effects will work when ERs with WAP27 are localized between plasma membrane and other cytoplasmic organelle membranes (e.g. cold-acclimated cells subjected to dehydration), rather than localized far from plasma membranes (e.g. non-acclimated cells subjected to dehydration).

Some transgenic plants showed significant increase in freezing tolerance, but sometimes the acquisition of freezing tolerance was associated with reduction in plant growth (Gilmour *et al.*, 2000). The *WAP27* transgenic plants did not show any visible changes in plant growth (results not shown) and no change in the ultrastructure of cells. The protective effects of WAP27 appear only after cold acclimation and after subjection to freezing-induced dehydration. Thus, WAP27 may be a useful gene to confer plant freezing tolerance without significant effects on the growth of transgenic plants.

In Vitro Functional Analysis of Purified Mulberry Winter-accumulating Proteins

Purification of mulberry winter-accumulating proteins

For characterizing *in vitro* functional roles of winter-accumulating mulberry proteins, recombinant proteins of WAP27, WAP20 and MbDHN were prepared using the pET vector system for the expression of proteins in bacteria (Novagen Corp., USA). Recombinant proteins of WAP20 and WAP27 (rWAP20 and rWAP27) corresponded to the mature forms in mulberry, and that of MbDHN corresponded to the mature form with histidine-tag. To obtain rWAP27, rWAP20 and rMbDHN, bacterial cells, in which the gene expression was induced by isopropyl-beta-D-thiogalactopyranoside, were disrupted by sonication. The recombinant proteins in the crude soluble fraction were concentrated by ammonium sulphate precipitation. Concentrated protein fractions were purified by combination of the hydrophobic anion exchange, gel filtration column chromatography or heating at 50°C. In the case of rMbDHN, protein fractions were further purified using the Ni^{2+} affinity column (Amersham Bioscience Corp., USA). Fractions containing the abundant recombinant proteins were collected and used for functional analyses since minor contamination by other proteins was deemed to be negligible for further analyses.

WAP18 for *in vitro* functional analysis was isolated directly from mulberry twigs in winter because the purification was relatively easier than other winter-inducible proteins (Ukaji *et al.*, 2004). After extraction of the crude soluble fraction from cortical tissues of mulberry, WAP18 was effectively purified from the soluble fraction by the addition of ammonium sulphate to the final concentration of 85% saturation. WAP18 was still soluble under 85%-saturated ammonium sulphate solution while most of the other proteins were precipitated. The fraction containing abundant WAP18 was further purified by hydrophobic column and subsequent gel filtration column chromatography. The resultant WAP18-enriched fraction, with almost homogeneity, was utilized for the functional analysis.

Effect of purified proteins as a cryoprotectant to freezing-labile protein

Four winter-accumulating mulberry proteins purified from bacterial homogenates or directly from cortical tissue homogenates, rWAP27, rWAP20, WAP18 and rMbDHN, were used for testing the protective effect to the activity of freezing-sensitive protein, lactate dehydrogenase (LDH) (Fig. 13.3A–D). LDH solution was frozen at −20°C for 1 day in the presence of purified winter-accumulating mulberry proteins or in the presence of other known cryoprotective substances including sucrose, bovine serum albumin (BSA) and ovalbumin (OVA). Among these four winter-accumulating mulberry proteins, the protective effects of rWAP27 and rMbDHN, in low concentration on the basis of protein concentration (less than 10 μg/ml), were much higher than those of known cryoprotective substances (Fig. 13.3A,D). The effect of WAP18 on the basis of protein concentration was not so high and almost similar to those of BSA and OVA, even in the range of low concentration (Ukaji *et al.*, 2004).

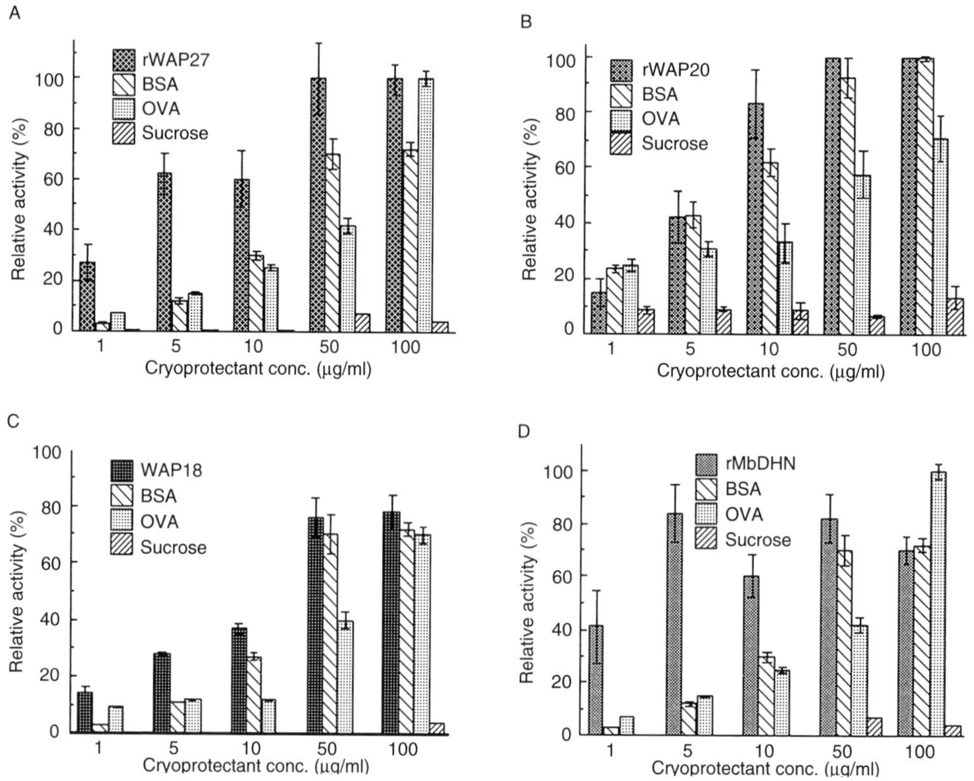

Fig. 13.3. *In vitro* functional analysis of purified winter-accumulating proteins, (A) rWAP27, (B) rWAP20, (C) WAP18 and (D) rMbDHN, from cortical parenchyma cells of mulberry as a cryoprotectant to freeze-labile lactate dehydrogenase (LDH). After various concentrations of the winter-accumulating proteins were added to LDH solution (at a final concentration of 2.5 g/ml of LDH), the mixtures were frozen at −20°C for 1 day. LDH activities in the mixtures were immediately measured after thawing. For comparison, cryoprotective effects of bovine serum albumin (BSA), ovalbumin (OVA) and sucrose were also measured.

The CP_{50} values, the concentration of cryoprotectant to protect 50% of LDH activity, between rWAP27, rWAP20 and rMbDHN were very similar to, and slightly higher than, that of WAP18 (Table 13.1). The CP_{50} values of rWAP27, rWAP20 and rMbDHN were also similar to those of other cryoprotective LEA proteins (Kazuoka and Oeda, 1992; Wisniewski *et al.*, 1999; Hara *et al.*, 2001; Bravo *et al.*, 2003; Lopez-Matas *et al.*, 2004), but higher than those of COR15a (Lin and Thomashow, 1992). It was noted that, compared with COR15a and DHN, winter-accumulating proteins from mulberry, especially WAP27 and MbDHN, might be more effective at lower concentrations (less than 5 μg/ml).

When the protective effects were shown as a molar basis of monomer, CP_{50} values of rMbDHN and rWAP27 were similar to each other, slightly lower than that of rWAP20 and far lower than that of WAP18. The protective effects of these three proteins, rWAP27, rWAP20 and rMbDHN, were also similar to those of other cryoprotective LEA proteins (Kazuoka and Oeda, 1992; Wisniewski *et al.*, 1999; Hara *et al.*, 2001; Bravo *et al.*, 2003; Lopez-Matas *et al.*, 2004; Table 13.1). In our system, rWAP20 and rWAP27 were found to form oligomeric structures similar to native WAP20

Table 13.1. Comparison of cryoprotective effects of cold-regulated proteins. Cryoprotective effects against lactate dehydrogenase (LDH) during freeze–thaw cycles are represented as CP_{50} values calculated by data or referred to in previous reports. The CP_{50} values are expressed on a basis of protein concentration (μg/ml) and on a molar basis of monomer (nM). Molecular masses of these proteins, except for COR85 and PCA60, were calculated from the predicted amino acid sequences. MM, molecular mass.

Proteins	MM ($\times 10^3$)	CP_{50} (μg/ml)	CP_{50} (nM)	References
rWAP27	22.5	3.7	164	This study
rWAP20	18.6	6	323	This study
rMbDHN	15.5	1.7	110	This study
WAP18	16.6	22	1.3×10^3	Ukaji *et al.* (2004)
COR15a	15.0	0.1	7	Lin and Thomashow (1992)
CsHSP17.5	17.5	1.2	69	Lopez-Matas *et al.* (2004)
DHN5	58.6	3	51	Bravo *et al.* (2003)
CuCOR19	19.0	7	368	Hara *et al.* (2001)
COR85	85.0	15	176	Kazuoka and Oeda (1992)
PCA60	60.0	20	333	Wisniewski *et al.* (1999)
BSA[a]	66.0	23	348	This study
OVA[a]	45.0	53	1.2×10^3	This study

[a]CP_{50} values were represented as an average of four independent experiments shown in Fig. 13.4A–D.

and WAP27, respectively, while the information about oligomeric formation of rMbDHN has not yet been determined. If the cryoprotective effects were detected under the conditions of the oligomeric structures, the effects of rWAP20 and rWAP27 shown as a molar basis of the oligomer may be much higher than those of the present data.

While mulberry winter-accumulating proteins have shown clear cryoprotection effects on freezing-labile proteins, the mechanisms of cryoprotection are unknown. However, it is suggested that group 3 LEA proteins could act as ion scavengers by binding to ions concentrated in the solution during freeze-induced dehydration (Dure, 1993). One LEA protein, CuCOR19, that was induced by chilling stress in *Citrus unshu*, has been shown to act as a possible radical-scavenging protein, since the recombinant CuCOR19 produced in a bacterial protein-expression system reduced peroxidation of liposomes of soybean *in vitro* (Hara *et al.*, 2003). It is also suggested that the predicted amphipathic alpha-helical structure of DHN proteins may interact with macromolecules to protect them from dehydration-induced damages (Close, 1997). It is well known that smHSPs have a function for refolding of denatured proteins as a mechanism of molecular chaperoning (Lee *et al.*, 1995; Collada *et al.*, 1997; Lopez-Matas *et al.*, 2004). Furthermore, in terms of membrane protection, COR15am in *Arabidopsis* (Steponkus *et al.*, 1998), cryoprotectin in cabbage (Hincha, 2002), osmotin-like protein in bittersweet nightshade (Newton and Duman, 2000), beta-1,3-glucanase in tobacco (Hincha *et al.*, 1997a) and lectins (Hincha *et al.* 1993; 1997b), have been shown to reduce freezing injury of membrane vesicles *in vitro*, possibly by direct interaction of these proteins with membrane lipids or membrane glycolipids. Since the structural features of mulberry winter-accumulating proteins are different to each other, their individual modes of cryoprotective action could be different.

Other effects of winter-accumulating proteins

Effects of WAP20 as a molecular chaperone

Among the proteins induced or upregulated in plants by low temperatures, there are HSPs. The members of the HSP70 family (Neven *et al.*, 1992; Anderson *et al.*, 1994) and also HSP90 (Krishna *et al.*, 1995; Pareek *et al.*, 1995) isoforms are induced by low temperature and exhibit molecular chaperone activity (Guy *et al.*, 1998). Low temperatures have also been shown to stimulate the accumulation of smHSPs in cold-stored potato (van Berkel *et al.*, 1994) and in tomato fruit after heat treatment (Sabehat *et al.*, 1998). Winter-specific accumulation of smHSPs has also been observed in cortical parenchyma cells of mulberry (Ukaji *et al.*, 1999), *Acer platanoides*, *Sambucus nigra* and *Aristolochia macrophylla* (Lubaretz and zur Nieden, 2002). Some of the plant smHSPs are also known to have molecular chaperone activity. The molecular chaperone activity of plant smHSP has been reported for two recombinant smHSPs, HSP18.1 and HSP17.7 in pea (Lee *et al.*, 1995), CssmHSP1 in chestnut seeds (Collada *et al.*, 1997) and CsHSP17.5 in chestnut stems (Lopez-Matas *et al.*, 2004).

WAP20 proteins, also belonging to the smHSP family, from cortical parenchyma cells of mulberry, have cryoprotective activity to LDH. Therefore, we tested the function of WAP20 as a molecular chaperone using rWAP20 proteins (Fig. 13.4A–D). Recovery of activity of citrate synthase (CS) by refolding, previously having been denatured in guanidine hydrochloride, was enhanced in the presence of rWAP20, while the absence of rWAP20 had little effect (Fig. 13.4A). Similarly, rWAP20 recovered CS from thermal inactivation more effectively than did the absence of rWAP20 (Fig. 13.4B). Moreover, rWAP20 suppressed thermal aggregation of CS while catalase, which possesses a similar molecular mass to the native form to rWAP20, had no effect (Fig. 13.4C). Most interestingly, WAP20 had molecular chaperone activity to recover previously freeze-denatured LDH (Fig. 13.4D). Thus, winter-accumulating mulberry WAP20 had dual roles as molecular chaperone and cryoprotectant. CsHSP17.5 in chestnut stems also shows a dual role as molecular chaperone on the refolding of chemically denatured endochitinase and as cryoprotectant to LDH (Lopez-Matas *et al.*, 2004).

Among the stresses characteristic of winter – except for freezing – low temperature may directly affect the stability and solubility properties of many globular proteins (Pace, 1990; Privalov, 1990). The presence of cold-labile proteins in plants has been confirmed in spinach leaves (Guy *et al.*, 1998). It is also indicated that low temperature caused increased association of some cell proteins with two chaperones of the HSP70 family (Guy *et al.*, 1998). It has also been reported that the smHSP, CsHSP17.5, enhanced cell survival at low temperature when expressed in *E. coli* (Soto *et al.*, 1999). The present study showed not only that WAP20 might have a function in protecting cold-labile proteins as molecular chaperone, as suggested in previous studies (Guy *et al.*, 1998), but also in the role of molecular chaperone to repair functions of freeze-denatured proteins.

Effects of WAP18 on pathogenic infection

WAP18 proteins have a high homology with the PR group 10/Bet v 1 family (Ukaji *et al.*, 2004). We have, therefore, examined, although it is still indirect, the functional role of WAP18 as PR proteins. It has been reported that PR proteins accumulate by pathogenesis-related stresses, including wounding treatment and exogeneous application of ethephon, salicylic acid (SA) or methyl jasmonate (MeJA) (Yu *et al.*, 2001; Kuwabara *et al.*, 2002). Northern blot analysis showed that *WAP18* gene expression increased within 1 h, reached maximum level between 6 h and 12 h, and then decreased in response to wounding treatment of the mulberry twigs harvested in August. The *WAP18* gene expression increased within 3 h, reached maximum level in 12 h and then decreased by application of ethephon or MeJA, also in twigs harvested in August. The *WAP18* transcripts increased after 72-h treatment by endogenous application of SA in the same twigs. The *WAP18* gene expression also increased by all these treatments indicated above in the twigs harvested in autumn. The RT-PCR analysis revealed that the transcripts of *WAP18A, WAP18B and WAP18C* genes were all accumulated in response to wounding or exogeneous application of ethephon, SA or MeJA. It is well established that ethylene, jasmonic acid (JA) and SA act as signalling molecules for the induction of PR genes followed by acquisition of resistance to pathogenic infection (Gaffney *et al.*, 1993). The role of SA for the production of PR genes differs from that of ethylene and JA. In general, ethylene and MeJA intermediate signals in response to wounding and pathogen infection,

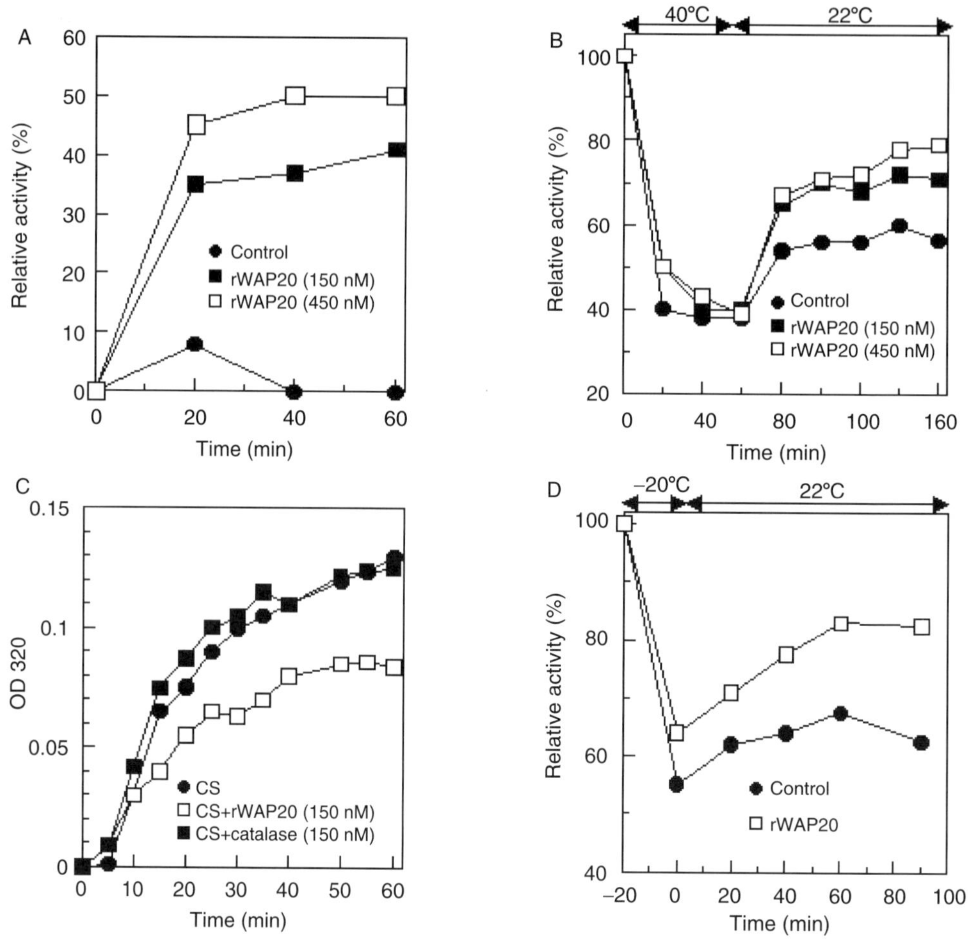

Fig. 13.4. Functional role of WAP20 as a molecular chaperone. (A) Recovery of activity of chemically denatured citrate synthase (CS) by addition of rWAP20. CS was previously denatured by 6 M guanidine hydrochloride, and the recovery of CS activities (at a final concentration of 150 nM) was measured by addition of rWAP20 at a final concentration of 150 nM (closed square), 450 nM (open square) or 0 nM (closed circle). (B) Recovery of thermally inactivated CS activity by addition of rWAP20. CS monomer solution (150 nM) was heated at 40°C for 1 h in the absence (closed circle) or in the presence of 150 nM (closed square) or 450 nM (open square) rWAP20. Treatment temperature was reduced to 22°C and samples were further incubated. (C) Prevention of thermal aggregation of CS. CS monomer solution (150 nM) was heated at 37°C for 1 h in the absence (closed circle) or in the presence of 150 nM catalase (closed square) or 150 nM rWAP20 (open square). Optical density (OD) at 320 nM was measured as an index of the formation of thermal aggregation. (D) Recovery of freeze-inactivated lactate dehydrogenase (LDH) by addition of rWAP20. LDH solution (150 nM) was frozen at –20°C. After thawing, samples were incubated at 22°C for 90 min, in the absence of rWAP20 (closed circle) or in the presence of 150 nM rWAP20 (open square).

whereas SA is required for the induction of broad-spectrum disease resistance in systemic tissues of plants previously infected with a pathogen (Gaffney *et al.*, 1993).

The distinct expression profiles of *WAP18* genes by these treatments were consistent with the general expression profile of PR genes in plants that were infected by pathogens. It is, thus, speculated that in addition to the roles of cryoprotectant, WAP18 may also have the function of PR proteins, although experiments to obtain the direct evidence are necessary. Additionally, because WAP18 is the most abundantly accumulating protein in cortical parenchyma cells of mulberry during winter, the role as storage proteins also cannot be ignored.

Conclusion

Among cold acclimation-induced diverse changes that are closely associated with the acquisition of freezing tolerance, the functional role of cold-induced proteins is generally unclear. In the present study, we focused on four proteins that are distinctly accumulated in winter in association with the acquisition of extremely high freezing tolerance in cortical parenchyma cells of mulberry. These winter-accumulating mulberry proteins seemed to exhibit different functional roles against winter-induced diverse stresses. The group 3 LEA protein, WAP27, had functional effects in protecting cells from inter-bilayer events as a main cause of freezing injury, and as an effective cryoprotectant. The smHSP, WAP20, had effects not only as a cryoprotectant, but also as a molecular chaperone. The PR-10 protein, WAP18, had functions not only as a cryoprotectant but also may work as a PR protein. MbDHDN had high cryoprotective activity. Thus, all mulberry winter-accumulating proteins examined in this study had some protective effects against diverse winter-induced stresses, such as low temperature, freezing or winter-specific infection of pathogens. It is also interesting to note that many of the winter-accumulating proteins analysed in this study had complex roles for protection against diverse winter-induced stresses. However, it is suggested that the cause of the extremely high freezing tolerance of mulberry in winter, as well as improved freezing tolerance of other plants by cold acclimation, will not be explained by only consideration of the total effects of winter-induced proteins as end products. In addition to the combination of protective effects of winter-induced proteins, plants can obtain total resistance against winter-induced biotic and abiotic stresses by combining these with other diverse cold-induced changes.

Acknowledgements

The present study was supported by a grant from PROBRAIN.

References

Anderson, J.V., Li, Q.B., Haskell, D.W. and Guy, C.L. (1994) Structural organization of the spinach endoplasmic reticulum-luminal 70-kilodalton heat-shock genes during cold acclimation. *Plant Physiology* 104, 1359–1370.

Artus, N.N., Uemura, M., Steponkus, P.L., Gilmour, S.J., Lin, C. and Thomashow,

M.F. (1996) Constitutive expression of the cold-regulated *Arabidopsis thaliana COR15a* gene affects both chloroplast and protoplast freezing tolerance. *Proceedings of National Academy of Science USA* 93, 13404–13409.

Bravo, L.A., Gallardo, J., Navarrete, A., Olave, N., Martinez, J., Aberdi, M., Close, T.J. and Corcuera L.J. (2003) Cryoprotective activity of a cold-induced dehydrin purified from barley. *Physiologia Plantarum* 118, 262–269.

Breiteneder, H., Pettenburger, K., Bito, A., Valenta, R., Kraft, D., Rumpold, H., Scheiner, O. and Breitenbach, M. (1989) The gene coding for the major birch pollen allergen Betv1, is highly homologous to a pea disease resistance response gene. *EMBO Journal* 8, 1935–1938.

Close, T.J. (1997) Dehydrins: a commonalty in the response of plants to dehydration and low temperature. *Physiologia Plantarum* 100, 291–296.

Clough, S.J. and Bent, A.F. (1998) Floral dip: a simplified method for *Agrobacterium*-mediated transformation of *Arabidopsis thaliana*. *Plant Journal* 16, 735–743.

Collada, C., Gomez, L., Casado, R. and Aragoncillo, C. (1997) Purification and *in vitro* chaperone activity of a class I small heat-shock protein abundant in recalcitrant chestnut seeds. *Plant Physiology* 115, 71–77.

Crowell, D.N., John, M.E., Russell, D. and Amasino, R.M. (1992) Characterization of a stress-induced, developmentally regulated gene family from soybean. *Plant Molecular Biology* 18, 459–466.

Dure III, L. (1993) A repeating 11-mer amino acid motif and plant desiccation. *Plant Journal* 3, 363–369.

Fujikawa, S. (1994) Seasonal ultrastructural alterations in the plasma membranes produced by slow freezing in cortical tissues of mulberry (*Morus bombycis* Koidz. cv. Goroji). *Trees* 8, 288–296.

Fujikawa, S. and Kuroda, K. (2000) Cryo-scanning electron microscopic study on freezing behavior of xylem ray parenchyma cells in hardwood species. *Micron* 31, 669–686.

Fujikawa, S. and Miura, K. (1986) Plasma membrane ultrastructural changes caused by mechanical stress in the formation of extracellular ice as a primary cause of slow freezing injury in fruit-bodies of Basidiomycetes (*Lyophyllum ulmarium* (Fr.) Kuhner). *Cryobiology* 23, 371–382.

Fujikawa, S. and Takabe, K. (1996) Formation of multiplex lamellae by equilibrium freezing of cortical parenchyma cells of mulberry and its possible relationship to freezing tolerance. *Protoplasma* 190, 198–203.

Fujikawa, S., Jitsuyama, Y. and Kuroda, K. (1999) Determination of the role of cold-acclimation-induced diverse changes in plant cells from the viewpoint of avoïdance of freezing injury. *Journal of Plant Research* 112, 237–244.

Gaffney, T., Friedrich, L., Vernooij, B., Negrotto, D., Nye, G., Uknes, S., Ward, E., Kessmann, H. and Ryals, J. (1993) Requirement of salicylic acid for the induction of systemic acquired resistance. *Science* 261, 754–756.

Gilmour, S.J., Sebolt, A.M., Salazar, M.P., Everard, J.D. and Thomashow, M.F. (2000) Overexpression of the *Arabidopsis CBF3* transcriptional activator mimics multiple biochemical changes associated with cold acclimation. *Plant Physiology* 124, 1854–1865.

Gordon-Kamm, W.J. and Steponkus, P.L. (1984) Lamellar-to-hexagonal II phase transitions in the plasma membrane of isolated protoplasts after freeze-induced dehydration. *Proceedings of the National Academy of Science USA* 81, 6373–6377.

Guy, C.L. (1990) Cold acclimation and freezing stress tolerance: role of protein metabolism. *Annual Review of Plant Physiology and Plant Molecular Biology* 41, 187–223.

Guy, C.L., Haskell, D. and Li, Q.-B. (1998) Association of proteins with the stress 70 molecular chaperones at low temperature: evidence for the existence of cold labile proteins in spinach. *Cryobiology* 36, 301–314.

Hara, M., Terashima, S. and Kuboi, T. (2001) Characterization and cryoprotective activity of cold-responsive dehydrin from *Citrus unshu*. *Journal of Plant Physiology* 158, 1333–1339.

Hara, M., Terashima, S., Fukaya, T. and Kuboi, T. (2003) Enhancement of cold tolerance and inhibition of lipid peroxidation by citrus dehydrin in transgenic tobacco. *Planta* 217, 290–298.

Hare, P.D., Cress, W.A. and van Staden, J. (1998) Dissecting the role of osmolyte accumulation during stress. *Plant, Cell and Environment* 21, 535–553.

Helm, K.W., LaFayette, P.R., Nagao, R.T., Key, J.L. and Vierling, E. (1993) Localization of small heat shock proteins to the higher plant endomembrane system. *Molecular and Cellular Biology* 13, 238–247.

Helm, K.W., Schmeits, J. and Vierling, E. (1995) An endomembrane-localized small heat-shock protein from *Arabidopsis thaliana*. *Plant Physiology* 107, 287–288.

Hincha, D.K. (2002) Cryoprotectin: a plant lipid-transfer protein homologue that stabilizes membranes during freezing. *Philosophical Transactions of the Royal Society of London. Series B* 357, 909–916.

Hincha, D.K., Bakaltcheva, I. and Schmitt, J.M. (1993) Galactose-specific lectins protect isolated thylakoids against freeze–thaw damage. *Plant Physiology* 103, 59–65.

Hincha, D.K., Meins, F. Jr and Schmitt, J.M. (1997a) Beta-1,3-glucanase is cryoprotective *in vitro* and is accumulated in leaves during cold acclimation. *Plant Physiology* 114, 1077–1083.

Hincha, D.K., Bratt, P.J. and Williams, W.P. (1997b) A cryoprotective lectin reduces the solute permeability and lipid fluidity of thylakoid membranes. *Cryobiology* 34, 193–199.

Hon, W.C., Griffith, M.M., Kowk, Y.C. and Yang, D.S.C. (1995) Antifreeze proteins in winter rye are similar to pathogenesis-related proteins. *Plant Physiology* 109, 879–889.

Iturriaga, G., Schneider, K., Salamini, F. and Bartles, D. (1994) Expression of desiccation-related proteins from the resurrection plant *Craterostigma plantagineum* in transgenic tobacco. *Plant Molecular Biology* 20, 555–558.

Jaglo-Ottosen, K.R., Gilmour, S.J., Zarka, D.G., Schabenberger, O. and Thomashow, M.F. (1998) *Arabidopsis CBF1* overexpression induces COR genes and enhances freezing tolerance. *Science* 280, 104–106.

Kasuga, M., Liu, Q., Miura, S., Yamaguchi-Shinozaki, K. and Shinozaki, K. (1999) Improving plant drought, salt, and freezing tolerance by gene transfer of a single stress-inducible transcriptional factor. *Nature Biotechnology* 17, 287–291.

Kawamura, Y. and Uemura, M. (2002) Changes in the plasma membrane from *Arabidopsis thaliana* within one week of cold acclimation. In: Li, P.H. and Palva, E.T. (eds) *Plant Cold Hardiness, Gene Regulation and Genetic Engineering*. Kluwer Academic/Plenum Publishers, New York, pp. 181–194.

Kaye, C., Neven, L., Hofig, A., Li, Q.B. and Guy, C. (1998) Characterization of a gene for spinach CAP160 and expression of two spinach cold-acclimated proteins in tobacco. *Plant Physiology* 116, 1367–1377.

Kazuoka, T. and Oeda, K. (1992) Heat-stable COR (cold-regulated) proteins associated with freezing tolerance in spinach. *Plant and Cell Physiology* 33, 1107–1114.

Krishna, P., Sacco, M., Cherutti, J.F. and Hill, S. (1995) Cold-induced accumulation of hsp90 transcripts in *Brassica napus*. *Plant Physiology* 107, 915–923.

Kuwabara, C., Takezawa, D., Shimada, T., Hamada, T., Fujikawa, S. and Arakawa, K. (2002) Abscisic acid- and cold-induced thaumatin-like protein in winter wheat has an antifungal activity against snow mould, *Microdochium nivale*. *Physiologia Plantarum* 115, 101–110.

Lauzon, L.M., Helm, K.W. and Vierling, E. (1990) A cDNA clone from *Pisum sativum* encoding a low molecular weight heat shock protein. *Nucleic Acids Research* 18(14), 4274.

Lee, J.B., Pokala, N. and Vierling, E. (1995) Structure and *in vitro* molecular chaperone activity of cytosolic small heat shock proteins from pea. *Journal of Biological Chemistry* 270, 10432–10438.

Lin, C. and Thomashow, M.F. (1992) A cold-regulated *Arabidopsis* gene encodes a polypeptide having potent cryoprotective activity. *Biochemical and Biophysical Research Communication* 183, 1103–1108.

Liu, Q., Kasuga, M., Sakuma, Y., Abe, H., Miura, S., Yamaguchi-Shinozaki, K. and Shinozaki, K. (1998) Two transcription factors, DREB1 and DREB2, with an EREBP/AP2 DNA binding domain separate two cellular signal transduction pathways in drought- and low temperature-responsive gene expression, respectively, in *Arabidopsis*. *Plant Cell* 10, 1391–1406.

Lopez-Matas, M.,-A., Nunez, P., Soto, A., Allona, I., Casado, R., Collada, C., Guevara, M.-A., Aragoncillo, C. and Gomez, L. (2004) Protein cryoprotective activity of a cytosolic small heat shock protein that accumulates constitutively in chestnut stems and is up-regulated by low and high temperatures. *Plant Physiology* 134, 1708–1717.

Lubaretz, O. and zur Nieden, U. (2002) Accumulation of plant small heat-stress proteins in storage organs. *Planta* 215, 220–228.

Neven, L.G., Haskell, D.W., Guy, C.L., Denslow, N., Klein, P.A., Green, L.G. and Silverman, A. (1992) Association of 70-kilodalton heat-shock cognate proteins with acclimation to cold. *Plant Physiology* 99, 1362–1369.

Newton, S.S. and Duman, J.D. (2000) An osmotin-like cryoprotective protein from the bittersweet nightshade *Solanum dulcamara*. *Plant Molecular Biology* 44, 581–589.

Niki, T. and Sakai, A. (1981) Ultrastructural changes related to frost hardiness in the cortical parenchyma cells from mulberry twigs. *Plant and Cell Physiology* 22, 171–183.

Pace, C.N. (1990) Conformational stability of globular proteins. *Trends in Biochemical Science* 15, 14–17.

Pareek, A., Singla, S.L. and Grover, A. (1995) Immunological evidence for accumulation of two high-molecular-weight (104 and 90 kDa) HSPs in response to different stresses in rice and in response to high temperature stress in diverse plant genera. *Plant Molecular Biology* 29, 293–301.

Privalov, P.L. (1990) Cold denaturation of proteins. *Critical Review of Biochemical and Molecular Biology* 25, 281–306.

Puhakainen, T., Hess, M.W., Makela, P., Svensson, J., Heino, P. and Palva, E.T. (2004) Overexpression of multiple dehydrin genes enhances tolerance to freezing stress in *Arabidopsis*. *Plant Molecular Biology* 54, 743–753.

Rajashekar, C.B. and Lafta, A. (1996) Cell-wall changes and cell tension in response to cold acclimation and exogenous abscisic acid in leaves and cell culture. *Plant Physiology* 111, 605–612.

Rinne, P.L.H., Kaikuraranta, P.L.M., van der Plas, L.H.W. and van der Shoot, C. (1999) Dehydrins in cold-acclimated apices of birch (*Betula pubescent* Ehrh.): production, localization and potential role in rescuing enzyme function during dehydration. *Planta* 209, 377–388.

Ristic, Z. and Ashworth, E.N. (1993) Changes in leaf ultrastructure and carbohydrates in *Arabidopsis thaliana* L. (Heyn) cv. Columbia during rapid cold acclimation. *Protoplasma* 172, 111–123.

Sabehat, A., Lurie, S. and Weiss, D. (1998) Expression of small heat-shock proteins at low temperature: a possible role in protecting against chilling injuries. *Plant Physiology* 117, 651–658.

Sakai, A. and Larcher, W. (1987) *Frost Survival of Plant Responses and Adaptation to Freezing Stress.* Springer-Verlag, Berlin, Germany.

Sakai, A. and Yoshida, S. (1968) The role of sugar and related compounds in variations of freezing resistance. *Cryobiology* 5, 160–174.

Scheurer, S., Metzner, K., Haustein, D. and Vieths, S. (1997) Molecular cloning, expression and characterization of Pru a 1, the major cherry allergen. *Molecular Immunology* 34, 619–629.

Somssich, I.E., Schmelzer, E., Bollmann, J. and Hahlbrock, K. (1986) Rapid activation by fungal elicitor of genes encoding 'pathogenesis-related' proteins in cultured parsley cells. *Proceedings of the National Academy of Sciences USA* 83, 2427–2430.

Soto, A., Allona, I., Collada, C., Guevara, M.A., Casado, R., Rodriguez-Cerezo, E., Aragoncillo, C. and Gomez, L. (1999) Heterologous expression of a plant small heat-shock protein enhances *Escherichia coli* viability under heat and cold stress. *Plant Physiology* 120, 5213–5218.

Steponkus, P.L. (1984) Role of the plasma membrane in freezing injury and cold acclimation. *Annual Review of Plant Physiology* 35, 543–584.

Steponkus, P.L. and Lynch, D.V. (1989) Freeze/thaw-induced destabilization of the plasma membrane and the effects of cold acclimation. *Journal of Bioenergy and Biomembrane* 21, 21–41.

Steponkus, P.L., Uemura, M. and Webb, M.S. (1993) A contrast of the cryostability of the plasma membrane of winter rye and spring oat; two species that widely differ in their freezing tolerance and plasma membrane lipid composition. In: Steponkus, P.L. (ed.) *Advances in Low Temperature Biology*, Vol. 2, JAI Press, London, pp. 211–312.

Steponkus, P.L., Uemura, M., Joseph, R.A., Gilmour, S.J. and Thomashow, M.F. (1998) Mode action of the *COR15a* gene on the freezing tolerance of *Arabidopsis thaliana*. *Proceedings of the National Academy of Science USA* 95, 14570–14575.

Thomashow, M.F. (1994) *Arabidopsis thaliana* as a model for studying mechanisms of plant cold tolerance. In: Meyerowitz, E. and Sommerville, C. (eds) *Arabidopsis*. Cold Spring Harbor Laboratory Press, Cold Spring Harbor, New York, pp. 807–834.

Thomashow, M.F. (1999) Plant cold acclimation: freezing tolerance genes and regulatory mechanisms. *Annual Review of Plant Physiology and Plant Molecular Biology* 50, 571–599.

Uemura, M., Joseph, R.A. and Steponkus, P.L. (1995) Cold acclimation of *Arabidopsis thaliana*. Effect on plasma membrane lipid composition and freeze-induced lesions. *Plant Physiology* 109, 15–30.

Ukaji, N., Kuwabara, C., Takezawa, D., Arakawa, K., Yoshida, S. and Fujikawa, S. (1999) Accumulation of small heat-shock protein homologs in the endoplasmic reticulum of cortical parenchyma cells in mulberry in association with seasonal cold acclimation. *Plant Physiology* 120, 481–489.

Ukaji, N., Fujikawa, S. and Yoshida, S. (2000) Isolation and characterization of endoplasmic reticulum from mulberry cortical parenchyma cells. In: Dashek, W.V. (ed.) *Methods in Plant Electron Microscopy and Cytochemistry*, Humana Press, Totowa, New Jersey, pp. 147–160.

Ukaji, N., Kuwabara, C., Takezawa, D., Arakawa, K. and Fujikawa, S. (2001) Cold acclimation-induced WAP27 localized in endoplasmic reticulum in cortical parenchyma cells of mulberry tree was homologous to group 3 late-embryogenesis abundant proteins. *Plant Physiology* 126, 1588–1597.

Ukaji, N., Kuwabara, C., Takezawa, D., Arakawa, K. and Fujikawa S. (2004) Accumulation of pathogenesis-related (PR) 10/Bet v 1 protein homologues in mulberry (*Morus bombycis* Koidz.) tree during winter. *Plant, Cell and Environment* 27, 1112–1121.

van Berkel, J., Salamini, F. and Gebhardt, C. (1994) Transcripts accumulating during cold storage of potato (*Solanum tuberosum* L.) tubers are sequence related to stress-responsive genes. *Plant Physiology* 104, 445–452.

Vanek-Krebitz, M., Hoffmann-Sommergruber, K., da Camara Machado, M.L., Susami, M., Ebner, C., Kraft, D., Scheiner, O. and Breiteneder, H. (1995) Cloning and sequencing of Mal d 1, the major allergen from apple (*Malus domestica*), and its immunological relationship to Bet v 1, the major birch pollen allergen. *Biochemical and Biophysical Research Communications* 214, 538–551.

Walter, M.H., Liu, J.-W., Wünn, J. and Hess, D. (1996) Bean ribonuclease-like pathogenesis-related protein genes (Ypr10) display complex patterns of developmental, dark-induced and exogenous-stimulus-dependent expression. *European Journal of Biochemistry* 239, 281–293.

Waters, E.R., Lee, G.J. and Vierling, E. (1996) Evolution, structure and function of the small heat shock proteins in plants. *Journal of Experimental Botany* 47, 325–338.

Wisniewski, M., Webb, R., Balsamo, R., Close, T.J., Yu, X.-M. and Griffith, M. (1999) Purification, immunolocalization, cryoprotective, and antifreeze activity of PCA60: a dehydrin from peach (*Prunus persica*). *Physiologia Plantarum* 105, 600–608.

Yamada, T., Kuroda, K., Jitsuyama, Y., Takezawa, D., Arakawa, K. and Fujikawa, S. (2002) Roles of the plasma membrane and the cell wall in the responses of plant cells to freezing. *Planta* 215, 770–778.

Yoshida, S. (1984) Chemical and biophysical changes in the plasma membrane during cold acclimation of mulberry bark cells (*Morus bombycis* Koidz. Cv. Goroji). *Plant Physiology* 76, 257–265.

Yu, X.-M., Griffith, M. and Wiseman, S.B. (2001) Ethylene induces antifreeze activity in winter rye leaves. *Plant Physiology* 126, 1232–1240.

Zhou, B.-L., Arakawa, K., Fujikawa, S. and Yoshida, S. (1994) Cold-induced alterations in plasma membrane proteins that are specifically related to the development of freezing tolerance in cold-hardy winter wheat. *Plant and Cell Physiology* 35, 175–182.

14 The Role of Compatible Solutes in Plant Freezing Tolerance: a Case Study on Raffinose

D.K. Hincha,[1,*] E. Zuther,[1] M. Hundertmark[1] and A.G. Heyer[1,2]

[1]*Max-Planck-Institut für Molekulare Pflanzenphysiologie, D-14424 Potsdam, Germany;* [2]*Biologisches Institut, Abteilung Botanik, Universität Stuttgart, Pfaffenwaldring 57, D-70550 Stuttgart, Germany*

Introduction

Compatible solutes are synthesized by many organisms ranging from bacteria to animals and plants, in response to desiccation, osmotic stress, salt stress or low temperature. This chemically heterogeneous group of substances comprises some amino acids (e.g. proline), quaternary ammonium compounds (e.g. betaine), many sugars, sugar alcohols and several others (see Yancey *et al.*, 1982; Somero, 1992, for reviews). The functional role of sugars in cellular desiccation tolerance, however, has recently been challenged (Tunnacliffe and Lapinski, 2003), since it has been shown that some organisms are extremely desiccation tolerant without producing elevated amounts of sugars such as sucrose (Suc) or trehalose (Lapinski and Tunnacliffe, 2003). Nevertheless, there is a wealth of information in the literature on the stabilizing effects of various sugars on biological molecules, cells and organisms, which indicates a functional role of sugars in the stress tolerance of many, although not all, organisms. In this chapter, we review the possible role of raffinose (Raf) and raffinose family oligosaccharides (RFO) in plant stress tolerance, and in particular in plant freezing tolerance. It should, however, be noted that natural stress tolerance in any organism can never be explained by the action of just one compound and that in addition to sugar synthesis, many other physiological adaptations have to take place to allow an organism to survive extremes of stress such as freezing or desiccation (see Oliver *et al.*, 2001, 2002, for recent reviews).

*Corresponding author (e-mail: hincha@mpimp-golm.mpg.de; Fax: +49-331-567-8250)

 Cold Hardiness in Plants: Molecular Genetics, Cell Biology and Physiology (eds T.H.H. Chen *et al.*)

The Role of Raffinose in Cellular Desiccation Tolerance

The seed stage is a phase in the life cycle of many plants that is naturally desiccation tolerant. During seed maturation, RFO are accumulated concurrently with the reduction of tissue water content and the development of desiccation tolerance in many species (see Obendorf, 1997, for a review). RFO are derived from Suc and galactinol, which is synthesized by galactinol synthase (GS). It has, however, been shown in maize seeds that there are other regulatory steps in Raf accumulation, in addition to GS gene expression and enzyme activity (Zhao *et al.*, 2004a,b). One such step may be the synthesis of *myo*-inositol, one of the substrates of GS (Karner *et al.*, 2004). The addition of galactose to the glucose moiety of Suc through an α(1–6) bond is catalysed by raffinose synthase (RS) and leads to Raf. Further galactose molecules can be added to the terminal galactose in the oligosaccharide, again through α(1–6) bonds, to form the tetrasaccharide stachyose, the pentasaccharide verbascose, and further higher oligomers. An enzyme catalysing these chain elongations has been identified in pea seeds (Peterbauer *et al.*, 2002).

It has been proposed that RFO and Suc are involved in cytoplasmic vitrification in dry seeds (Sun and Leopold, 1997), thereby stabilizing sensitive macromolecular structures (see Crowe *et al.*, 1998, for a review). However, while the protective role of RFO for proteins and membranes *in vitro* is well established (see below), the quantitative contribution of RFO to the desiccation tolerance of seeds is not clear. RFO content is not related to intracellular glass stability in seeds (Buitink *et al.*, 2000a; Buitink and Leprince, 2004) and, in fact, seeds may show low desiccation tolerance even in the presence of large amounts of sugars (Hoekstra *et al.*, 2001). Also, quantitative genetic studies, quantitative trait loci (QTL) mapping, in *Arabidopsis thaliana* did not reveal a significant coincidence of QTL for seed storage stability and Suc or RFO content (Bentsink *et al.*, 2000; Clerkx *et al.*, 2004). This does not mean that sugars in general, or RFO in particular, are irrelevant for seed desiccation tolerance, but only that the differences in desiccation tolerance between the investigated genotypes are not strongly related to their sugar content and that other factors are more important in these cases. However, a strong correlation between RFO content and desiccation tolerance was found in soybean seeds under different experimental treatments (Blackman *et al.*, 1992). Therefore, the exact role and quantitative contribution of RFO to seed desiccation tolerance under different storage conditions are still open questions.

The Effects of Raffinose on the Stability of Proteins and Membranes During Drying

As mentioned above, cytoplasmic vitrification is considered to be an important mechanism in cellular desiccation tolerance. Many sugars form glasses (vitrify) during drying at ambient temperatures (see Crowe *et al.*, 1998; Buitink and Leprince, 2004, for reviews). The melting temperature of a dry sugar glass (glass transition temperature, T_g) is a convenient measure of glass stability, although some recent work indicates that it may not be the only important parameter when the stabilizing effects of a carbohydrate glass are considered (Buitink *et al.*, 2000b; Hincha and Hagemann, 2004). The collapse temperature or critical temperature (T_c) can be significantly above T_g and may be more closely related to the stability of biological structures in the dry state than T_g. T_c, however, is more difficult to measure and therefore the calorimetrically determined T_g is more commonly used to compare the vitrification properties of biological glass formers such as sugars. In general, T_g increases with the molecular weight of the solute (Levine and Slade, 1988). This is also true for RFO, where T_g increases from Suc (69°C) to Raf (106°C), and stachyose (123°C) (Buitink *et al.*, 2000a). In addition, it has been suggested that mixtures of Suc and Raf are particularly potent glass formers (Kajiwara *et al.*, 1999), because Raf would effectively prevent Suc from crystallizing during drying (Caffrey *et al.*, 1988) or long-term storage in the dry state (Saleki-Gerhardt and Zografi, 1994).

The activity of isolated enzymes can be preserved during drying in the presence of sugars (Crowe *et al.*, 1990). As a mechanistic explanation of this effect it has been proposed that the OH groups of sugars can H-bond to proteins in the dry state, thereby replacing water molecules and stabilizing the three-dimensional structure of proteins (Carpenter and Crowe, 1989). In addition, vitrification is also necessary to stabilize proteins in the dry state (Crowe *et al.*, 1998; Buitink and Leprince, 2004). A positive effect of mixtures between Suc and Raf on protein stability, as suggested from the glass transition behaviour, however, was not found (Davidson and Sun, 2001).

A stabilizing effect of Raf on biological membranes during drying has been reported for microsomes from lobster muscle (Crowe *et al.*, 1984) and for protoplasts from desiccation-tolerant pea seed embryos (Xiao and Koster, 2001). In the latter case, however, pure Raf was not effective, but mixtures of Raf and Suc provided slightly better protection than Suc alone.

In liposomes, the superior glass-forming ability of longer-chain RFO results in reduced liposome fusion in the dry state, especially at elevated temperatures (Hincha *et al.*, 2003). This, however, is only evident for stachyose and verbascose in comparison to Suc, but not for Raf. Since it has been suggested that mixtures of Suc and Raf may be particularly potent protectants for liposomes during drying (Caffrey *et al.*, 1988), and that a specific ratio of Suc to Raf is necessary for optimum stability of dry seeds, we tested the effectiveness of such mixtures in the preservation of liposomes. Figure 14.1 shows results from air-drying experiments with phosphatidylcholine liposomes that encapsulate the fluorescent dye carboxyfluorescein (CF) as a marker for membrane damage during drying and rehydration. As shown before, both Suc and Raf protect liposomes from loss of internal solutes, but there is no clear difference in the effects of the two sugars. The same is also true for the two different mixtures of the sugars. This would suggest that, similar to the case of soluble proteins, such mixtures of Suc and Raf have no superior protective properties, when compared to the pure compounds. If such mixtures of Suc and Raf are indeed required for optimum stability of dry seeds, as suggested by the experiments with protoplasts described above (Xiao and Koster, 2001), we would hypothesize that the effect is not due to the stabilization of proteins or membranes, but rather involves other, so far unidentified, cellular structures.

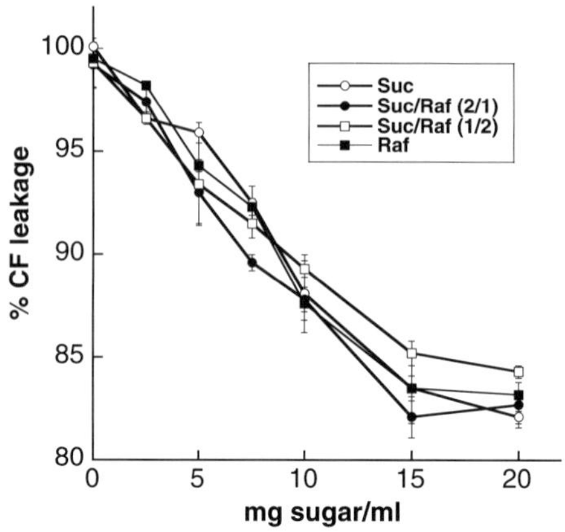

Fig. 14.1. Stability of liposomes during drying and rehydration in the presence of sucrose (Suc), raffinose (Raf), or mixtures of the two sugars. Liposomes encapsulating the soluble fluorescent dye carboxyfluorescein (CF) were made from egg phosphatidylcholine. They were air-dried in the presence of the indicated amounts of sugars. After rehydration, leakage of CF from the liposomes was determined as a measure of membrane damage.

In addition to glass formation and the concomitant effects on membrane fusion, RFO also interact with membranes in the dry state by replacing water molecules in the hydration shell of the lipid-head groups, thereby preventing deleterious lipid phase transitions (Crowe *et al.*, 1996; Hincha *et al.*, 2003). In this case, however, the effect slightly decreases with increasing chain length of the oligosaccharides. This is due to a reduced ability of the longer-chain RFO to H-bond to the phosphate moiety of the phospholipid-head group, as shown by Fourier-transform infrared spectroscopy (Hincha *et al.*, 2003).

In conclusion, the available data on the effects of Raf on dry proteins and membranes suggest that Raf is a potent stabilizer of these biological structures, but that it is not superior to Suc. Also, in model system experiments, Raf does not seem to have any special properties in mixtures with Suc, which would explain why Raf should be necessary in addition to Suc for the stability of cells in the dry state.

The Effects of Raffinose on the Stability of Membranes during Freezing *In Vitro*

Unlike the situation during drying, there is surprisingly little information on the effects of Raf or other RFO on membrane stability during freezing. To the best of our knowledge, there are no published data available on the effects of RFO on model membranes during freezing. There are, however, some data that were obtained with chloroplast thylakoid membranes (Lineberger and Steponkus, 1980; Hincha, 1990). This may actually have some relevance to the *in vivo* situation, since it has been shown that Raf may be localized in the chloroplasts after plant cold acclimation (Santarius and Milde, 1977).

When thylakoids are frozen to −20°C and damage after thawing is assessed as the loss of the soluble electron-transport protein, plastocyanin (see Hincha *et al.*, 1996, for a review), it can be clearly seen that Raf is a superior cryoprotectant, when compared to Suc (Fig. 14.2; compare also Hincha, 1990). A similar result was also obtained when functional assays were used to determine freezing damage to thylakoids (Lineberger and Steponkus, 1980). A mechanistic explanation of this difference between Suc and Raf will have to come from detailed physico-chemical studies using defined model membranes. These results are, nevertheless, in accord with a role of Raf in plant freezing tolerance.

Using Natural Genetic Diversity to Create Variability in Leaf Raffinose Content

In many plant species, RFO, together with other compatible solutes, are accumulated during cold acclimation, when plants acquire increased freezing tolerance

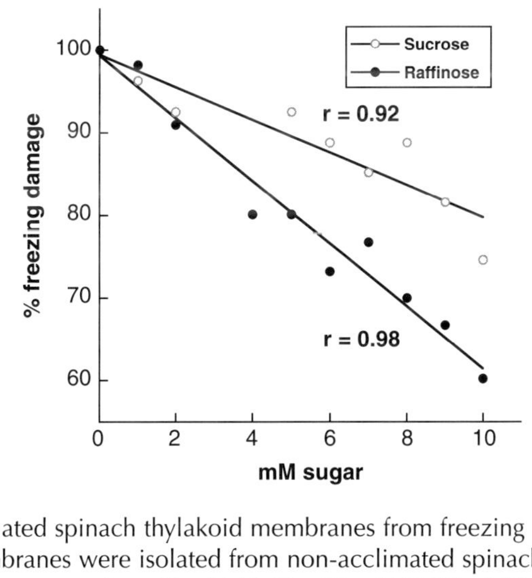

Fig. 14.2. Protection of isolated spinach thylakoid membranes from freezing damage by sucrose and raffinose. Membranes were isolated from non-acclimated spinach plants and were suspended in an artificial stroma medium (5 mM $MgCl_2$, 10 mM K_2SO_4, 150 mM K-glutamate, 50 mM sucrose; compare Hincha and Schmitt, 1988). The samples were frozen for 3 h at −20°C and rapidly thawed in a water bath at room temperature. The membranes were then removed from the samples by centrifugation and the amount of plastocyanin in the supernatants was determined immunologically (Hincha *et al.*, 1985). Plastocyanin release in the absence of additional sugar was taken as 100% freezing damage. The lines were fitted to the data by linear regression analysis and the correlation coefficients (r) are shown in the figure. (Figure redrawn from data presented in Hincha, 1990.)

(Koster and Lynch, 1992; Bachmann *et al.*, 1994; Castonguay *et al.*, 1995; Castonguay and Nadeau, 1998; Gilmour *et al.*, 2000). In leaves, RFO are synthesized from Suc by subsequent addition of activated galactose moieties donated by galactinol (Peterbauer and Richter, 2001). The synthesis of galactinol from *myo*-inositol and UDP-galactose is considered a key regulatory step in RFO synthesis (Keller and Pharr, 1996), and enzyme activity of GS is increased at low temperatures (Castonguay and Nadeau, 1998; Sprenger and Keller, 2000). In *Arabidopsis*, this increase is correlated with increased transcript abundance (Liu *et al.*, 1998), which results from cold induction of one of the seven GS genes (Taji *et al.*, 2002). This induction is mediated by CBF/DREB1 transcription factors, which control a complex set of plant responses to low temperature (Fowler and Thomashow, 2002; Maruyama *et al.*, 2004). Overexpression of CBF3/DREB1A causes increased frost tolerance and the accumulation of Raf in transgenic *Arabidopsis* plants, indicating physiological relevance of the GS induction (Gilmour *et al.*, 2000). Raf is synthesized by RS from Suc and galactinol (Lehle and Tanner, 1973; Zuther *et al.*, 2004). In some plants, longer-chain RFO are synthesized either by a galactan:galactan galactosyltransferase (GGT) (Tapernoux-Lüthi *et al.*, 2004), or by stachyose synthase (Peterbauer *et al.*, 2002). Both RS (Zuther *et al.*, 2004) and GGT (Tapernoux-Lüthi *et al.*, 2004) are cold inducible in leaves.

There is evidence for sufficient genetic variability among accessions of *A. thaliana* to allow investigation of genotype × environment interactions (see Alonso-Blanco and Koorneef, 2000, for a review). The laboratory strains C24 and Columbia (Col) have been shown, by several methods, to be genetically distinct accessions (Barth *et al.*, 2002; Törjek *et al.*, 2003). Fig. 14.3 shows that Col is more freezing tolerant than C24, when the accessions are compared under identical experimental conditions. Both in the non-acclimated and in the cold-acclimated state, leaves from Col plants are significantly more freezing tolerant than leaves from C24 plants (compare also Klotke *et al.*, 2004; Rohde *et al.*, 2004). Also, the cold-acclimation capacity, i.e. the ability to increase freezing tolerance under acclimating conditions, is higher for Col (3.5°C) than for C24 (1.9°C).

In addition, crosses of *Arabidopsis* accessions can be used to create additional genetic and phenotypic variation through heterosis effects (e.g. Griffing and Scholl, 1991; Narang and Altmann, 2001; Barth *et al.*, 2003; Meyer *et al.*, 2004). The term heterosis was introduced into the literature in 1914 (Shull, 1914) to describe the phenomenon of increased physiological performance of many crosses in comparison to their parents. Such crosses have been used extensively by breeders to increase the performance of crop plants. Our understanding of heterosis on a molecular level is, nevertheless, still rudimentary (see Birchler *et al.*, 2003, for a recent review). Heterosis has been studied mainly for traits such as biomass accumulation or grain yield, but it has also been described for plant freezing tolerance in winter wheat (Sutka, 1981; Parodi *et al.*, 1983). It can be distinguished between mid-parent heterosis, i.e. the offspring being superior to the mean of the parents, and best-parent heterosis, i.e. the offspring being superior to the better parent. Obviously, for plant-breeding purposes, only best-parent heterosis is of interest. For genetic studies, however, the deviation of the offspring from the parental mean is more relevant (Griffing, 1990).

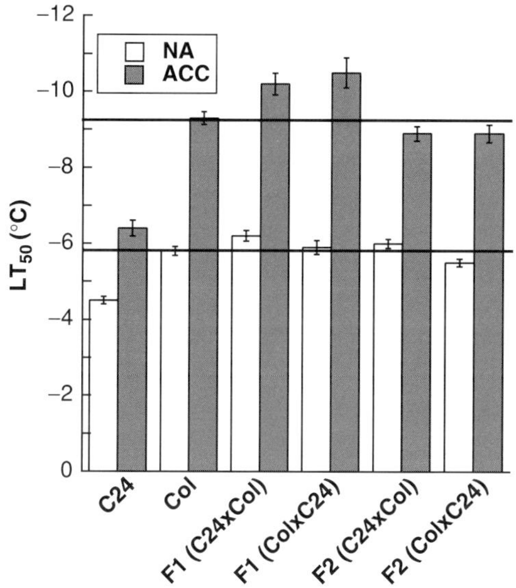

Fig. 14.3. Freezing tolerance of *Arabidopsis thaliana* leaves determined as the temperature that causes 50% electrolyte leakage after a freeze–thaw cycle (LT_{50}). Leaves were frozen at a rate of 2°C/h to different temperatures. After thawing, electrolyte leakage was determined in the bathing solution before and after boiling the samples. The figure shows a comparison of the freezing tolerance of the *A. thaliana* accessions C24 and Columbia-0 (Col) with the F1 and F2 generations of their reciprocal crosses. The horizontal lines indicate the non-acclimated (NA) and acclimated (ACC) freezing tolerance of the better parent, Col. The SEM is indicated for each bar. (Figure redrawn from data published in Rohde *et al.*, 2004.)

In an almost exclusively selfing species like *A. thaliana*, accessions can be expected to be largely homozygous. This can lead to inbreeding depression. Conversely, crossing such accessions would lead to increased heterozygosity and heterosis. To investigate possible heterosis in *Arabidopsis* freezing tolerance, we performed reciprocal crosses between the accessions C24 and Col. The resulting F1 plants were allowed to self-fertilize to create two reciprocal F2 populations. Figure 14.3 summarizes the results of freezing experiments with the different genotypes, both before and after cold acclimation. The two horizontal lines indicate the LT_{50} values of the better parent (Col) in the non-acclimated and cold-acclimated state.

Mid-parent heterosis occurs in the F1 populations (Fig. 14.3), both in the non-acclimated and cold-acclimated state. These effects are reduced in the F2 populations. Significant best-parent heterosis is observed in the F1 populations after cold acclimation. In addition to the effects on acclimated and non-acclimated freezing tolerance *per se*, heterosis is also evident in a cold-acclimation capacity (i.e. the difference between these values), which is 4°C and 4.6°C for the reciprocal F1 populations, and 2.9°C and 3.4°C for the respective F2 populations. The mean acclimation capacity of the parents is 2.5°C and the acclimation capacity of the best parent is 3.5°C. The fact that the heterosis observed in the F1 generation is reduced in the F2 generation is in accordance with the classical definition of heterosis (Shull, 1914). There are no significant differences in freezing tolerance between reciprocal crosses, indicating that maternal effects play no strong role in leaf freezing tolerance as assayed by the electrolyte-leakage method.

During plant cold acclimation, the content of compatible solutes in leaf cells increases (see Xin and Browse, 2000; Smallwood and Bowles, 2002, for reviews). We have, therefore, measured the amounts of soluble sugars (Fru, Glc, Suc, Raf) in leaf samples from all genotypes (C24, Col, F1, F2), both before and after cold acclimation. The data show clearly that the content of all sugars increases dramatically in all genotypes during cold acclimation (Rohde *et al.*, 2004). A similar behaviour has been observed in lucerne (Castonguay *et al.*, 1995) and in saltgrass (Shahba *et al.*, 2003).

Although the parental accessions C24 and Col show significant differences in their freezing tolerance, of the measured sugars only the Raf content of Col is higher than that of C24, both before and after cold acclimation (Klotke *et al.*, 2004; Rohde *et al.*, 2004). The content of all sugars is higher in the leaves of F1 plants than in either parent in the cold-acclimated state. Only the level of Raf follows a pattern similar to changes in leaf freezing tolerance of the F2, i.e. it is significantly higher than the mid-parent value, but below the level of the better parent (Col) (Rohde *et al.*, 2004).

The exceptional behaviour of Raf is also apparent in an analysis of correlations between Raf content and leaf freezing tolerance (Fig. 14.4). The analysis shows strong correlations (r = 0.97 and 0.96) for acclimated and non-acclimated

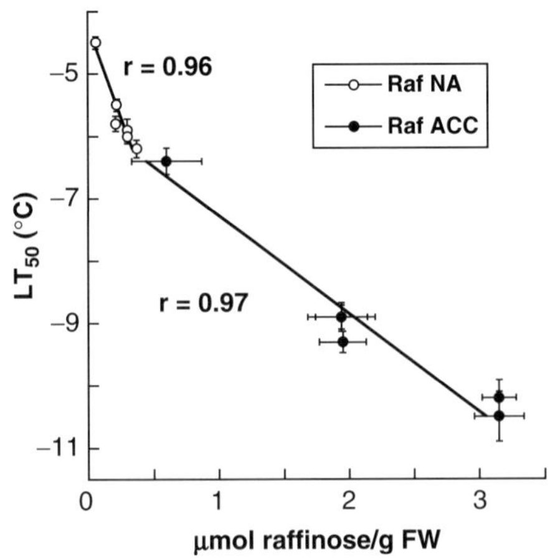

Fig. 14.4. Correlation between the raffinose content of *Arabidopsis* leaves and leaf freezing tolerance expressed as the LT_{50} values (compare Figure 14.3). Linear regression analysis was performed separately for data from non-acclimated (NA) and acclimated (ACC) plants. The correlation coefficients (r) are shown next to the curve. (Figure redrawn from data presented in Rohde *et al.*, 2004.)

plants. For all other sugars, the correlation coefficients are much smaller (Rohde *et al.*, 2004). Similar correlations of freezing tolerance and Raf content have also been shown for different cultivars of lucerne and saltgrass (Castonguay *et al.*, 1995; Shahba *et al.*, 2003).

Transgenic Approaches to Influence the Raffinose Content of Leaves

Different approaches have been used to manipulate the Raf content of plant leaves. Increased Raf content was obtained by constitutive overexpression of a GS gene in *Arabidopsis* (Taji *et al.*, 2002; Zuther *et al.*, 2004) or by antisense inhibition of the α-galactosidase gene in petunia plants (Pennycooke *et al.*, 2003). A reduction of the leaf Raf content was achieved in petunia by overexpression of an α-galactosidase gene (Pennycooke *et al.*, 2003), and a complete loss of Raf synthesis in *Arabidopsis* leaves, even under cold-acclimating conditions, was found in a RS knockout mutant (Zuther *et al.*, 2004).

Leaf Raf content was indirectly influenced in CBF3/DREB1A-overexpressing plants, in which multiple physiological processes associated with cold acclimation are already operating under non-acclimating conditions. In these transgenic plants, Raf levels are elevated about fourfold and further increase during cold acclimation to a level sevenfold higher than in wild-type (WT) plants (Gilmour *et al.*, 2000). Plants constitutively expressing CBF transcription factors show enhanced freezing tolerance including a higher acclimation capacity. This, however, also correlates with transcriptional induction of pyrroline-5-carboxylate synthase, a key enzyme in the proline biosynthesis pathway, and an induction of a large array of other genes (Fowler and Thomashow, 2002; Seki *et al.*, 2002; Maruyama *et al.*, 2004). Therefore, from these plants no conclusions about the specific contribution of Raf to freezing tolerance can be drawn.

To create *Arabidopsis* lines with specifically elevated Raf content, we cloned a GS cDNA from cucumber (*Cucumis sativus* L.) and expressed it under the control of the constitutive CMV 35S promoter (Zuther *et al.*, 2004). The *A. thaliana* plants used for transformation were from the accessions Columbia-0 (Col) and Cape Verde Islands (Cvi). While the Col WT plants acclimate by about 3.5–4°C (see above), Cvi only acclimates by 1.9°C, very similar to the accession C24, which we investigated previously (Klotke *et al.*, 2004; Rohde *et al.*, 2004).

The GS-overexpressing lines in the Col background contain between 2 and 3.5 times as much Raf as the WT under non-acclimating conditions and 1.6 times more after cold acclimation. Galactinol is significantly elevated in some lines under non-acclimating conditions, but not after cold acclimation, presumably due to an increased expression of RS during acclimation, which converts galactinol into Raf. The relative increases in Raf content are even higher in transformants in the Cvi background. Raf content is 20-fold higher than in the WT under non-acclimating conditions, and 2.3-fold higher in acclimated plants. The absolute values, however, are always much lower in the Cvi plants than in the Col plants (Fig. 14.5; Zuther *et al.*, 2004).

Surprisingly, although the WT and transgenic plants differ widely in their Raf content, in both the non-acclimated and the acclimated state, no significant

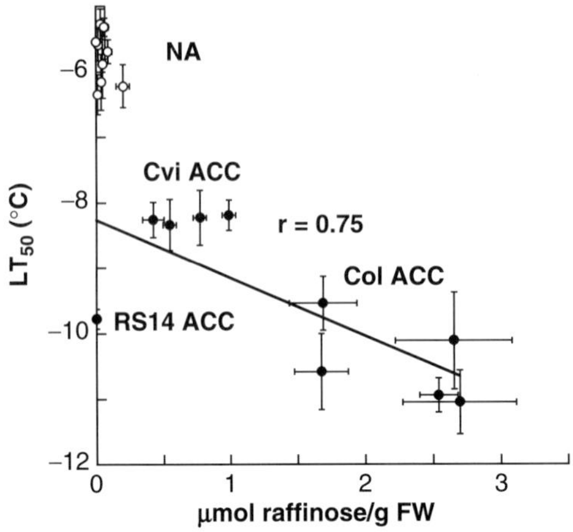

Fig. 14.5. Correlation between the raffinose content of *Arabidopsis* leaves and leaf freezing tolerance expressed as the LT_{50} values. Linear regression analysis was performed only for data from acclimated (ACC) plants, and not for non-acclimated (NA). The correlation coefficient (r) is shown next to the curve. These samples were taken from plants of the accession Col-0, comprising wild type and plants overexpressing the galactinol synthase gene from cucumber (Col ACC) and a raffinose synthase knockout mutant (RS14 ACC) carrying a T-DNA insertion in the RS gene. In addition, plants from the accession Cvi were included in the analysis, which also comprised wild type and plants overexpressing the galactinol synthase gene from cucumber (Cvi ACC). The SEM is indicated for both LT_{50} and raffinose content. (Figure redrawn from data presented in Zuther *et al.*, 2004.)

differences in freezing tolerance are detectable (Fig. 14.5; Zuther *et al.*, 2004). In contrast, increased drought tolerance has been demonstrated for *Arabidopsis* plants overexpressing an endogenous GS gene (Taji *et al.*, 2002). This could indicate that Raf is only important for drought, but not for freezing tolerance of plants. In transgenic petunia, however, the accumulation of Raf is associated with increased freezing tolerance of the leaves (Pennycooke *et al.*, 2003). Increased Raf and stachyose accumulation was induced by antisense inhibition of the expression of the α-galactosidase gene. These data could be interpreted in different ways. Either freezing tolerance in petunia and *Arabidopsis* is affected differently by Raf; or the stachyose that is also accumulated in petunia, but not in *Arabidopsis*, is the active agent in petunia freezing tolerance; or α-galactosidase has additional effects on freezing tolerance that are unrelated to effects on RFO accumulation. The last assumption seems reasonable, since it has been shown that α-galactosidases have a variety of substrates besides RFO (see Peterbauer and Richter, 2001, for a review). A conclusive answer, however, could only be provided by manipulating the α-galactosidase gene expression in *Arabidopsis* and/or overexpressing the GS gene in petunia.

Since a moderate increase in Raf content after overexpression of the GS gene in transgenic *Arabidopsis* does not result in increased freezing tolerance of the leaf tissue, it was of interest to determine whether Raf makes a significant contribution to freezing tolerance and cold acclimation at all. For this purpose, we used a knockout mutant of the RS gene of *A. thaliana* that is not able to accumulate Raf (Zuther *et al.*, 2004). The RS mutant RS14 in the Col background was obtained from the GABI-Kat collection and carries a 5.8 kb T-DNA insert in the RS gene. Homozygous mutants were identified by polymerase chain reaction (PCR). To establish that line RS14 really carries a knockout mutation in the RS gene, RS transcript levels were determined in mutant and WT plants. The RS gene is cold induced in the WT Col plants, but the transcript is not detectable in the mutant RS14 plants, under either non-acclimating or acclimating conditions (Zuther *et al.*, 2004).

Raf is completely absent in the mutant, both before and after cold acclimation for 14 days at 4°C. Nevertheless, neither the basic freezing tolerance of non-acclimated leaves nor their ability to cold acclimate is influenced in the RS mutant (Fig. 14.5), demonstrating that Raf accumulation is dispensable for both components of freezing tolerance (Zuther *et al.*, 2004). In the case of non-acclimated plants, this may be expected, because of the low concentration of Raf. The lack of an effect of the RS mutation on the ability of the plants to cold acclimate, however, is astonishing.

It therefore seemed possible that the RS mutant would compensate for its lack of Raf by increasing levels of other sugars or proline. Measurements showed that this is not the case for proline, or sugars other than Raf, where Suc, Glc and Fru show a 2- to 3.5-fold reduction in the acclimated RS14 (Zuther *et al.*, 2004). The galactinol content, on the other hand, is 8.7-fold higher in RS14 than in the WT plants after cold acclimation, while it increases 4.8-fold in non-acclimated plants. Reports on a possible role of galactosyl cyclitols in plant abiotic stress tolerance are rare. They are accumulated, sometimes together with RFO, in the seeds of many plant species (Peterbauer and Richter, 2001), but their functional significance has not been investigated. Other cyclitols, such as ononitol, pinitol and quercitol, however, show a degree of protection for isolated thylakoid membranes during freezing that is similar to that provided by sucrose (Orthen and Popp, 2000).

Again, these data on the RS knockout mutant, like the data on the GS overexpressors, are contrary to the report on the manipulation of the expression and activity of α-galactosidase (Pennycooke *et al.*, 2003). Overexpression of this gene resulted in reduced Raf and stachyose content and reduced freezing tolerance of the leaves. We think that this is additional evidence that the effects reported in the study by Pennycooke *et al.* are not a direct result of changes in Raf content, but result from other effects of the α-galactosidase, that remain to be analysed.

Conclusions and Perspectives

Taken together, our data demonstrate that Raf is not an essential component of basic freezing tolerance and cold acclimation in *A. thaliana*. Since overexpression of GS did not lead to the large increase in Raf concentration observed in CBF3/DREB1A-overexpressing plants, it will be interesting to study whether upregulating RS gene expression could further increase Raf levels and whether this would improve acclimation capacity. In addition, it will be interesting to determine whether galactinol can substitute for Raf during cold acclimation, or whether other, so far unidentified, solutes are accumulated in the mutant.

More generally, most of the results on plant freezing tolerance obtained over the last years emphasize the fact that freezing tolerance is a multigenic trait that depends on the coordinate action of many different cellular components. This point is also illustrated in Fig. 14.5. While there is no correlation between Raf content and LT_{50} within the Cvi or Col lines, there is a clear correlation when lines from both accessions in the acclimated state are included in the analysis. This may indicate that Raf, in conjunction with other factors that are different between the accessions, can be one determinant of freezing tolerance. It will therefore be necessary to employ appropriate multiparallel profiling technologies to unravel the complex physiological adaptations that take place during cold acclimation on the gene expression, protein and metabolite levels. It will then be necessary to take these results back to a critical evaluation of the quantitative contributions of the single components, through, e.g. the use of carefully targeted knockout mutations.

References

Alonso-Blanco, C. and Koorneef, M. (2000) Naturally occurring variation in *Arabidopsis*: an underexploited resource for plant genetics. *Trends in Plant Science* 5, 22–29.

Bachmann, M., Matile, P. and Keller, F. (1994) Metabolism of the raffinose family oligosaccharides in leaves of *Ajuga reptans* L. *Plant Physiology* 105, 1335–1345.

Barth, S., Melchinger, A.E. and Lübberstedt, T. (2002) Genetic diversity in *Arabidopsis thaliana* L. Heynh. investigated by cleaved amplified polymorphic sequences (CAPS) and inter-simple sequence repeat (ISSR) markers. *Molecular Ecology* 11, 495–505.

Barth, S., Busimi, A.K., Utz, H.F. and Melchinger, A.E. (2003) Heterosis for biomass yield and related traits in five hybrids of *Arabidopsis thaliana* L. Heynh. *Heredity* 91, 36–42.

Bentsink, L., Alonso-Blanco, C., Vreugdenhil, D., Tesnier, K., Groot, S.P.C. and Koornneef, M. (2000) Genetic analysis of seed-soluble oligosaccharides in relation to seed storability of *Arabidopsis*. *Plant Physiology* 124, 1595–1604.

Birchler, J.A., Auger, D.L. and Riddle, N.C. (2003) In search of the molecular basis of heterosis. *Plant Cell* 15, 2236–2239.

Blackman, S.A., Obendorf, R.L. and Leopold, A.C. (1992) Maturation proteins and sugars in desiccation tolerance of developing soybean seeds. *Plant Physiology* 100, 225–230.

Buitink, J. and Leprince, O. (2004) Glass formation in plant anhydrobiotes: survival in the dry state. *Cryobiology* 48, 215–228.

Buitink, J., Hemminga, M.A. and Hoekstra, F.A. (2000a) Is there a role for oligosaccharides in seed longevity? An assessment of intracellular glass stability. *Plant Physiology* 122, 1217–1224.

Buitink, J., van den Dries, I.J., Hoekstra, F.A., Alberda, M. and Hemminga, M.A. (2000b) High critical temperature above T_g may contribute to the stability of biological systems. *Biophysical Journal* 79, 1119–1128.

Caffrey, M., Fonseca, V. and Leopold, A.C. (1988) Lipid-sugar interactions. *Plant Physiology* 86, 754–758.

Carpenter, J.F. and Crowe, J.H. (1989) An infrared spectroscopic study of the interactions of carbohydrates with dried proteins. *Biochemistry* 28, 3916–3922.

Castonguay, Y. and Nadeau, P. (1998) Enzymatic control of soluble carbohydrate accumulation on cold-acclimated crowns of alfalfa. *Crop Science* 38, 1183–1189.

Castonguay, Y., Nadeau, P., Lechasseur, P. and Chouinard, L. (1995) Differential accumulation of carbohydrates in alfalfa cultivars of contrasting winterhardiness. *Crop Science* 35, 509–516.

Clerkx, E.J.M., El-Lithy, M.E., Vierling, E., Ruys, G.J., Blankensteijn-De Vries, H., Groot, S.P.C., Vreugdenhil, D. and Koornneef, M. (2004) Analysis of natural allelic variation of *Arabidopsis* seed germination and seed longevity traits between the accessions Landsberg *erecta* and Shakdara, using a new recombinant inbred line population. *Plant Physiology* 135, 432–443.

Crowe, J.H., Carpenter, J.F., Crowe, L.M. and Anchordoguy, T.J. (1990) Are freezing and dehydration similar stress vectors? A comparison of modes of interaction of stabilizing solutes with biomolecules. *Cryobiology* 27, 219–231.

Crowe, J.H., Hoekstra, F.A., Nguyen, K.H.N. and Crowe, L.M. (1996) Is vitrification involved in depression of the phase transition temperature in dry phospholipids? *Biochimica et Biophysica Acta* 1280, 187–196.

Crowe, J.H., Carpenter, J.F. and Crowe, L.M. (1998) The role of vitrification in anhydrobiosis. *Annual Reviews in Physiology* 60, 73–103.

Crowe, L.M., Mouradian, R., Crowe, J.H., Jackson, S.A. and Womersley, C. (1984) Effects of carbohydrates on membrane stability at low water activities. *Biochimica et Biophysica Acta* 769, 141–150.

Davidson, P. and Sun, W.Q. (2001) Effect of sucrose/raffinose mass ratios on the stability of co-lyophilized protein during storage above T_g. *Pharmaceutical Research* 18, 474–479.

Fowler, S. and Thomashow, M.F. (2002) *Arabidopsis* transcriptome profiling indicates that multiple regulatory pathways are activated during cold acclimation in addition to the CBF cold response pathway. *Plant Cell* 14, 1675–1690.

Gilmour, S.J., Sebolt, A.M., Salazar, M.P., Everard, J.D. and Thomashow, M.F. (2000) Overexpression of the *Arabidopsis CBF3* transcriptional activator mimics multiple biochemical changes associated with cold acclimation. *Plant Physiology* 124, 1854–1865.

Griffing, B. (1990) Use of a controlled-nutrient experiment to test heterosis hypotheses. *Genetics* 126, 753–767.

Griffing, B. and Scholl, R.L. (1991) Qualitative and quantitative genetic studies of *Arabidopsis thaliana*. *Genetics* 129, 605–609.

Hincha, D.K. (1990) Differential effects of galactose containing saccharides on mechanical freeze–thaw damage to isolated thylakoid membranes. *CryoLetters* 11, 437–444.

Hincha, D.K. and Hagemann, M. (2004) Stabilization of model membranes during drying by compatible solutes involved in the stress tolerance of plants and microorganisms. *Biochemical Journal* 383, 277–283.

Hincha, D.K. and Schmitt, J.M. (1988) Mechanical freeze–thaw damage and frost hardening in leaves and isolated thylakoids from spinach. I. Mechanical freeze–thaw damage in an artificial stroma medium. *Plant Cell and Environment* 11, 41–46.

Hincha, D.K., Heber, U. and Schmitt, J.M. (1985) Antibodies against individual thylakoid membrane proteins as molecular probes to study chemical and mechanical freezing damage *in vitro*. *Biochimica et Biophysica Acta* 809, 337–344.

Hincha, D.K., Sieg, F., Bakaltcheva, I., Köth, H. and Schmitt, J.M. (1996) Freeze–thaw damage to thylakoid membranes: specific protection by sugars and proteins. In: Steponkus, P.L. (ed.) *Advances in Low-Temperature Biology*, Vol. 3, JAI Press, London, pp. 141–183.

Hincha, D.K., Zuther, E. and Heyer, A.G. (2003) The preservation of liposomes by raffinose family oligosaccharides during drying is mediated by effects on fusion and lipid phase transitions. *Biochimica et Biophysica Acta* 1612, 172–177.

Hoekstra, F.A., Golovina, E.A., Tetteroo, F.A.A. and Wolkers, W.F. (2001) Induction of desiccation tolerance in plant somatic embryos:

how exclusive is the role of sugars? *Cryobiology* 43, 140–150.

Kajiwara, K., Franks, F., Echlin, P. and Greer, A.L. (1999) Structural and dynamic properties of crystalline and amorphous phases in raffinose-water mixtures. *Pharmaceutical Research* 16, 1441–1448.

Karner, U., Peterbauer, T., Raboy, V., Jones, D.A., Hedley, C.L. and Richter, A. (2004) *myo*-inositol and sucrose concentrations affect the accumulation of raffinose family oligosaccharides in seeds. *Journal of Experimental Botany* 55, 1981–1987.

Keller, F. and Pharr, D.M. (1996) Metabolism of carbohydrates in sinks and sources: galactosyl-sucrose oligosaccharides. In: Zamski, E. and Schaffer A.A. (eds) *Photoassimilate Distribution in Plants and Crops.* Marcel Dekker, New York, pp. 115–184.

Klotke, J., Kopka, J., Gatzke, N. and Heyer, A.G. (2004) Impact of soluble sugar concentrations on the acquisition of freezing tolerance in accessions of *Arabidopsis thaliana* with contrasting cold acclimation – evidence for a role of raffinose in cold acclimation. *Plant Cell and Environment*, 27,1395–1404.

Koster, K.L. and Lynch, D.V. (1992) Solute accumulation and compartmentation during the cold acclimation of Puma rye. *Plant Physiology* 98, 108–113.

Lapinski, J. and Tunnacliffe, A. (2003) Anhydrobiosis without trehalose in bdelloid rotifers. *FEBS Letters* 553, 387–390.

Lehle, L. and Tanner, W. (1973) The function of *myo*-inositol in the biosynthesis of raffinose. *European Journal of Biochemistry* 38, 103–110.

Levine, H. and Slade, L. (1988) Principles of "cryostabilization" technology from structure/property relationships of carbohydrate/water systems – a review. *CryoLetters* 9, 21–63.

Lineberger, R.D. and Steponkus, P.L. (1980) Cryoprotection by glucose, sucrose, and raffinose to chloroplast thylakoids. *Plant Physiology* 65, 298–304.

Liu, J.-J.J., Krenz, D.C., Galvez, A.F. and de Lumen, B.O. (1998) Galactinol synthase (GS): increased enzyme activity and levels of mRNA due to cold and desiccation. *Plant Science* 134, 11–20.

Maruyama, K., Sakuma, Y., Kasuga, M., Ito, Y., Seki, M., Goda, H., Shimada, Y., Yoshida, S., Shinozaki, K. and Yamaguchi-Shinozaki, K. (2004) Identification of cold-inducible downstream genes of the *Arabidopsis* DREB1A/CBF3 transcriptional factor using two microarray systems. *Plant Journal* 38, 982–993.

Meyer, R.C., Törjek, O., Becher, M. and Altmann, T. (2004) Heterosis of biomass production in *Arabidopsis.* Establishment during early development. *Plant Physiology* 134, 1813–1823.

Narang, R.A. and Altmann, T. (2001) Phosphate acquisition heterosis in *Arabidopsis thaliana*: a morphological and physiological analysis. *Plant Soil* 134, 91–97.

Obendorf, R.L. (1997) Oligosaccharides and galactosyl cyclitols in seed desiccation tolerance. *Seed Science Research* 7, 63–74.

Oliver, A.E., Leprince, O., Wolkers, W.F., Hincha, D.K., Heyer, A.G. and Crowe, J.H. (2001) Non-disaccharide-based mechanisms of protection during drying. *Cryobiology* 43, 151–167.

Oliver, A.E., Hincha, D.K. and Crowe, J.H. (2002) Looking beyond sugars: the role of amphiphilic solutes in preventing adventitious reactions in anhydrobiotes at low water contents. *Comparative Biochemistry and Physiology* 131A, 515–525.

Orthen, B. and Popp, M. (2000) Cyclitols as cryoprotectants for spinach and chickpea thylakoids. *Environmental and Experimental Botany* 44, 125–132.

Parodi, P.C., Nyquist, W.E., Patterson, F.L. and Hodges, H.F. (1983) Traditional combining-ability and Gardner-Eberhart analyses of a diallel for cold resistance in winter wheat. *Crop Science* 23, 314–318.

Pennycooke, J.C., Jones, M.L. and Stushnoff, C. (2003) Down-regulating α-galactosidase enhances freezing tolerance in transgenic petunia. *Plant Physiology* 133, 901–909.

Peterbauer, T. and Richter, A. (2001) Biochemistry and physiology of raffinose family oligosaccharides and galactosyl cyclitols in seeds. *Seed Science Research* 11, 185–197.

Peterbauer, T., Mucha, J., Mach, L. and Richter, A. (2002) Chain elongation of raffinose in pea seeds. *Journal of Biological Chemistry* 277, 194–200.

Rohde, P., Hincha, D.K. and Heyer, A.G. (2004) Heterosis in the freezing tolerance of crosses between two *Arabidopsis thaliana* accessions (Columbia-0 and C24) that show differences in non-acclimated and acclimated freezing tolerance. *Plant Journal* 38, 790–799.

Saleki-Gerhardt, A. and Zografi, G. (1994) Non-isothermal and isothermal crystallization of sucrose from the amorphous state. *Pharmaceutical Research* 11, 1166–1173.

Santarius, K.A. and Milde, H. (1977) Sugar compartmentation in frost-hardy and partially dehardened cabbage leaf cells. *Planta* 136, 163–166.

Seki, M., Narusaka, M., Ishida, J., Nanjo, T., Fujita, M., Oono, Y., Kamiya, A., Nakajima, M., Enju, A., Sakurai, T., Satou, M., Akiyama, K., Taji, T., Yamaguchi-Shinozaki, K., Carninci, P., Kawai, J., Hayashizaki, Y. and Shinozaki, K. (2002) Monitoring of the expression profiles of 7000 *Arabidopsis* genes under drought, cold, and high-salinity stresses using a full-length cDNA microarray. *Plant Journal* 31, 279–292.

Shahba, M.A., Qian, Y.L., Hughes, H.G., Koski, A.J. and Christensen, D. (2003) Relationships of soluble carbohydrates and freeze tolerance in saltgrass. *Crop Science* 43, 2148–2153.

Shull, G.H. (1914) Duplicate genes for capsule-form in *Bursa bursa-pastoris*. *Zeitschrift für induktive Abstammungs- und Vererbungslehre* 12, 97–149.

Smallwood, M. and Bowles, D.J. (2002) Plants in a cold climate. *Philosophical Transactions of the Royal Society of London. Series B* 357, 831–847.

Somero, G.N. (1992) Adapting to water stress: convergence on common solutions. In: Somero, G.N., Osmond, C.B. and Bolis, C.L. (eds) *Water and Life*. Springer, Berlin, pp. 3–18.

Sprenger, N. and Keller, F. (2000) Allocation of raffinose family oligosaccharides to transport and storage pools in *Ajuga reptans*: the roles of two distinct galactinol synthases. *Plant Journal* 21, 249–258.

Sun, W.Q. and Leopold, A.C. (1997) Cytoplasmic vitrification of anhydrobiotic organisms. *Comparative Biochemistry and Physiology* 117A, 327–333.

Sutka, J. (1981) Genetic studies of frost resistance in wheat. *Theoretical and Applied Genetics* 59, 145–152.

Taji, T., Ohsumi, C., Iuchi, S., Seki, M., Kasuga, M., Kobayashi, M., Yamaguchi-Shinozaki, K. and Shinozaki, K. (2002) Important roles of drought- and cold-inducible genes for galactinol synthase in stress tolerance in *Arabidopsis thaliana*. *Plant Journal* 29, 417–426.

Tapernoux-Lüthi, E.M., Böhm, A. and Keller, F. (2004) Cloning, functional expression, and characterization of the raffinose oligosaccharide chain elongation enzyme, galactan: galactan galactosyltransferase, from common bugle leaves. *Plant Physiology* 134, 1377–1387.

Törjek, O., Berger, D., Meyer, R.C., Müssig, C., Schmid, K.J., Rosleff Sörensen, T., Weisshaar, B., Mitchell-Olds, T. and Altmann, T. (2003) Establishment of a high-efficiency SNP-based framework marker set for *Arabidopsis*. *Plant Journal* 36, 122–140.

Tunnacliffe, A. and Lapinski, J. (2003) Resurrecting Van Leeuwenhoek's rotifers: a reappraisal of the role of disaccharides in anhydrobiosis. *Philosophical Transactions of the Royal Society of London. Series B* 358, 1755–1771.

Xiao, L. and Koster, K.L. (2001) Desiccation tolerance of protoplasts isolated from pea embryos. *Journal of Experimental Botany* 52, 2105–2114.

Xin, Z. and Browse, J. (2000) Cold comfort farm: the acclimation of plants to freezing temperatures. *Plant Cell and Environment* 23, 893–902.

Yancey, P.H., Clark, M.E., Hand, S.C., Bowlus, R.D. and Somero, G.N. (1982) Living with water stress: evolution of osmolyte systems. *Science* 217, 1214–1222.

Zhao, T.-Y., Martin, D., Meeley, R.B. and Downie, B. (2004a) Expression of the maize *GALACTINOL SYNTHASE* gene family: (II) kernel abscission, environmental stress and *myo*-inositol influences accumulation of transcript in developing seeds and callus cells. *Physiologia Plantarum* 121, 647–655.

Zhao, T.-Y., Thacker, R., Corum, J.W. III, Snyder, J.C., Meeley, R.B., Obendorf, R.L. and Downie, B. (2004b) Expression of the maize *GALACTINOL SYNTHASE* gene family: (I) expression of two different genes during seed development and germination. *Physiologia Plantarum* 121, 634–646.

Zuther, E., Büchel, K., Hundertmark, M., Stitt, M., Hincha, D.K. and Heyer, A.G. (2004) The role of raffinose in the cold acclimation response of *Arabidopsis thaliana. FEBS Letters* 576, 169–173.

15 Dehydration in Model Membranes and Protoplasts: Contrasting Effects at Low, Intermediate and High Hydrations

K.L. Koster[1,*] and G. Bryant[2]

[1]Department of Biology, The University of South Dakota, Vermillion, SD 57069, USA; [2]Department of Applied Physics, RMIT University, Melbourne 3001, Australia

Introduction

Freezing of plant tissues can cause lethal injury to sensitive species, and much of the damage can be ascribed to the deleterious effects of water loss from the cells. Although moderate dehydration can disrupt metabolic processes within tissues and cells, which may harm the plant, the lethal effects of extensive dehydration are thought primarily to result from physical stresses to vulnerable cellular structures, in particular, cell membranes. The physical forces that arise during dehydration have been thoroughly characterized using model membranes. The knowledge gained from these *in vitro* systems can be applied to more complex living cells in order to understand some of the mechanisms of damage and tolerance that lead to injury or survival, respectively. It is apparent that these mechanisms differ according to the extent of dehydration, and that there is not a single type of lesion that can account for all dehydration injury experienced by cells.

The Spectrum of Dehydration

For purposes of discussion, it is convenient to categorize the extent of dehydration according to the amount of water remaining in the system. Thus, we can discuss damage that occurs at high, intermediate and low hydrations. While this categorization is somewhat arbitrary, with few clear physically defined boundaries, there

*Corresponding author (e-mail: kkoster@usd.edu; Fax: +1-605-677-6557)

is broad correspondence to the behaviour of model membranes and to damage that has been observed in some plant species. Moreover, quantification of the extent of dehydration associated with lethal damage enables comparison to the measurable physical forces that arise during drying and, thus, facilitates development of a physical model to explain dehydration injury. This physical approach has been used successfully to explain freezing injury to leaf mesophyll cells, and one of our goals is to determine whether this model can be extended to other biological systems, such as desiccating seeds.

Cellular dehydration during freezing occurs when ice forms in extracellular spaces and creates a gradient for the diffusion of water from the intracellular solution to the extracellular ice. Dehydration causes physical stresses within the membranes and macromolecules that form the cell, and these stresses may be lethal (Bryant and Wolfe, 1992; Wolfe and Bryant, 1999). The magnitude of the physical stress depends on the extent of cellular dehydration during freezing, which in turn depends on the temperature of the extracellular ice. Therefore, freezing tolerance to some extent can be equated with dehydration tolerance. Plant freezing tolerance can differ among species and among developmental stages within a species. During the process of cold acclimation, plant freezing tolerance increases and, for many species, this has been linked to increasing tolerance of cellular dehydration (e.g. Levitt, 1972; Siminovitch and Cloutier, 1982; Steponkus and Lynch, 1989).

Just as various plant species differ in their freezing tolerance, the desiccation tolerance of different seeds can vary from species to species. The developmental stage of the embryo also plays a role in desiccation tolerance; tolerance generally increases as the embryo matures and decreases during germination (Vertucci and Farrant, 1995). Even when fully mature, however, seeds of some species sustain lethal damage at relatively high water contents, while others can survive the almost complete loss of cellular water. The reasons underlying the wide range of dehydration tolerances measured among seed species are not completely known. The quantitative nature of desiccation tolerance has prompted the suggestion that different types of injury develop at different hydration levels (Vertucci and Farrant, 1995; Walters, 1999; Walters *et al.*, 2002). Both metabolic and physical stresses have been implicated in desiccation damage to seeds, with metabolic dysfunction contributing more to damage at high hydrations and physical stresses being more injurious at low hydrations. It is thought that seeds that survive only mild dehydration have not acquired all of the protective mechanisms that enable complete desiccation tolerance (Vertucci and Farrant, 1995; Walters *et al.*, 2002).

When discussing relative levels of hydration, we consider cells at water potentials between 0 and approximately –6 MPa to be at high hydration. These high hydration levels are found in tissues frozen to temperatures greater than about –5°C, or in seeds equilibrated to relative humidities greater than 96%. Dehydration within this level may result in the loss of up to 80% of the so-called 'bulk' water in the cell; the remaining solution is highly viscous, but still liquid at temperatures above the freezing point. Although this represents a relatively mild dehydration, it can be lethal to leaves of non-freezing-tolerant species (Levitt, 1972; Steponkus *et al.*, 1993) and to seeds of some desiccation-sensitive species, such as *Avicennia marina* (Farrant *et al.*, 1993; Walters, 1999). The metabolic disruption that injures air-dried seeds at these relatively high hydrations is probably less important during freezing, when low temperature effectively dampens metabolism. Instead, physical and osmotic effects resulting from extracellular ice formation typically cause the most significant damage to freezing-sensitive species at high hydrations (Steponkus and Lynch, 1989; Steponkus *et al.*, 1993), as described further below.

Intermediate hydration levels are found between about –6 MPa and –60 MPa, where –60 MPa represents the average upper inflection point on water-sorption isotherms for numerous seeds (C. Walters, unpublished results) and their constituents such as phospholipids, soluble proteins and polysaccharides (K.L. Koster, unpublished results). Dehydration to these levels is brought about by freezing to temperatures between –6°C and –65°C or by equilibration to relative humidities between 96% and 65%. At these levels, embryo water contents may be reduced down to 0.15 g H_2O/g dry weight (DW), resulting in the removal of the remaining bulk water and some of the interfacial water that coats the surfaces of membranes and macromolecules in the cell. Most herbaceous species will succumb to freezing injury within this range of temperatures, while many woody plants survive (Levitt, 1972). Seeds of many subtropical species, such as *Aesculus hippocastanum* and *Coffea arabica*, sustain lethal damage at intermediate hydrations (Sun, 1999; Walters, 1999).

Low hydration is attained at water potentials less than –60 MPa and can be achieved by slow freezing to temperatures below –65°C or equilibration to relative humidities less than 65%. In this range, embryo water contents are less than about 0.15 g H_2O/g DW, and not only the bulk water, but also most of the interfacial water, has been removed. Cells that survive drying to low-hydration levels generally are considered to be desiccation tolerant, and the seeds of most temperate crop species fall into this category. Tolerance of these low-water potentials may also be required for successful cryopreservation using slow cooling. However, in many protocols, freezing of the sample is so rapid that water does not diffuse out of the cells, but instead vitrifies within the cell. Under these highly non-equilibrium conditions, cellular hydration is maintained at much higher levels than would be the case in equilibrium, thus avoiding the damage associated with dehydration.

Complete desiccation may be considered as a distinct level, when water contents are less than 0.02 g H_2O/g DW and essentially all of the interfacial water has been removed from the cell. Complete desiccation may be attained at water potentials on the order of –500 MPa, as estimated by extrapolation of hydration plots to determine the water potential at which separation between adjacent membranes and biomolecules is zero (Bryant and Koster, 2004). Equilibration to a relative humidity less than about 2% may bring samples to this very low hydration, and some careful freeze-drying protocols may also enable the complete desiccation of the cells or tissues. Although complete desiccation is an extreme situation that is probably seldom encountered in nature, seeds of many temperate crop species can survive this stress (Walters *et al.*, 2002), as do some vegetative plant tissues such as the moss *Tortula ruralis* (Bewley and Krochko, 1982).

Membranes as a Site of Dehydration Damage

For many years, membranes have been considered a primary site of damage due to cellular dehydration; however, it is clear that the mechanism of injury varies with the extent of dehydration. At high hydrations, membrane damage often manifests as lysis, when the osmotic contraction and expansion of the cell cannot be supported by the existing membrane structure. Membrane rupture has been observed in freezing-sensitive mesophyll protoplasts when the extent of osmotic expansion during thawing and rehydration exceeded the available membrane surface area (Dowgert and Steponkus, 1984; Dowgert *et al.*, 1987). In this example, lysis occurred during expansion of the protoplasts. In contrast, membrane rupture has also been detected in rapidly dried pea embryo protoplasts (Xiao and Koster, 2001), and we speculate, in this case, that the hydraulic conductivity of the plasma membrane was not sufficient to sustain the rapid rate of water loss.

Damage resulting from intermediate and low hydrations has been detected as the leakage of electrolytes and other solutes from injured cells and tissues (e.g. Senaratna and McKersie, 1983; Koster and Leopold, 1988; Reisdorph and Koster, 1999) and as the loss of osmotic responsiveness in protoplasts damaged by freezing (Gordon-Kamm and Steponkus, 1984; Uemura and Steponkus, 1989; Uemura *et al.*, 1995). Both these phenomena suggest that the semipermeability of the plasma membrane has been compromised, and electron micrographs of damaged cells reveal areas where membrane structure has been altered after moderate dehydration. Aparticulate domains, in which integral membrane proteins have been excluded from the lipid bilayer matrix, appear in some damaged membranes, while regions in which the lipids have undergone a transition to the non-bilayer hexagonal II phase can be seen in membranes injured at somewhat lower hydration (Gordon-Kamm and Steponkus, 1984). Membrane damage at very low hydration is visible in electron micrographs as regions of fused and disrupted membranes (Crowe and Crowe, 1982, 1988), representing a more extensive version of the injuries seen at low and intermediate hydrations.

Membrane Damage and Protection at Intermediate and Low Hydrations – the Hydration Forces Explanation

The importance of structural damage to membranes resulting from dehydration is apparent from both physiological data and microscopic observations, and one cause of this damage is thought to be the close approach of membranes during progressive water loss (Wolfe, 1987; Bryant and Wolfe, 1989, 1992). At high hydrations, membranes are kept separated by both interfacial and bulk water; however, as water molecules are removed at decreasing water potentials (lower

Fig. 15.1. *(Opposite.)* Schematic of the effects of dehydration on membranes. When two membranes are brought into close proximity, there may be demixing of lipid species and exclusion of intrinsic proteins. Cytoplasmic solutes (e.g. proteins and other large polymers) may be excluded from the appressed region. As further dehydration takes place, phase transitions may occur in the membranes, including fluid–gel transitions of some lipids. When continued dehydration brings the membranes into closer apposition, non-bilayer structures, such as the

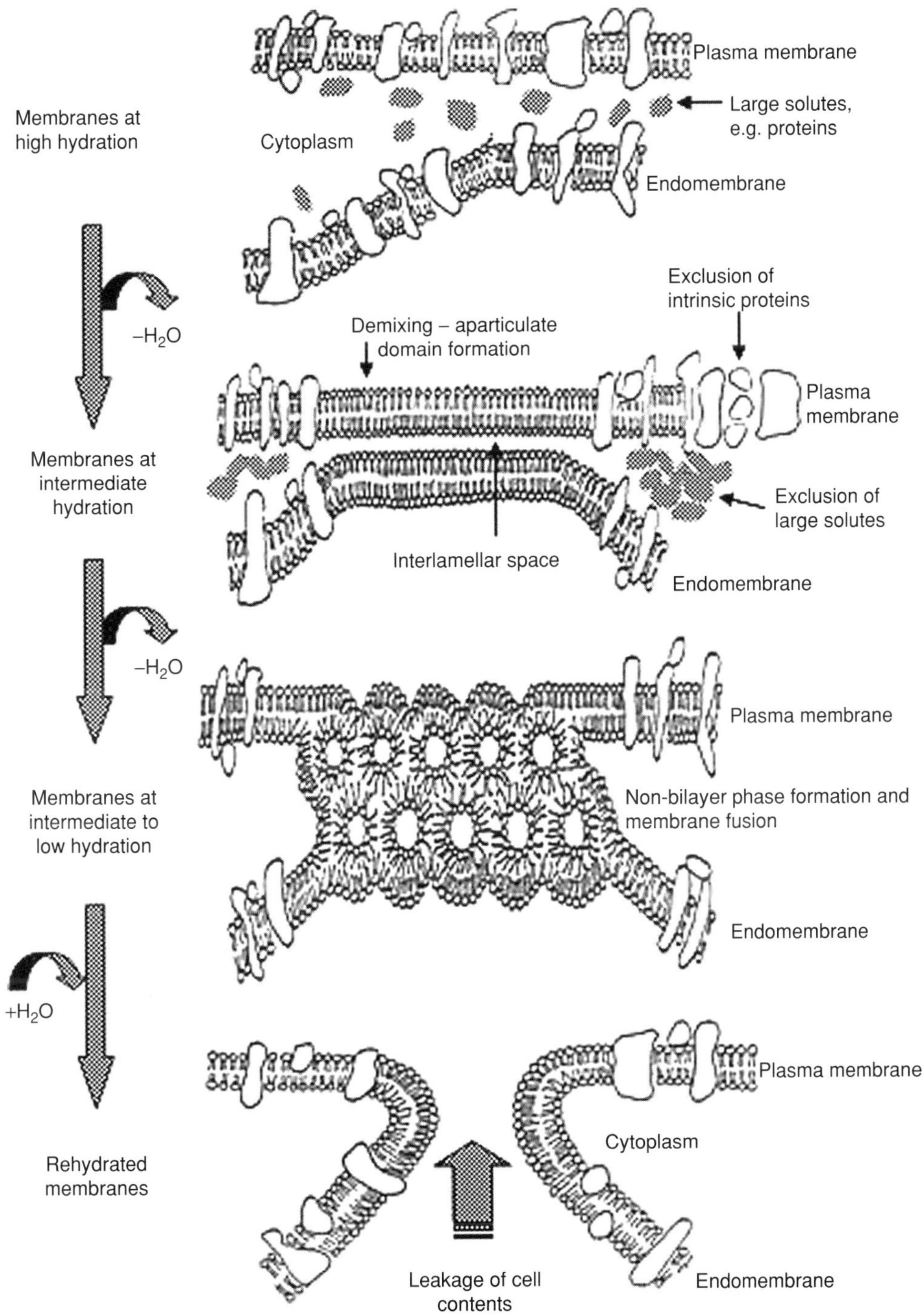

Fig. 15.1. *(Continued.)* inverse hexagonal II phase shown, may form between the two membranes. These non-bilayer structures lead to membrane lysis and leakage upon rehydration. (Adapted with permission from Walters *et al.*, 2002, based on an original diagram by Steponkus *et al.*, 1993.)

temperatures, in the case of freezing), membrane surfaces move closer together. At intermediate and low hydrations, hydration forces become increasingly important. The hydration force is a repulsive force that acts in a normal direction to opposing hydrophilic surfaces and resists their further close approach (Rand and Parsegian, 1989). Hydration forces have an exponential decay length, so they become stronger as the surfaces are brought closer together. As the water potential drops, the suction required to overcome the strong hydration force and remove interfacial water gives rise to a compressive stress (π) in the plane of the membrane (Bryant and Wolfe, 1992). At intermediate water contents, the hydration force and resultant lateral compression can cause the demixing of membrane components, leading to the exclusion of membrane proteins from regions of close approach (Fig. 15.1; Bryant and Wolfe, 1989, 1992). This phenomenon can explain the aparticulate domains seen in electron micrographs of membranes in cells injured by dehydration. As dehydration progresses, the membranes move closer together and the hydration force becomes stronger. The lateral compression π increases correspondingly and squeezes the membrane lipids together. This facilitates the transition of some phospholipids, such as phosphatidylcholine, from the fluid phase to the more tightly packed gel phase (Bryant and Wolfe, 1989; Koster *et al.*, 1994). The development of the compressive stress is reflected in the increased fluid–gel phase transition temperature, T_m, measured for phosphatidylcholines during progressive dehydration (Fig. 15.2).

Often, demixing of the membrane components leads to the localized enrichment of non-bilayer-forming lipids, such as phosphatidylethanolamine (Bryant *et al.*, 1992). Continued close approach and lateral compression can induce fusion of these domains in opposing bilayers, resulting in the formation of regions of the non-bilayer hexagonal II phase (Fig. 15.1). This is an interbilayer event, requiring two facing membranes, such as the plasma membrane and a subtending endomembrane, or perhaps two proximal regions of a highly folded membrane. Hexagonal II phase formation disrupts the semipermeability of the membrane, which can explain the high rates of leakage and the loss of osmotic responsiveness characteristically seen in dehydration-damaged tissues and protoplasts (Gordon-Kamm and Steponkus, 1984; Webb and Steponkus, 1993).

The presence of relatively high concentrations of di- and oligosaccharides has long been associated with tolerance of dehydration resulting from freezing (e.g. Levitt, 1972; Sauter and Kloth, 1987; Koster and Lynch, 1992) and equilibration to low relative humidities (e.g. Crowe *et al.*, 1984a; Koster and Leopold, 1988; Smirnoff, 1993; Vertucci and Farrant, 1995). In model membrane systems, sugars can prevent leakage and fusion during freeze–thaw events, freeze-drying and rehydration, and air-drying followed by rehydration (Crowe *et al.*, 1986; Strauss *et al.*, 1986; Crowe and Crowe, 1988; Sun *et al.*, 1996; Hincha *et al.*, 2000). The mechanism by which sugars accomplish this protection can be explained through an understanding of the importance of hydration forces in causing dehydration injury, as described above. According to this model, injury results from the close approach of membrane surfaces during dehydration; therefore, hindering the close approach can prevent or limit damage. Soluble sugars that remain between the membranes serve this function. They act as volumetric and osmotic spacers that physically limit the close approach of membrane surfaces and, thus, limit the compressive

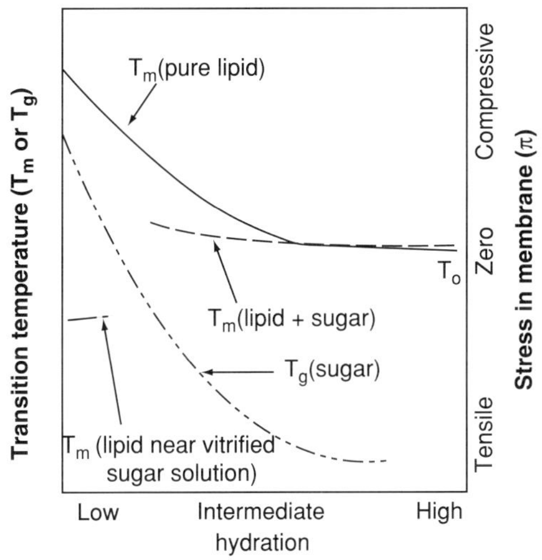

Fig. 15.2. Phase diagram demonstrating the effects of small solutes on the fluid–gel transition temperature, T_m, of phosphatidylcholine membranes. At high hydration the interlamellar separation is large, and the transition temperature is a constant value (T_o). For a system without solutes, the T_m increases as the hydration is reduced (bold line). For a system with solutes that remain between the bilayers (e.g. small sugars), the increase in T_m is significantly reduced, by an amount related to the size and number of solutes present (dashed line). As the water is removed, the solute–water mixture between the bilayers may form a glass, and the temperature at which this occurs (T_g) is indicated by the dot-dash line. Where T_g is greater than T_o (in the presence of solutes), the T_m is depressed due to the mechanical resistance provided by the glass, which hinders the transition to the gel phase. In this case, T_m is less than T_o. (Adapted from data in Koster *et al.*, 2003a.)

stress in the bilayer (Bryant and Wolfe, 1992; Bryant *et al.*, 2001). This protective mechanism is relatively non-specific; it depends upon the molecular volume of the sugars to keep the membranes separated and on the polarity of small sugars to confer osmotic properties. These osmotic properties resist the dehydrating effect of the external environment; thus, at any given water potential, an interlamellar solution that contains sugars will have more water than an interlamellar solution without sugars (Wolfe and Bryant, 1999). The polarity of small sugars also helps to keep them dissolved as the interlamellar solution becomes more concentrated.

Numerous experimental data support this explanation. Phosphatidylcholines, such as 1-palmitoyl-2-oleoylphosphatidylcholine (POPC), dried in the absence of sugars, display a gradual increase in their fluid–gel phase transition temperature, T_m (Fig. 15.2). If, however, the lipid bilayers are suspended in a sugar solution prior to drying, the rise in T_m is diminished and the phase transition temperature, T_m, remains near to that of the fully hydrated lipid (Koster *et al.*, 1994, 2000, 2003a). We have documented this effect using a number of different phosphatidylcholines and a range of sugars, including a sugar alcohol (sorbitol), monosaccharide (glucose), disaccharides (sucrose, trehalose, maltose), trisaccharides (raffinose, maltotriose), and larger maltodextrins up to and including dextran 1000 (MW = 1000). To first order, all these sugars were able to limit the rise in lipid T_m during progressive dehydration of the model membranes, as predicted by the model.

The protective mechanism proposed for sugars is non-specific, as long as the molecular volume of the sugars is not too large, in which case they may be squeezed out from the interlamellar space into discrete bulk phases during dehydration (Fig. 15.1; Koster *et al.*, 2000, 2003a). For example, we found that maltodextrins with a molecular mass of 5000 or greater did not prevent the increase in T_m of phosphatidylcholine during drying, so we infer that they were excluded from the interlamellar space (Koster *et al.*, 2003a). When excluded into bulk phases, large polymers may actually contribute to dehydration damage to membranes. We found that dextran (MW = 40,000) was excluded from the narrow interlamellar spaces and into a bulk phase, where it exerted an osmotic effect that dehydrated the membranes more than drying to the same water potential in the absence of the polymer (Koster *et al.*, 2000). Similar drying effects of large carbohydrate polymers have been reported by Rand and Parsegian (1989), Arnold and Gawrisch (1993), Hincha *et al.* (2000) and Vereyken *et al.* (2003). Crystallization of sugars during drying may also lead to their exclusion from interlamellar spaces and, thereby, remove their protection from the membranes. Caffrey *et al.* (1988) reported that crystallization of sucrose eliminated its ability to prevent the dehydration-induced rise in the T_m of DMPC. Although it is unclear whether solute crystallization is likely in a complex cellular environment, this facet of sugar behaviour can be an important consideration in model membrane systems and in protoplasts. Thus, not all carbohydrates exert similar stabilizing effects on membranes during dehydration; only those that are small enough and polar enough to remain between the membranes can diminish the interlamellar hydration force and the compressive stress that results from it.

At low hydrations, an additional effect of sugars becomes important to membrane stability. Highly concentrated sugar solutions may vitrify to form an amorphous solid in the interlamellar space. We first noted a decade ago that the presence of the glassy sugar solution depressed the fluid–gel T_m of POPC by 20 degrees below the T_m of the fully hydrated lipid (T_o) (Fig. 15.2; Koster *et al.*, 1994). Subsequent studies have confirmed this effect for a wide range of phosphatidylcholines and sugars (Koster *et al.*, 2000, 2003a; Bryant *et al.*, 2001). Zhang and Steponkus (1996) proposed that the solid glassy solution at the membrane surface resists the lateral compression of the membrane lipids that accompanies the fluid–gel phase transition. Effectively, the presence of the glass exerts a tensile stress ($^-\pi$) on the membranes that resists the phase transition and results in the depression of T_m (Fig. 15.2). Ultrasonic measurements of the elastic properties of sugar glasses confirm that vitrified sugar solutions can easily sustain the tension required to depress T_m below the fully hydrated T_o (Koster *et al.*, 2000). Thus, vitrification of the sugar solutions can confer an additional mechanical stability to membranes and cellular structures.

The stabilization of membranes by sugars extends to complete desiccation. During dehydration, the osmotic and volumetric effects of sugars keep membranes separated and limit the compressive stress that causes damage. At very low hydrations, when almost all water molecules are removed (approximately 0.02 g H_2O/g DW), sugars can also form hydrogen bonds with the polar groups of membrane lipids (Crowe *et al.*, 1984a,b), and if the sugar solution vitrifies at very low hydrations, the glass confers an additional mechanical stability to the membranes (Koster *et al.*, 2000).

Protoplast Desiccation Tolerance

Most studies of desiccation tolerance to date entail the use of either whole organisms, such as seeds (e.g. Senaratna and McKersie, 1983; Koster and Leopold, 1988; Farrant *et al.*, 1993; Reisdorph and Koster, 1999), or model systems, such as liposomes (e.g. Crowe *et al.*, 1986; Caffrey *et al.*, 1988; Koster *et al*, 1994, 2000, 2003a; Sun *et al.*, 1996; Hincha *et al.*, 2000; Vereyken *et al.*, 2003). The relative simplicity of the liposome model membrane system allows the development of mechanistic models for membrane stabilization; however, the complexity of the entire organism may make it difficult to obtain unambiguous results when testing these models. We recently developed a system to study seed desiccation tolerance at the cellular level using protoplasts derived from pea embryonic axes (Xiao and Koster, 2001). The protoplast system has the advantage of simplicity in that individual cells can be manipulated and their membranes viewed microscopically, while it retains the complexity of a living system. Protoplasts have been widely used to study the effects of freeze-induced dehydration on membranes (e.g. Dowgert and Steponkus, 1984; Dowgert *et al.*, 1987; Uemura and Steponkus, 1989, 2003; Steponkus *et al.*, 1993; Murai and Yoshida, 1998), and much of our understanding of membrane behaviour at high and intermediate hydrations comes from these studies.

The work of Steponkus, Uemura and their colleagues has drawn a detailed picture of the dynamic behaviour of the plasma membrane of leaf mesophyll cells during freeze-induced dehydration. This research has enabled a mechanistic understanding of how leaf freezing tolerance is influenced by membrane behaviours at high and intermediate hydrations, and how changes in membrane behaviour are brought about during cold acclimation. These studies have revealed a progression of injuries that occur as the ice temperature and, therefore, cellular water contents decrease. In protoplasts from non-cold-acclimated leaves, a relatively mild dehydration resulting from freezing at temperatures > −5°C results in the deletion of plasma membrane surface area during osmotic contraction of the protoplasts. During the rehydration that accompanies thawing, expansion-induced lysis occurs, as the membrane cannot expand to its original surface area (Dowgert and Steponkus, 1984; Dowgert *et al.*, 1987; Uemura *et al.*, 1995). Changes to plasma membrane composition brought about by cold acclimation diminish the loss of plasma membrane surface area in protoplasts from acclimated leaves, and expansion-induced lysis is therefore precluded (Steponkus and Lynch, 1989; Uemura *et al.*, 1995).

As the temperature of extracellular ice drops, more extensive cellular dehydration occurs to intermediate hydrations, causing membrane injury that manifests as the loss of osmotic responsiveness (LOR) (reviewed in Steponkus, 1984; Uemura and Steponkus, 1989; Steponkus *et al.*, 1993). LOR results when the plasma membrane and internal membranes come into close proximity during dehydration, as described above. In protoplasts from non-acclimated leaves, regions of the adjacent bilayers undergo a transition to a non-bilayer, hexagonal II phase that disrupts the membranes (Fig. 15.1; Gordon-Kamm and Steponkus, 1984; Webb and Steponkus, 1993). During cold acclimation, the temperature at which LOR occurs decreases, suggesting that biochemical and physical modifications hinder the close approach of membranes in the more freezing-tolerant cells (Steponkus and Lynch, 1989; Fujikawa and Steponkus, 1990; Webb and Steponkus, 1993; Uemura *et al.*, 1995).

The membrane behaviours and injuries that result from freeze-induced dehydration have been described in mesophyll protoplasts isolated from several species, including winter rye, oats and *Arabidopsis* (Webb *et al.*, 1994; Uemura *et al.*, 1995). Insight into the nature of these injuries has led to improved understanding of the mechanisms of freezing tolerance and the adaptive changes that occur during cold acclimation. One of our goals in our recent work was to determine whether similar types of membrane damage occur during dehydration of seeds that are becoming sensitive to desiccation as they germinate, and thus, whether changes in membrane behaviour contribute to desiccation tolerance.

Protoplasts isolated from pea embryonic axes retain some of the features of the tissue from which they were isolated. Just as the desiccation tolerance of the embryos decreases during germination, the tolerance of the protoplasts derived from the embryos also decreases (Xiao and Koster, 2001; Koster *et al.*, 2003b). Pea embryos progressively become less tolerant of desiccation after 18 h of imbibition, around the time that the radicle begins to emerge from the seed coat (Reisdorph and Koster, 1999). At 24 h of imbibition, embryos are intermediate in their tolerance, and longer germination times result in even less tolerance of dehydration (Fig. 15.3). By analogy to the LT_{50} values often used to quantify the freezing tolerance of a tissue, we can quantify the desiccation tolerance of protoplasts as the WC_{50}, the water content that results in the death of 50% of the protoplast population (Koster *et al.*, 2003b). Thus, protoplasts isolated from pea radicles prior to or at 18 h of imbibition had WC_{50} values of 0 g H_2O/g DW, as most protoplasts survived desiccation (Fig. 15.3). Protoplasts isolated from radicles at later germination times had correspondingly greater WC_{50} values, suggesting that they were less desiccation tolerant. By comparing the mechanisms of damage among protoplasts from various stages of germination that differ in their desiccation tolerance, we hope to gain additional insight into mechanisms of desiccation tolerance in seed embryos.

Protoplasts isolated from pea radicles at 12 h of imbibition are considered to be desiccation tolerant, as more than 50% survived drying to hydrations less than 0.15 g H_2O/g DW (Xiao and Koster, 2001). At this point in germination, radicle cells do not contain large vacuoles; instead, most of the cellular volume is occupied by the nucleus, protein bodies, lipid bodies and the dense cytoplasm (Ramsay, 2002). In order to observe membrane behaviour in these dried protoplasts, samples were fixed using an anhydrous freeze-substitution technique, then prepared for transmission electron microscopy. The dried protoplasts appeared shrunken, with the plasma membrane folded around the clustered lipid bodies

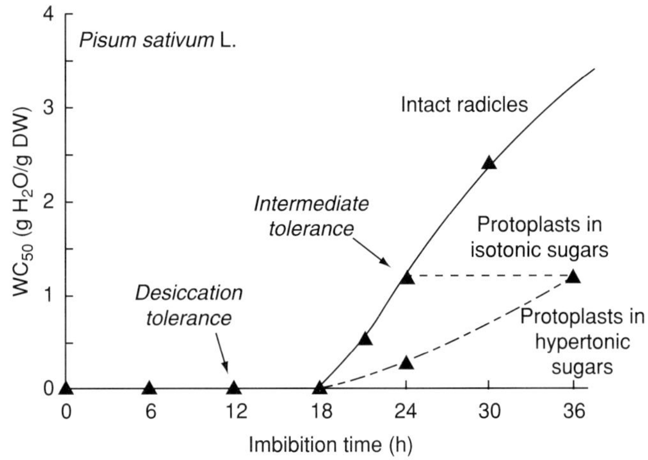

Fig. 15.3. Graph showing the desiccation tolerance (as measured by the WC_{50}, the water content that results in the death of 50% of the population) as a function of imbibition time for intact pea radicles (solid line), and protoplasts isolated from pea radicles in an isotonic sucrose/raffinose solution (dotted line) and in a hypertonic sucrose/raffinose solution (dot-dash line). (Data for radicles from Reisdorph and Koster, 1999.)

and protein bodies within the dense cytoplasm (see Fig. in Xiao and Koster, 2001). We conclude that this folding enables the protoplasts to conserve their surface area and, thus, preclude expansion-induced lysis during rehydration (Xiao and Koster, 2001). Other types of membrane damage resulting from their close approach may be prevented by the high sugar content of embryo cells at this stage of germination. The sugar content of pea embryos at 12 h of imbibition is about 14% of the dry matter, comprised primarily of sucrose and raffinose series oligosaccharides (Koster and Leopold, 1988). Given the desiccation tolerance of these protoplasts, we speculate that the high intracellular sugar concentrations keep membranes and macromolecules separated to minimize physical stresses during dehydration.

Protoplasts isolated at 24 h of imbibition have intermediate desiccation tolerance, which can be manipulated somewhat by altering the osmotic strength of the suspending solution, as discussed further below (Fig. 15.3). At this stage, germination is almost complete and the cells are becoming vacuolated as the protein in the storage bodies is broken down (Ramsay, 2002). Average sizes of protoplasts isolated at this time are larger than they are at 12 h of imbibition; however, after drying, the protoplasts from both imbibition times shrink to the same size (Ramsay, 2002). This suggests that the protoplasts isolated at 24 h of imbibition undergo larger volume changes during drying. Although some of these protoplasts do survive drying to intermediate and low hydrations, our data suggest that many lyse after relatively mild dehydration (S.J. Halperin and K.L. Koster, unpublished results).

Protoplast desiccation tolerance depends not only on the developmental stage of the radicle tissue (i.e. how long germination has progressed prior to isolation), but also on the osmotic strength and sugar composition of the suspending solution. At both 12 h and 24 h of imbibition, desiccation tolerance of the protoplasts can be manipulated by changing the sugar composition of the suspension buffers. At 12 h, protoplast desiccation tolerance was maximal (76%) when they were isolated and dried in a solution containing a mixture of sucrose and raffinose (85:15 w/w) that somewhat mimics the endogenous sugar composition of the embryos (Xiao and Koster, 2001). Protoplast survival after drying to water contents less than 0.2 g H_2O/g DW was successively lower when the protoplasts were dried in pure sucrose (56% survival), trehalose (44%), glucose (27%), raffinose (20%), sorbitol (11%) or mannitol (7%) (Xiao and Koster, 2001). In the case of raffinose and mannitol, the sugars frequently crystallized during drying, and when this occurred, all the suspended protoplasts were physically destroyed.

Desiccation tolerance of protoplasts isolated at 24 h of imbibition was also influenced by the osmotic strength and composition of the suspending solution (Koster *et al.*, 2003b). Protoplasts isolated and dried in an isotonic solution containing the same sucrose and raffinose mixture as above (85:15 w/w) had intermediate desiccation tolerance, with a WC_{50} of 1.2 g H_2O/g DW (Koster *et al.*, 2003b). However, if the protoplasts were isolated and dried in hypertonic sucrose/raffinose solutions, tolerance was significantly improved, with the WC_{50} at 0.3 g H_2O/g DW (Halperin *et al.*, 2004). Increasing the osmotic strength was, however, not the only requirement for improving the survival of these protoplasts. When hypertonic sorbitol solutions were used, the WC_{50} was high, about 2.7 g H_2O/g DW. Substituting sorbitol for the sucrose/raffinose mixture in the solution used for either isolation or drying resulted in diminished desiccation tolerance of the protoplasts (Halperin *et al.*, 2004). Thus, it seems clear that the presence of sucrose, and perhaps raffinose, is important for desiccation tolerance of the pea radicle protoplasts.

It is not known to what extent sugars may exchange across the protoplast plasma membrane during isolation and desiccation, though it seems likely, given the existence of sucrose transporters in legume seeds (Tegeder *et al.*, 1999; Patrick and Offler, 2001) that this sugar might be transported into or out of radicle cells. Therefore, it is possible that the beneficial effect of extracellular sucrose on desiccation tolerance in the complex protoplast system results not from any specific effects of sucrose on membrane lipids at low hydration, but rather from the fact that sucrose can enter and remain in the protoplasts during dehydration.

In summary, sugars can exert several different protective effects in dehydrating systems. At high hydrations, sugars have osmotic effects that enable cells to resist water loss in the face of dehydrating environments. Sugars also contribute to the freezing-point depression of cellular solutions, which may facilitate survival of mild subzero temperatures. There is also evidence that sugars can act as scavengers of free radicals, which may be generated during episodes of dehydration stress (Smirnoff, 1993). Such scavenging may be an important buffer against damage caused by disruption of metabolic processes. Sugars can also act as metabolic substrates to provide energy and carbon skeletons to cells. These substrates may facilitate the adaptive alterations in cellular architecture that accompany processes such as acclimation (Uemura and Steponkus, 2003). Finally, sugars can act as signalling molecules and may play an important role in mediating metabolic responses to dehydration (Finkelstein and Gibson, 2001).

At intermediate and low hydrations, the protective role of sugars is likely to be more physical than physiological. At these lower water contents, metabolism is constrained by the limited diffusion possible in the viscous cellular environment (Vertucci and Farrant, 1995; Walters *et al.*, 2002), and the damage that occurs results, instead, from the physical forces engendered as hydrophilic surfaces come into close apposition (Bryant and Wolfe, 1989, 1992; Wolfe and Bryant, 1999). The protection afforded by sugars is due to their ability to act as volumetric spacers that hinder the close approach of membranes and other surfaces and, thereby, diminish the physical stresses that would otherwise damage cellular structures. At low hydrations, if the sugar solutions vitrify, an additional mechanical stabilization is conferred (Koster *et al.*, 2000) that may be important to the long-term stability of the system.

Acknowledgements

The authors thank Dr Christina Walters for the use of Fig. 15.1, unpublished results, and many fruitful discussions. Karen L. Koster also thanks Dr Stephen Halperin for the use of unpublished results. Karen L. Koster acknowledges funding from USDA-NRI-CSREES award #2003-35100-13364, from NSF-EPSCoR Grant #OSR-9108773, and from the South Dakota Future Fund.

References

Arnold, K. and Gawrisch, K. (1993) Effects of fusogenic agents in membrane hydration. In: Düzgumes, N. (ed.) *Methods in Enzymology*, Vol. 220. Academic Press, New York, pp. 143–157.

Bewley, J.D. and Krochko, J.E. (1982) Desiccation-tolerance. In: Lange, O.L., Nobel, P.S., Osmond, C.B., Ziegler, H. (eds) *Physiological Plant Ecology II Water Relations and Carbon Assimilation. Encyclopedia of Plant Physiology* (New Series), Vol. 12B. Springer-Verlag, New York, pp. 325–378.

Bryant, G. and Koster, K.L. (2004) Dehydration of solute-lipid systems: hydration forces analysis. *Colloids & Surfaces B: Biointerfaces* 35, 73–79.

Bryant, G. and Wolfe, J. (1989) Can hydration forces induce lateral phase separations in lamellar phases? *European Biophysics Journal* 16, 369–374.

Bryant, G. and Wolfe, J. (1992) Interfacial forces in cryobiology and anhydrobiology. *Cryo-Letters* 13, 23–36.

Bryant, G., Pope, J. and Wolfe, J. (1992) Low hydration phase properties of phospholipid mixtures. Evidence for dehydration-induced fluid-fluid separations. *European Biophysics Journal* 21, 223–232.

Bryant, G., Koster, K.L. and Wolfe, J. (2001) Membrane behaviour in seeds and other systems at low water content: the various effects of solutes. *Seed Science Research* 11, 17–25.

Caffrey, M., Fonseca, V. and Leopold, A.C. (1988) Lipid-sugar interactions. Relevance to anhydrous biology. *Plant Physiology* 86, 754–758.

Crowe, J.H., Crowe, L.M. and Chapman, D. (1984a) Preservation of membranes in anhydrobiotic organisms: the role of trehalose. *Science* 223, 701–703.

Crowe, J.H., Crowe, L.M. and Chapman, D. (1984b) Infrared spectroscopic studies on interactions of water and carbohydrates with a biological membrane. *Archives of Biochemistry and Biophysics* 232, 400–407.

Crowe, L.M. and Crowe, J.H. (1982) Hydration-dependent hexagonal phase lipid in a biological membrane. *Archives of Biochemistry and Biophysics* 217, 582–587.

Crowe, L.M. and Crowe, J.H. (1988) Trehalose and dry dipalmitoylphosphatidylcholine revisited. *Biochimica et Biophysica Acta* 946, 193–201.

Crowe, L.M., Womersley, C., Crowe, J.H., Reid, D., Appel, L. and Rudolph, A. (1986) Prevention of fusion and leakage in freeze-dried liposomes by carbohydrates. *Biochimica et Biophysica Acta* 861, 131–140.

Dowgert, M.F. and Steponkus, P.L. (1984) Behavior of the plasma membrane of isolated protoplasts during a freeze-thaw cycle. *Plant Physiology* 75, 1139–1151.

Dowgert, M.F., Wolfe, J. and Steponkus, P.L. (1987) The mechanics of injury to isolated protoplasts following osmotic contraction and expansion. *Plant Physiology* 83, 1001–1007.

Farrant, J.M., Pammenter, N.W. and Berjak, P. (1993) Seed development in relation to desiccation tolerance: a comparison between desiccation-sensitive (recalcitrant) seeds of *Avicennia marina* and desiccation-tolerant types. *Seed Science and Technology* 3, 1–13.

Finkelstein, R.R. and Gibson, S.I. (2001) ABA and sugar interactions regulating development: cross-talk or voices in a crowd? *Current Opinion in Plant Biology* 5, 26–32.

Fujikawa, S. and Steponkus, P.L. (1990) Freeze-induced alterations in the ultrastructure of the plasma membrane of rye protoplasts isolated from cold-acclimated leaves. *Cryobiology* 27, 665–666.

Gordon-Kamm, W.J. and Steponkus, P.L. (1984) Lamellar-to-hexagonal II phase transitions in the plasma membrane of isolated protoplasts after freeze-induced dehydration. *Proceedings of the National Academy of Sciences USA* 81, 6373–6377.

Halperin, S.J., Rowen, R. and Koster, K.L. (2004) Desiccation damage and tolerance in pea embryo protoplasts related to oxidative stress and sugar content. *Plant Biology* 2004, Abstract # 991.

Hincha, D.K., Hellwege, E.M., Heyer, A.G. and Crowe, J.H. (2000) Plant fructans stabilize phosphatidylcholine liposomes during freeze-drying. *European Journal of Biochemistry* 267, 535–540.

Koster, K.L. and Leopold, A.C. (1988) Sugars and desiccation tolerance in seeds. *Plant Physiology* 88, 829–832.

Koster, K.L. and Lynch, D.V. (1992) Solute accumulation and compartmentation during the cold acclimation of Puma rye. *Plant Physiology* 98, 108–113.

Koster, K.L., Webb, M.S., Bryant, G. and Lynch, D.V. (1994) Interactions between soluble sugars and POPC (1-palmitoyl-2-oleoylphosphatidylcholine) during dehydration: vitrification of sugars alters the phase behavior of the phospholipid. *Biochimica et Biophysica Acta* 1193, 143–150.

Koster, K.L., Lei, Y.P., Anderson, M., Martin, S. and Bryant, G. (2000) Effects of vitrified and non-vitrified sugars on phosphatidylcholine fluid-to-gel phase transitions. *Biophysical Journal* 78, 1932–1946.

Koster, K.L., Maddocks, K. and Bryant, G. (2003a) Exclusion of maltodextrins from phosphatidylcholine multilayers during dehydration: effects on membrane phase behaviour. *European Biophysics Journal* 32, 96–105.

Koster, K.L., Reisdorph, N. and Ramsay, J.L. (2003b) Changing desiccation tolerance of pea embryo protoplasts during germination. *Journal of Experimental Botany* 54, 1607–1614.

Levitt, J. (1972) *Responses of Plants to Environmental Stresses.* Academic Press, New York.

Murai, M. and Yoshida, S. (1998) Evidence for the cell wall involvement in temporal changes in freezing tolerance of Jerusalem artichoke (*Helianthus tuberosus* L.) tubers during cold acclimation. *Plant and Cell Physiology* 39, 97–105.

Patrick, J.W. and Offler, C.E. (2001) Compartmentation of transport and transfer events in developing seeds. *Journal of Experimental Botany* 52, 551–564.

Ramsay, J.L. (2002) Ultrastructure of pea embryonic axes (*Pisum sativum* L. cv. Alaska) and the associated loss of desiccation tolerance during germination. PhD dissertation, The University of South Dakota, Vermillion, South Dakota.

Rand, R.P. and Parsegian, V.A. (1989) Hydration forces between phospholipid bilayers. *Biochimica et Biophysica Acta* 988, 351–376.

Reisdorph, N.A. and Koster, K.L. (1999) Progressive loss of desiccation tolerance in germinating pea (*Pisum sativum*) seeds. *Physiologia Plantarum* 105, 266–271.

Sauter, J.J. and Kloth, S. (1987) Changes in carbohydrates and ultrastructure in xylem ray cells of *Populus* in response to chilling. *Protoplasma* 137, 45–55.

Senaratna, T. and McKersie, B.D. (1983) Characterization of solute efflux from dehydration injured soybean (*Glycine max* L. Merr.) seeds. *Plant Physiology* 72, 911–914.

Siminovitch, D. and Cloutier, Y. (1982) Twenty-four hour induction of freezing and drought tolerance in plumules of winter rye seedlings by desiccation stress at room temperature in the dark. *Plant Physiology* 69, 250–255.

Smirnoff, N. (1993) The role of active oxygen in the response of plants to water deficit and desiccation. *New Phytologist* 125, 27–58.

Steponkus, P.L. (1984) Role of the plasma membrane in freezing injury and cold acclimation. *Annual Review of Plant Physiology* 35, 543–584.

Steponkus, P.L. and Lynch, D.V. (1989) Freeze/thaw induced destabilization of the plasma membrane and the effects of cold acclimation. *Journal of Bioenergetics and Biomembranes* 21, 21–41.

Steponkus, P.L., Uemura, M. and Webb, M.S. (1993) A contrast of the cryostability of the plasma membrane of winter rye and spring oat – two species that widely differ in their freezing tolerance and plasma membrane lipid composition. In: Steponkus, P.L. (ed.) *Advances in Low Temperature Biology*, Vol. 2. JAI Press, London, pp. 211–312.

Strauss, G., Schurtenberger, P. and Hauser, H. (1986) The interaction of saccharides with lipid bilayer vesicles: stabilization during freeze-thawing and freeze-drying. *Biochimica et Biophysica Acta* 858, 169–180.

Sun, W.Q. (1999) Desiccation sensitivity of recalcitrant seeds and germinated orthodox seeds: Can germinated orthodox seeds serve as a model system for studies of recalcitrance? In: Marzalina, M., Khoo, K.C. and Jayanthi, N. (eds) *IUFRO Seed Symposium 1998 Recalcitrant Seeds*. Forest Research Institute, Kuala Lumpur, Malaysia, pp. 29–42.

Sun, W.Q., Leopold, A.C., Crowe, L.M. and Crowe, J.H. (1996) Stability of dry liposomes in sugar glasses. *Biophysical Journal* 70, 1769–1776.

Tegeder, M., Wang, X.D., Frommer, W.B., Offler, C.E. and Patrick, J.W. (1999) Sucrose transport into developing seeds of *Pisum sativum* L. *The Plant Journal* 18, 151–161.

Uemura, M. and Steponkus, P.L. (1989) Effect of cold acclimation on the incidence of two forms of freezing injury in protoplasts isolated from rye leaves. *Plant Physiology* 91, 1131–1137.

Uemura, M. and Steponkus, P.L. (2003) Modification of the intracellular sugar content alters the incidence of freeze-induced membrane lesions of protoplasts isolated from *Arabidopsis thaliana* leaves. *Plant, Cell & Environment* 26, 1083–1096.

Uemura, M., Joseph, R.A. and Steponkus, P.L. (1995) Cold acclimation of *Arabidopsis thaliana*. Effect on plasma membrane lipid composition and freeze-induced lesions. *Plant Physiology* 109, 15–30.

Vereyken, I.J., Chupin, V., Islamov, A., Kuklin, A., Hincha, D.K. and de Kruijff, B. (2003) The effect of fructan on the phospholipid organization in the dry state. *Biophysical Journal* 85, 3058–3065.

Vertucci, C.W. and Farrant, J.M. (1995) Acquisition and loss of desiccation tolerance. In: Kigel, J. and Galili, G. (eds) *Seed Development and Germination*. Marcel Dekker, New York, pp. 237–271.

Walters, C. (1999) Levels of recalcitrance in seeds. In: Marzalina, M., Khoo, K.C. and Jayanthi, N. (eds) *IUFRO Seed Symposium 1998 Recalcitrant Seeds*. Forest Research Institute, Kuala Lumpur, Malaysia, pp. 1–13.

Walters, C., Farrant, J.M., Pammenter, N.W. and Berjak, P. (2002) Desiccation stress and damage. In: Black, M. and Pritchard, H.W. (eds) *Desiccation and Plant Survival: Drying without Dying.* CABI Publishing, Wallingford, UK, pp. 263–291.

Webb, M.S. and Steponkus, P.L. (1993) Freeze-induced membrane ultrastructural alterations in rye (*Secale cereale*) leaves. *Plant Physiology* 101, 955–963.

Webb, M.S., Uemura, M. and Steponkus, P.L. (1994) A comparison of freezing injury in oat and rye: two cereals at the extremes of freezing tolerance. *Plant Physiology* 104, 467–478.

Wolfe, J. (1987) Lateral stresses in membranes at low water potential. *Australian Journal of Plant Physiology* 14, 311–318.

Wolfe, J. and Bryant, G. (1999) Freezing, drying, and/or vitrification of membrane-solute-water systems. *Cryobiology* 39, 103–129.

Xiao, L. and Koster, K.L. (2001) Desiccation tolerance of protoplasts isolated from pea embryos. *Journal of Experimental Botany* 52, 2105–2114.

Zhang, J. and Steponkus, P.L. (1996) Proposed mechanism for depression of the liquid-crystalline to gel phase transition temperature of phospholipids in dehydrated sugar-phospholipid mixtures. *Cryobiology* 33, 21A.

16 Effect of Plasma Membrane-associated Proteins on Acquisition of Freezing Tolerance in *Arabidopsis thaliana*

Y. TOMINAGA,[1,2] C. NAKAGAWARA,[2] Y. KAWAMURA[2] AND M. UEMURA[2,*]

[1]Département des Sciences Biologiques, Université du Québec à Montréal, Montréal, Québec, Canada H3C 3P8; [2]Cryobiosystem Research Center, Iwate University, Morioka 020-8550, Japan

Introduction

Plasma membrane as the primary site of freezing injury

Many plants growing in temperate regions can acquire tolerance to freezing through exposure to low, non-freezing temperatures, a phenomenon known as cold acclimation. For example, freezing tolerance in *Arabidopsis* leaves substantially increases during the first day of cold acclimation and reaches a maximum after 7 days (Gilmour *et al.*, 1988; Uemura *et al.*, 1995). Similarly, the survival of protoplasts isolated from *Arabidopsis* leaves after a freeze–thaw cycle significantly increases after 1 day of cold acclimation and gradually increases to a maximum following 7 days of cold acclimation (Kawamura and Uemura, 2003). These results indicate that the acquisition of freezing tolerance in *Arabidopsis* occurs during a short period of cold acclimation, i.e. within 1 day.

During the dynamic process of cold acclimation in plants, numerous physiological and biochemical changes have been reported. Such changes include growth reduction, changes in water content, altered levels of abscisic acid (ABA), accumulation of compatible solutes (proline, glycinebetaine and soluble sugars) and changes in membrane lipid composition (for review, see Guy, 1999; Pearce, 1999; Thomashow, 1999). From the large number of studies on these changes, it is no exaggeration to state that the majority of the physiological changes occurring

*Corresponding author (e-mail: uemura@iwate-u.ac.jp; Fax: +81-19-621-6253)

during the process of cold acclimation are induced to avoid the structural and functional damage to the plasma membrane that is known as the primary site of freezing injury in plants (Steponkus, 1984).

The aforementioned statement can be altered to state that the apparently diverse responses occurring during cold acclimation ultimately increase the cryostability of membranes during freeze-induced dehydration (Levitt, 1980; Steponkus, 1984). For example, compatible solutes (highly soluble organic compounds of low molecular weights) accumulate to increase internal osmotic concentration and, hence, prevent the loss of water from the cell upon ice formation. In addition, it is considered that compatible solutes contribute to increase freezing tolerance by reducing the membrane damage derived from freeze-induced dehydration and to protect macromolecules against dehydration-induced denaturation (Steponkus, 1984). In fact, both *in vivo* and *in vitro* studies have clearly indicated a cryoprotective function of compatible solutes on biomembranes against both dehydration and freezing stresses (Crowe *et al.*, 1990; Uemura and Steponkus, 2003). Wanner and Junttila (1999) determined the freezing tolerance of *Arabidopsis* leaves during cold acclimation under various light conditions. They concluded that soluble sugar accumulation was a fundamental component in the acquisition of freezing tolerance that functioned to maintain membrane stability at low temperatures.

Transgenic approaches with genes encoding cold-regulated proteins demonstrated the importance of membrane cryostability for acquiring freezing tolerance in plants. Constitutive expression of the *COR15a* gene that encodes COR15am, which is a chloroplastic stromal protein and accumulates during cold acclimation in *Arabidopsis* (Lin and Thomashow, 1992), increased the freezing tolerance of both chloroplasts and protoplasts (Artus *et al.*, 1996). Analyses on the occurrence of membrane-associated lesions during a freeze–thaw cycle demonstrated that the accumulation of the COR15am protein in chloroplasts increased cryostability of the plasma membrane and the chloroplast envelopes. Specifically, this overexpression resulted in a decrease in the incidence of a specific membrane lesion, the hexagonal II phase transition associated with the plasma membranes.

Plasma membrane alterations associated with cold acclimation

Plasma membrane lipid composition affects the cryostability of the plasma membrane during freeze-induced dehydration (Steponkus *et al.*, 1993). Early studies on plasma membrane lipid composition (Uemura and Yoshida, 1984; Yoshida and Uemura, 1984; Lynch and Steponkus, 1987) clearly showed that there were considerable changes in plasma membrane lipid classes (such as phospholipids, sterols and glucocerebrosides) and molecular species in two major phospholipid classes (phosphatidylcholine and phosphatidylethanolamine) during cold acclimation. Furthermore, membrane engineering studies in which plasma membrane lipid composition of isolated protoplasts was artificially altered by protoplast-liposome fusion revealed that there was a large decrease in the incidence of plasma membrane-associated lesions and, hence, an increase in the freezing tolerance of protoplasts (Steponkus *et al.*, 1988; Uemura *et al.*, 1995). Although lipid changes cannot solely explain the mechanism of increased stability of the plasma membrane during freeze-induced dehydration, these results clearly indicate that the acquisition of freezing tolerance requires considerable alterations of plasma membrane lipids during the cold-acclimation process.

In contrast to the detailed studies on the role of the plasma membrane lipids in freezing tolerance, plasma membrane proteins have not been thoroughly studied in relation to the ability of cold acclimation in plants. However, it is generally considered that plasma membrane proteins play a key role in many important biological functions such as signal transduction, transport regulation and the recognition of various stimuli. These functions are also involved in the determination of responses to low temperatures (Shinozaki and Yamaguchi-Shinozaki, 2000). Due to the lack of research in this particular area, we focused our efforts on analysing changes in plasma membrane proteins associated with cold acclimation of *Arabidopsis* and thereby set out to determine the function of cold-responsive plasma membrane proteins during the process of cold acclimation. By utilizing a peptide mass finger printing method that was coupled with MALDI-TOF MS analysis, we have recently succeeded in identifying a number of cold acclimation-responsive plasma membrane-associated proteins in *Arabidopsis* (Kawamura and Uemura, 2003). With a purified plasma membrane fraction, 30 proteins were identified as cold-responsive proteins. Eighteen of these proteins were found to exist in the solubilized fraction after the plasma membrane was treated with a urea-thiourea-CHAPS buffer for isoelectric point electrophoresis and the remaining 12 proteins were in the insolubilized fraction. A protein database search further revealed that the proteins in the solubilized fraction lacked transmembrane domains, suggesting that they are the peripheral membrane proteins that may be associated with integral membrane proteins or hydrophilic surfaces of the membrane bilayers. On the other hand, the proteins in the insolubilized fraction were predicted to bind to the plasma membrane directly via transmembrane domains or by myristoylation. Functions of these cold-responsive plasma membrane proteins described from previous studies and protein databases suggested that they may be involved in membrane repair, protection of the membrane against osmotic stress or proteolysis, and so on.

In this study, several cold-responsive plasma membrane proteins were selected and their functions in direct relation to cold acclimation were examined. The proteins included dehydrins (ERD10 and ERD14), outer-membrane lipoprotein-like protein (lipocalin-like: AtLCN), and DREPP-like protein and nodulin VfNOD18-like protein; all of these five proteins were found in the solubilized fraction. The genes encoding this subset of five proteins were initially characterized and their effects on the freezing tolerance and freeze-induced lesions of *Arabidopsis in planta* with independent transgenic plants that constitutively express one of these genes were determined subsequently.

Cold-responsive Plasma Membrane Proteins

Protein characteristics analysed in the present study

ERD10 and ERD14, both of which belong to the dehydrin family (group II late embryogenic abundant (LEA) protein family), were originally identified as the products encoded in genes that rapidly respond to dehydration stress in *Arabidopsis* (Kiyosue *et al.*, 1994). The expression profiles determined by microarray analysis showed that *ERD10* was the target gene of DREB1A, and was strongly induced by cold, drought or high salinity stress (Seki *et al.*, 2001, 2002a). Although *ERD14* was also induced by drought and cold stress, the level of induction was not significant (Seki *et al.*, 2001). Under non-stress conditions, *ERD14* expression was observed in vascular tissue, bordering parenchyma cells and root tips. However, under stressed conditions, *ERD14* transcript localization was not restricted to certain tissues or cell types (Nylander *et al.*, 2001). At this time, there is no structural evidence available to indicate the interaction between ERD10 or ERD14 and the plasma membrane. However, it has been shown that the acidic dehydrin, WCOR410, accumulated in the vicinity of the plasma membrane in wheat during cold acclimation, suggesting that WCOR410 may have a role in minimizing the destabilization of the plasma membrane caused by freeze-induced dehydration (Danyluk *et al.*, 1998). Because ERD10 and ERD14 are classified to the acidic SK-type dehydrins (Nylander *et al.*, 2001), it is feasible that both proteins accumulate in the vicinity of the plasma membrane during cold acclimation and are related to acquisition of freezing tolerance by protecting against freeze-induced membrane injury. Further immunocytochemical studies are warranted and necessary to confirm their putative intracellular localization.

It is important to note that AtLCN was one of the proteins that showed a remarkable increase during cold acclimation in *Arabidopsis* (Kawamura and Uemura, 2003). Over the last ten years, many publications have been devoted to describe the common features and characteristics of the lipocalin superfamily. The results collectively indicated that all of the proteins in the lipocalin family have a similar structure, which includes a conserved ligand-binding pocket (Akerstrom *et al.*, 2000). In plants, two xanthophyll cycle enzymes, violaxanthin de-epoxidase and zeaxanthin epoxidase, were initially identified as members of the lipocalin family (Bugos *et al*, 1998). Frenette-Charron *et al.* (2002) subsequently identified *TaTIL* as a novel temperature stress-induced gene from wheat that encodes a lipocalin-like protein. They also reported that *TaTIL* and its orthologue from *Arabidopsis* (referred to as *AtLCN* in the present study) were induced by low temperatures, suggesting that lipocalin-like proteins may contribute to the acquisition of freezing tolerance in both monocots and dicots. Nevertheless, the function of lipocalin-like proteins in the cold acclimation process in plants is still to be determined.

The protein termed DREPP-like was originally identified as a developmentally regulated plasma membrane polypeptide in tobacco (Logan *et al.*, 1997). A Glu-rich region at the C-terminus of the protein showed sequence homology with the vacuolar Ca^{2+}-binding protein (Yuasa and Maeshima, 2000), suggesting that the protein may function in a Ca^{2+} signal transduction pathway. However, there are no reports for this protein in relation to the cold-acclimation process.

VfNOD18-like protein is homologous to the ATP-binding protein, MJ0577, from *Mechanococcus jannaschii* (Hohnjec *et al.*, 2000). MJ0577 binds to ATP and mediates an ATP-dependent function such as a molecular switch. VfNOD18-like protein is structurally similar to UspA (universal stress protein A of *Escherichia coli*), which was originally identified as the protein that mediates cell survival under severe stress conditions such as osmotic stress, UV-light or toxic chemical exposure (Nystrom and Neidhardt, 1993). Thus, it is reasonable to consider that the VfNOD18-like protein in *Arabidopsis* may respond to the osmotic stress that is commonly associated with freeze-induced dehydration.

Expression of the genes encoding cold-responsive plasma membrane proteins

Transcripts of the genes encoding ERD10, ERD14, AtLCN and VfNOD18-like proteins were found to accumulate rapidly upon the exposure of plants to low temperature (Fig. 16.1). mRNA expression patterns of these four genes induced rapidly (within 30 min) and reached the maximum level after 24-h exposure to low temperature. On the contrary, a prolonged low-temperature exposure for 48 h or longer resulted in a gradual reduction of transcript levels for all genes. We also found that all of these genes were responsive to 1 h of heat-shock and the transcripts of two dehydrins (*ERD10* and *ERD14*) were increased in response to drought treatment (data not shown). In contrast, mRNA expression of *DREPP-like* was constitutive at the early stage of cold acclimation but slightly decreased after 3 days of low-temperature exposure.

In positive correlation to the rapid upregulation of *ERD10*, *ERD14*, *AtLCN* and *VfNOD18-like* genes within 24 h of low-temperature exposure, protein levels also increased during the first day of cold acclimation (Kawamura and Uemura, 2003). In the plasma membrane fraction, ERD10 and ERD14 proteins newly appeared after 24 h of cold acclimation. AtLCN and VfNOD18-like proteins increased during the first day of cold acclimation. Thus, these data are in good agreement between mRNA expression levels and the kinetics of protein accumulation. However, when plants were exposed to low temperature for a longer period (2–7 days), the protein levels remained as much as or greater than that after 1-day acclimation. On the contrary, mRNA expression of the genes encoding the proteins decreased gradually during this same time period. These results collectively suggest that these proteins may be regulated by several factors.

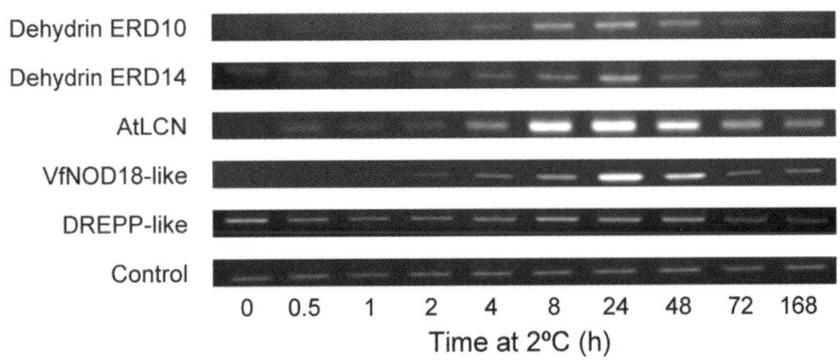

Fig. 16.1. Accumulation of transcripts of the genes encoding ERD10, ERD14, DREPP-like protein, AtLCN and VfNOD18-like protein during cold acclimation. A gene that is not cold responsive (At3g17020) was used as a positive control.

Using the PLACE database (http://www.dna.affrc.go.jp/PLACE/signalscan.html; Higo *et al.*, 1999), we determined putative *cis*-element regulatory regions in the promoter region (1000 bases) of the genes encoding the plasma membrane-associated proteins. In particular, we looked for the presence/absence for the core motifs of the low temperature-responsive element (C/DRE) of the *COR15* gene (Baker *et al.*, 1994) and for the ABA and osmotic stress-responsive element (ABRE) of the rab28 gene (Pla *et al.*, 1993). Four copies of the C/DRE were found in the promoter region of *ERD10*. In fact, the expression profiles of genes under various stress- and hormone-treatment conditions or with various mutants and transgenic plants have shown that *ERD10* was regulated by the CBF/DREB1A cold-responsive pathway (Seki *et al.*, 2001, 2002a; Fowler and Thomashow, 2002). In contrast, in the promoter regions of *ERD14*, *DREPP-like* and *AtLCN*, one or two copies of the C/DRE were identified even though these genes are not indicated as the target genes of CBF/DREB1A.

The core motif (-CCGAC-), which is an integral component of the C/DRE *cis*-acting element, has been shown to be essential for the transcriptional activation in ABA-independent pathways responsive to cold, drought and/or salt stress (Yamaguchi-Shinozaki and Shinozaki, 1994). To evaluate the effects of C/DRE, Kim *et al.* (2002) constructed a vector containing multiple copies of the synthetic C/DRE fused to a GUS reporter gene and then generated transgenic *Arabidopsis* with this vector. GUS staining patterns of the transgenic plants showed that four copies of C/DRE were sufficient to enhance the expression of a downstream gene in response to cold treatment. However, in transgenic plants that contained only two copies of C/DRE, gene expression was not enhanced by cold treatment. Thus, it is possible that the C/DREs that are present in the upstream region of *ERD14*, *DREPP-like* and *AtLCN* may not function in response to low temperature. It is possible that other elements are responsible for the induction of gene expression in the promoters for these three genes under low temperatures.

One of the cold-responsive plasma membrane proteins studied in the present study, DREPP-like protein, behaved differently from other proteins. DREPP-like protein was detected in the plasma membrane fraction of non-acclimated leaves and increased transiently during the first day of cold acclimation (Kawamura and Uemura, 2003). The mRNA expression of *DREPP-like*, however, barely changed during cold treatment (Fig. 16.1). Thus, it is apparently suggested that DREPP-like protein accumulation in the plasma membrane upon exposure to low temperatures may be regulated by post-transcriptional means such as modifications of the protein, interaction with other proteins or via other factors.

It was shown that *TaTIL* expression in wheat, the orthologue of *AtLCN*, was induced by low temperatures (Frenette-Charron *et al.*, 2002). The accumulation of *TaTIL* transcript was detected even after 36 days of cold acclimation. In contrast, *AtLCN* transcript decreased gradually after 1 day of cold acclimation (Fig. 16.1). In monocot species, the accumulation of the *TaTIL* transcript was lower in the less tolerant spring cultivars than in the more tolerant winter cultivars. The maximum freezing tolerance of *Arabidopsis* leaves after cold acclimation is reported to be around −10°C to −12°C, the extent of which is similar to that of spring cultivars of cereals (Uemura and Steponkus, 1994; Uemura *et al.*, 1995). Thus, the kinetic patterns of the transcript accumulation of the lipocalin-like proteins may be correlated to the genetically determined levels of freezing tolerance. Further quantitative analyses are warranted to validate this observation.

Although *AtLCN* is a cold-responsive gene in *Arabidopsis*, the expression of the gene may not be regulated by DREB/CBF transcription factors. Microarray analysis revealed that, in the *sfr6* mutant that is deficient in the COR gene expression and cannot acquire freezing tolerance after cold acclimation, *AtLCN* was downregulated after cold treatment (Boyce *et al.*, 2003). They also reported that *AtLCN* does not contain CRT/DRE elements in the promoter. In addition, it was also shown that transcriptional levels of *DREB1* and *DREB2* did not change in the *sfr6* mutant (Knight *et al.*, 1999). These results collectively suggest that *AtLCN* is a gene that is not a member of the DREB/CBF regulon. However, the introduction of the synthetic promoter, comprised of the C/DRE motif into the *sfr6* mutant, affects the activation of the expression via the C/DRE element itself and regulates DREB1/DREB2 action via a post-transcriptional mechanism. Thus, it is still possible that the *sfr6* mutation may be coupled with the DREB/CBF regulation pathway for cold-regulated gene expression.

In this study, the *VfNOD18-like* gene was apparently cold inducible (Fig. 16.1), which is consistent with results of the protein analysis reported previously (Kawamura and Uemura, 2003). VfNOD18-like protein is known to have a structure that is similar to the highly conserved domain found in '1MJH-like' small UspA proteins (Nystrom and Neidhardt, 1993). UspA proteins are known to be associated with the survival of bacterial cells under exposure to toxic chemicals, osmotic stress or UV-light. Thus, it is possible that the *VfNOD18-like* gene potentially functions as a stress gene and augments the freezing tolerance of plants. In *Arabidopsis*, Kawamura and Uemura (2003) reported that there were two different VfNOD18-like proteins that increased during cold acclimation and their deduced amino acid sequences were highly similar. However, it is interesting to note that one of these two *VfNOD-like* genes (At3g53990) responded to cold, whereas the other (At3g17020) did not (data not shown). Thus, it is necessary to further determine the expression profiles of these two genes in order to obtain conclusive evidence on the regulation for the differential cold response of the two *VfNOD-like* genes.

Functional analysis of the cold-responsive plasma membrane proteins on freezing tolerance *in planta*

To determine the effect of the quantitative changes of the aforementioned cold-responsive plasma membrane proteins on the acquisition of freezing tolerance *in planta*, transgenic plants that constitutively overexpress the respective genes were produced by the floral dip method with *Arabidopsis* (Clough and Bent, 1998). Overexpression of these genes was confirmed by real-time polymerase chain reaction (RT-PCR). Freezing tolerance of leaves of non-acclimated wild-type (WT) and transgenic plants was determined by the electrolyte-leakage assay as a method to evaluate the degree of freezing injury associated with the plasma membrane (Uemura *et al.*, 1995). The temperature at which 50% leakage occurred was determined as the T_{EL50} and used as an index of the degree of freezing tolerance. The genes used for these studies (*ERD10*, *DREPP-like* and *AtLCN*) were amplified by

PCR, and ones for *ERD14* and *VfNOD18-like* were obtained as cDNA clones from RIKEN Bioresource Centre (Seki *et al.*, 1998; 2002b).

To determine the freezing tolerance of transgenic plants, at least five independent lines for each gene were compared with that of WT (data not shown). Although freezing tolerance of leaves from ERD10 and DREPP-like transgenic plants were not obviously different from that of WT, some lines of ERD14, AtLCN and VfNOD18-like transgenic plants exhibited greater levels of freezing tolerance in comparison to WT. One line of ERD14 (ERD14-1) and two lines of AtLCN (AtLCN-1 and AtLCN-5), which were the best performers among the transgenic plants, were selected for further analysis.

Although there were no apparent differences in the degree of injury associated with the plasma membrane between WT and transgenic plants when frozen to the temperature over the range of −2°C to −4°C, electrolyte leakage after freezing at temperatures of −5°C to −7°C was significantly higher in WT than in transgenic plants ($P \leq 0.05$). Freezing injury at −8°C in transgenic AtLCN-1 was less than that in WT ($P \leq 0.01$), while this same difference was not significant between WT and AtLCN-5. These results collectively indicate that overexpression of the *AtLCN* gene results in an increase in freezing tolerance of non-acclimated *Arabidopsis* plants. ERD14-1 showed a similar result to AtLCN-5. Therefore, it was hypothesized that ERD14 is capable of functioning to increase freezing tolerance in *Arabidopsis*.

Because the overexpression of *AtLCN* or *ERD14* gene resulted in a decrease in the extent of electrolyte leakage of intact leaves, we next determined the effect of the *AtLCN* or *ERD14* overexpression on the plasma membrane-associated lesions with protoplasts that were isolated from leaves according to the method of Uemura *et al.* (1995). Briefly, freezing tolerance of isolated protoplasts was determined after a freeze–thaw cycle on a temperature-controlled stage attached with an inverted microscope. Protoplasts during the process of freeze–thawing were observed with a CCD camera and were video recorded. Survival rates were calculated from the number of uninjured and damaged cells observed in a field under the microscope. The occurrence of intracellular freezing (IF) was determined as blackening (flashing) of the protoplast and was scored during the freezing process. After thawing, the number of uninjured cells, which appeared as a smooth, round shape, was counted. The survival rate of protoplasts after a freeze–thaw cycle determined by FDA staining was nearly identical to that determined by visual assessment of the apparent intactness of protoplasts (data not shown). The protoplasts that were not intracellularly frozen but suffered damage were considered to be injured by dehydration-induced membrane lesions as a result of water migration from the cell to the growing extracellular ice crystal. We employed four different rates of cooling to distinguish the occurrence of the types of injury of protoplasts. Typically ten to 50 protoplasts were observed in a field of recorded images and survival rate was expressed as a percentage of normal cells counted after and before freeze–thawing.

When frozen to −10°C, the survival rate of protoplasts isolated from non-acclimated AtLCN transgenic plants was significantly greater than that of WT, especially at the cooling rates of 0.25°C/min and 0.5°C/min (Fig. 16.2). Over the cooling rates of 0.25°C/min to 1.0°C/min, the survival of protoplasts isolated from

AtLCN-1 was approximately twice as much as that of WT. For AtLCN-5 protoplasts, survival was considerably greater than WT at a rate of 0.5°C/min or 1.0°C/min and was somewhat greater at a rate of 0.25°C/min. At 1.0°C/min, both of these two transgenic lines were similar in survival rate and the survival was approximately 1.8-fold greater than that of WT. For ERD14-1, the survival of protoplasts was also greater than that of WT at all cooling rates ranging from 0.25°C/min to 2.5°C/min, and was most distinct when protoplasts were frozen at a rate of either 1.0°C/min or 2.5°C/min.

When the type of freezing injury was determined, the overexpression of *ERD14* had an apparent effect on the incidence of IF. In general, the occurrence of IF increased when cooling rates increased. For example, the incidence of IF in protoplasts of WT plants was 0.0%, 2.2%, 16.5% and 48.0% when frozen at a rate of 0.25°C/min, 0.5°C/min, 1.0°C/min and 2.5°C/min, respectively. At a rate of 2.5°C/min, the incidence of IF in protoplasts of ERD14-1 plants (26%) was significantly lower than that of WT, AtLCN-1 or AtLCN-5 plants (48.0%, 70.1% and 62.6%, respectively). This tendency was similarly observed at a rate of 1.0°C/min. With rates of 0.25°C/min or 0.5°C/min, there were only a few incidences of IF occurring in all samples. Overexpression of the *AtLCN* gene was not apparently effective on the incidence of IF at any of the tested cooling rates.

Under conditions that resulted in few or no occurrences of IF (i.e. when cooled at the rates of 0.25°C/min or 0.5°C/min), it seems likely that the overexpression of *AtLCN* resulted in a decrease in the incidence of dehydration-induced injury in protoplasts. For example, the incidence of dehydration-induced injury in protoplasts of AtLCN-1 and WT, which was calculated as [100% – (the incidence of IF + survival rate)], was 37.7% versus 70.6%, 60.5% versus 83.9%, 66.2% versus 68.6% and 22.2% versus 41.4% when frozen at 0.25°C/min, 0.5°C/min, 1.0°C/min

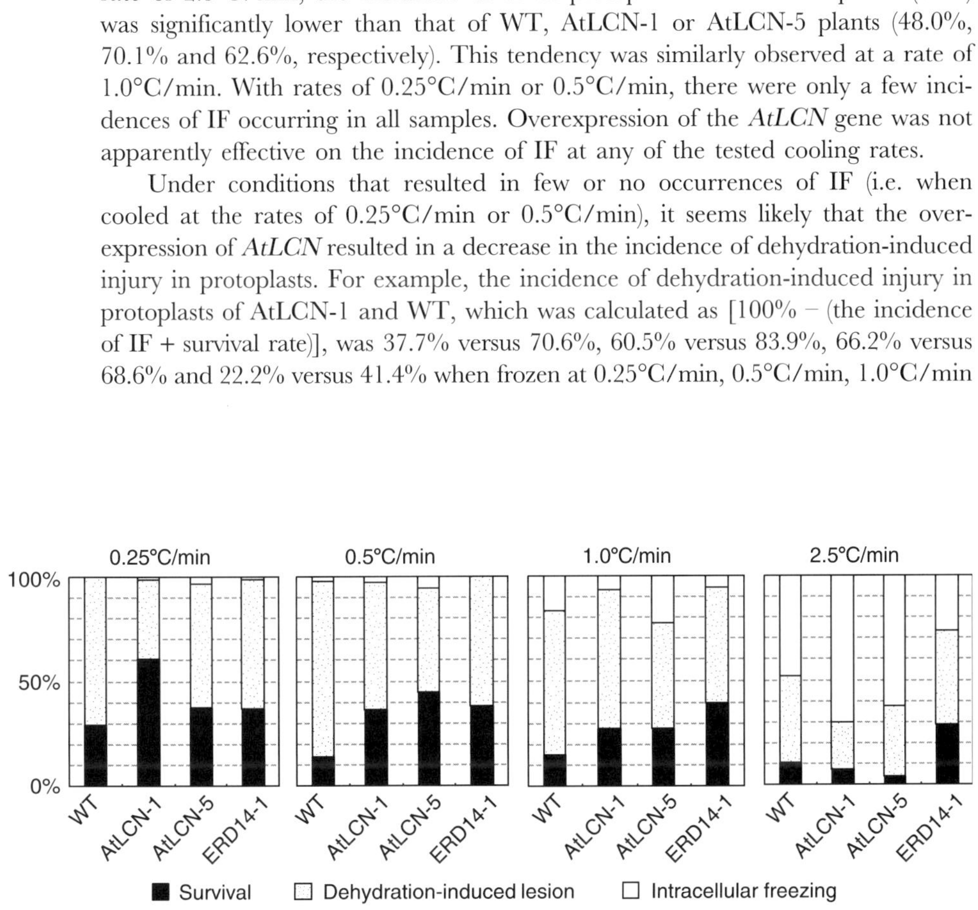

Fig. 16.2. Survival and the occurrence of two types of injury of protoplasts isolated from leaves of non-acclimated wild-type plants (WT) and transgenic plants (AtLCN-1, AtLCN-5 and ERD14-1) after a freeze–thaw cycle. Four different cooling rates (0.25°C/min, 0.5°C/min, 1.0°C/min and 2.5°C/min) were employed.

and 2.5°C/min, respectively. With another *AtLCN*-overexpressing line (AtLCN-5), the effect of *AtLCN* on dehydration-induced injury was not apparent at 0.25°C/min and 2.5°C/min, while, at 0.5°C/min and 1.0°C/min, the incidence (49.3% and 50.4%, respectively) was the lowest among the plants examined. In ERD14-1 protoplasts, the incidence was apparently less than that in WT plants, although the effect was not as great as those observed with AtLCN protoplasts.

Uemura *et al.* (1995) suggested that there are two dehydration-induced lesions associated with the plasma membrane: expansion-induced lysis (EIL) and loss of osmotic responsiveness (LOR). EIL is primarily observed when non-acclimated protoplasts are frozen at relatively high freezing temperatures (e.g. above or at the LT_{50}). When protoplasts are frozen, endocytotic vesiculation of the plasma membrane during freeze-induced osmotic contraction of protoplasts reduces the surface area of the membrane. Although endocytotic vesiculation of the plasma membrane is not injurious *per se*, the vesicles formed during freezing cannot be reincorporated into the plasma membrane during osmotic expansion upon thawing and, hence, protoplasts lyse during thawing of the suspending medium before it regains the original size. On the other hand, LOR occurs when freezing continues to lower temperatures (i.e. severe dehydration). During severe freeze-induced dehydration, membrane bilayers are brought into a close apposition and become mixed. This interaction subsequently results in the formation of non-bilayer structures such as hexagonal II phase transitions or the fracture-jump lesions (Steponkus *et al.*, 1993). These lesions affect the loss of semipermeability of the mixed membranes and, hence, osmotic responsiveness of cells is eliminated. To determine the incidence of EIL and LOR, thawing of the protoplasts in a hypertonic solution has been used and survival can be compared to protoplasts that are thawed in an isotonic solution (Steponkus *et al.*, 1988; Uemura and Steponkus, 1989). Hypertonic thawing minimizes the osmotic expansion of protoplasts during thawing, which minimizes the occurrence of EIL of the protoplasts.

From a preliminary experiment, the incidence of EIL appeared to be suppressed in both AtLCN and ERD14 transgenic plants when compared with that of WT plants. In contrast, there was little difference in the incidence of LOR between the transgenic and WT plants. Thus, the difference in the occurrence of the dehydration-induced injury between the transgenic and WT plants observed in a previous section was suggested to be primarily a result of the suppression of the incidence of EIL. Further experiments are necessary to elucidate the potential effect of *AtLCN* and *ERD14* on the incidence of EIL of isolated protoplasts in detail.

Subcellular localization of AtLCN in leaf cells

The results obtained in the present study indicate that overexpression of *AtLCN* results in an increase in freezing tolerance of excised leaves and isolated protoplasts as a consequence of minimization of freeze-induced lesions associated with the plasma membrane. We previously reported that accumulation of AtLCN in the solubilized fraction of the plasma membrane was detected after as little as 1 day of cold acclimation, and the accumulation was maintained for 7 days

(Kawamura and Uemura, 2003). However, there is no information available on whether the localization of the protein is restricted to the plasma membrane or not. As a result, we determined the localization of AtLCN using 2D-PAGE, Western blot analysis and immunoelectron microscopy (Y. Tominaga *et al.*, unpublished results).

Anti-AtLCN antiserum was raised against a peptide at a C-terminal region of AtLCN to determine the subcellular localization of the protein. After leaves of WT, AtLCN-1 or AtLCN-5 were roughly fractionated into cytosolic and microsomal fractions, Western blot analysis was carried out using the anti-AtLCN antiserum in order to detect AtLCN protein accumulation. The accumulation of AtLCN was observed in the microsomal fraction of non-acclimated *AtLCN* overexpression transgenic plants. However, no signal was detected in the cytosolic fraction of both WT and transgenic plants. Interestingly, the level of protein accumulation in the non-acclimated transgenic plants was much higher than that of the cold-acclimated WT. Furthermore, 2D-PAGE analysis of plasma membrane proteins from the leaves of AtLCN transgenic plants showed the accumulation of AtLCN only in the solubilized fraction of the plasma membrane.

Subcellular localization of AtLCN was next determined in detail with immunoelectron microscopy. AtLCN was indeed confirmed to localize in the vicinity of the plasma membrane, whereas no, or just a few, proteins were observed in cytoplasm, nucleus or other intracellular organelles. Thus, taken together with the results that freezing tolerance of AtLCN transgenic plants showed freezing tolerance higher than that of WT, it is concluded that AtLCN localized in or near the plasma membrane plays a role in acquisition of freezing tolerance of *Arabidopsis*.

Some lipocalin-like proteins are known to be stress inducible and suggested to have a protective effect on membranes under certain stress conditions. Human ApoD, *E. coli* outer membrane lipoprotein Blc precursor and American grasshopper Lazarillo precursor, all of which have significant sequence homology with the AtLCN protein and its orthologue (TaTIL) in wheat (Frenette-Charron *et al.*, 2002), are known to be anchored to biological membranes (Bishop, 2000). All of these membrane-anchored lipocalins appear to be expressed in association with the conditions of membrane stress or remodelling. For example, ApoD accumulated in re-myelinating peripheral nerve cells following nerve damage. *Blc* expression was observed under conditions of environmental stresses. These results suggest that these membrane-anchored lipocalins have a role in membrane biogenesis and/or repair (Bishop, 2000). The lipoprotein-like proteins in plants, AtLCN and TaTIL, contain the three structurally conserved regions that are considered to be characteristic domains of lipoproteins (Frenette-Charron *et al.*, 2002). However, they lack the N-terminal signal peptide that has been identified in ApoD, Blc and Lazarillo. Thus, it is necessary to characterize the plant lipocalin-like proteins in the stress-responsive reactions, especially from the viewpoint of the occurrence of membrane-associated lesions.

It is interesting to determine the way that lipocalin-like proteins associate with the membrane. Frenette-Charron *et al.* (2002) suggested the hydrophobic C-terminal tail of the Lazarillo protein enables it to receive a glycosylphosphatidylinositol (GPI) anchor from the membrane. However, the Big-PI Predictor program which predicts GPI modification (http://mendel.imp.univie.ac.at/gpi/

gpi_server.html; Eisenhaber *et al.*, 2003) indicated that there were no potential GPI-modification sites in the amino acid sequence of AtLCN. Nevertheless, it would be possible to argue that there is the physical attachment by the hydrophobic loop of the lipocalin-like proteins to the membrane in plants.

In Gram-negative bacteria, the localization of the lipoprotein on the membrane is determined by the lipoprotein-sorting signals. In *E. coli*, the lipoproteins were anchored to either the inner or the outer membrane through acyl chains that were covalently attached to an N-terminal cysteine of the protein (Terada *et al.*, 2001). They reported that outer membrane-specific lipoproteins were released from the inner membrane by the ATP binding-cassette transporter LolCDE complex and formed a complex with the periplasmic molecular chaperone LolA. Furthermore, it has been suggested that the lipoprotein-LolA complex is transported to the outer membrane and the lipoprotein is subsequently transferred from LolA to LolB and is subsequently anchored to the membrane (Matsuyama *et al.*, 1997; Miyamoto *et al.*, 2002; Narita *et al.*, 2003). However, no proteins that are similar to the lipoprotein-sorting system have been identified in plants. Thus, the interaction between lipocalin-like proteins and the plasma membrane is still unknown and should be determined in detail in the future.

Concluding Remarks

In order to reach the goal of improving tolerance to low-temperature stress, transgenic plants have been generated which overexpress particular genes, such as those that are associated with active oxygen species scavenging activity, compatible solute accumulation or membrane lipid alterations (for review, see Iba, 2002). With regard to the plasma membrane proteins, Li *et al.* (2004) have recently revealed that an integral plasma membrane protein, phospholipase Dδ, which is thought to be involved in membrane-lipid hydrolysis and cell signalling, enhances freezing tolerance in coordination with expression of cold-regulated genes such as *COR47* and *COR78*. In the present study, we clearly revealed that the accumulation of peripheral membrane proteins, AtLCN and ERD14, increased the freezing tolerance of non-acclimated *Arabidopsis* plants over the specific temperature ranges due to the mitigation of plasma membrane-associated lesions. In addition, *AtLCN* overexpression did not result in changes in the expression of several cold-regulated genes, including *DREB1a*, *COR15a* and *COR78* (data not shown). Therefore, it seems plausible that the accumulation of AtLCN protein results in an increase in the freezing tolerance of *Arabidopsis* independent from the CBF/DREB regulons. Nevertheless, further analysis on whether *AtLCN* overexpression is solely responsible for the increase in cryostability of the plasma membrane is still necessary to clarify the function of AtLCN on freezing tolerance in plants.

Acknowledgements

This work was, in part, supported by a grant from Bio-oriented Technology Research Advancement Institution, the Ministry of Education, Sports, Science and Technology of Japan (the 21st Century Centre of Excellence Program), and the Ministry of Agriculture, Forestry and Fisheries of Japan (Bio-Design Project). We also thank many students who worked on obtaining many transgenic *Arabidopsis* plants.

References

Akerstrom, B., Flower, D.R. and Salier, J.P (2000) Lipocalins: unity in diversity. *Biochimica et Biophysica Acta* 1482, 1–8.

Artus, N.N, Uemura, M., Steponkus, P.L., Gilmour, S.J., Lin, C.M. and Thomashow, M.F. (1996) Constitutive expression of the cold-regulated *Arabidopsis thaliana COR15a* gene affects both chloroplast and protoplast freezing tolerance. *Proceedings of the National Academy of Sciences USA* 93, 13404–13409.

Baker, S.S., Wilhelm, K.S. and Thomashow, M.F. (1994) The 5′-region of *Arabidopsis thaliana cor15a* has *cis*-acting elements that confer cold-, drought- and ABA-regulated gene expression. *Plant Molecular Biology* 24, 701–713.

Bishop, R.E. (2000) The bacterial lipocalins. *Biochimica et Biophysica Acta* 1482, 73–83.

Boyce, J.M., Knight, H., Deyholos, M., Openshaw, M.R., Galbraith, D.W., Warren, G. and Knight, M.R. (2003) The *sfr6* mutant of *Arabidopsis* is defective in transcriptional activation via CBF/DREB1 and DREB2 and shows sensitivity to osmotic stress. *Plant Journal* 34, 395–406.

Bugos, R.C., Hieber, A.D. and Yamamoto, H.Y. (1998) Xanthophyll cycle enzymes are members of the lipocalin family, the first identified from plants. *Journal of Biological Chemistry* 273, 15321–15324.

Clough, S.J. and Bent, A.F. (1998) Floral dip: a simplified method for *Agrobacterium*-mediated transformation of *Arabidopsis thaliana*. *Plant Journal* 16, 735–743.

Crowe, J.H., Carpenter, J.F., Crowe, L.M. and Anchordoguy, T.J. (1990) Are freezing and dehydration similar stress vectors? A comparison of modes of interaction of stabilizing solutes with biomolecules. *Cryobiology* 27, 219–231.

Danyluk, J., Perron, A., Houde, M., Limin, A., Fowler, B., Benhamou, N. and Sarhan, F. (1998) Accumulation of an acidic dehydrin in the vicinity of the plasma membrane during cold acclimation of wheat. *Plant Cell* 10, 623–638.

Eisenhaber, B., Wildpaner, M., Schultz, C.J., Borner, G.H., Dupree, P. and Eisenhaber, F. (2003) Glycosylphosphatidylinositol lipid anchoring of plant proteins. Sensitive prediction from sequence- and genome-wide studies for *Arabidopsis* and rice. *Plant Physiology* 133, 1691–1701.

Fowler, S. and Thomashow, M.F. (2002) *Arabidopsis* transcriptome profiling indicates that multiple regulatory pathways are activated during cold acclimation in addition to the CBF cold response pathway. *Plant Cell* 14, 1675–1690.

Frenette-Charron, J.B., Breton, G., Badawi, M. and Sarhan, F. (2002) Molecular and structural analyses of a novel temperature stress-induced lipocalin from wheat and *Arabidopsis*. *FEBS Letters* 517, 129–132.

Gilmour, S.J., Hajela, R.K. and Thomashow, M.F. (1988) Cold acclimation in *Arabidopsis thaliana*. *Plant Physiology* 87, 745–750.

Guy, C. (1999) Molecular responses of plants to cold shock and cold acclimation. *Journal of Molecular Microbiology and Biotechnology* 1, 231–242.

Higo, K., Ugawa, Y., Iwamoto, M. and Korenaga, T. (1999) Plant *cis*-acting regulatory DNA elements (PLACE) database: 1999. *Nucleic Acids Research* 27, 297–300.

Hohnjec, N., Kuster, H., Albus, U., Frosch, S.C., Becker, J.D., Puhler, A., Perlick, A.M. and Fruhling, M. (2000) The broad bean nodulin VfENOD18 is a member of a novel family of plant proteins with homologies to the bacterial MJ0577 superfamily. *Molecular Genetics and Genomics* 264, 241–250.

Iba, K. (2002) Acclimative response to temperature stress in higher plants: approaches of gene engineering for temperature tolerance. *Annual Review of Plant Physiology and Plant Molecular Biology* 53, 225–245.

Kawamura, Y. and Uemura, M. (2003) Mass spectrometric approach for identifying putative plasma membrane proteins of *Arabidopsis* leaves associated with cold acclimation. *Plant Journal* 36, 141–154.

Kim, H.J., Kim, Y.K., Park, J.Y. and Kim, J. (2002) Light signalling mediated by phytochrome plays an important role in cold-induced gene expression through the C-repeat/dehydration responsive element (C/DRE) in *Arabidopsis thaliana*. *Plant Journal* 29, 693–704.

Kiyosue, T., Yamaguchi-Shinozaki, K. and Shinozaki, K. (1994) Characterization of two cDNAs (*ERD10* and *ERD14*) corresponding to genes that respond rapidly to dehydration stress in *Arabidopsis thaliana*. *Plant and Cell Physiology* 35, 225–231.

Knight, H., Veale, E.L., Warren, G.J. and Knight, M.R. (1999) The *sfr6* mutation in *Arabidopsis* suppresses low-temperature induction of genes dependent on the CRT/DRE sequence motif. *Plant Cell* 11, 875–886.

Levitt, J. (1980) Chilling, freezing and high temperature stresses. In: *Responses of Plants to Environmental Stresses*. Vol. 1, 2nd edn. Academic Press, New York, pp. 407–414.

Li, W., Li, M., Zhang, W., Welti, R. and Wang, X. (2004) The plasma membrane-bound phospholipase D δ enhances freezing tolerance in *Arabidopsis thaliana*. *Nature Biotechnology* 22, 427–433.

Lin, C. and Thomashow, M.F. (1992) DNA sequence analysis of a complementary DNA for cold-regulated *Arabidopsis* gene *cor15* and characterization of the COR15 polypeptide. *Plant Physiology* 99, 519–525.

Logan, D.C., Domergue, O., Teyssendier de la Serve, B. and Rossignol, M. (1997) A new family of plasma membrane polypeptides differentially regulated during plant development. *Biochemistry and Molecular Biology International* 43, 1051–1062.

Lynch, D.V. and Steponkus, P.L. (1987) Plasma membrane lipid alterations associated with cold acclimation of winter rye seedlings (*Secale cereale* L. cv. Puma). *Plant Physiology* 83, 761–767.

Matsuyama, S., Yokota, N. and Tokuda, H. (1997) A novel outer membrane lipoprotein, LolB (HemM), involved in the LolA (p20)-dependent localization of lipoproteins to the outer membrane of *Escherichia coli*. *EMBO Journal* 16, 6947–6955.

Miyamoto, A., Matsuyama, S. and Tokuda, H. (2002) Dominant negative mutant of a lipoprotein-specific molecular chaperone, LolA, tightly associates with LolCDE. *FEBS Letters* 528, 193–196.

Narita, S., Kanamaru, K., Matsuyama, S. and Tokuda, H. (2003) A mutation in the membrane subunit of an ABC transporter LolCDE complex causing outer membrane localization of lipoproteins against their inner membrane-specific signals. *Molecular Microbiology* 49, 167–177.

Nylander, M., Svensson, J., Palva, E.T. and Welin, B.V. (2001) Stress-induced accumulation and tissue-specific localization of dehydrins in *Arabidopsis thaliana*. *Plant Molecular Biology* 45, 263–279.

Nystrom, T. and Neidhardt, F.C. (1993) Isolation and properties of a mutant of *Escherichia coli* with an insertional inactivation of the *uspA* gene, which encodes a universal stress protein. *Journal of Bacteriology* 175, 3949–3956.

Pearce, R.S. (1999) Molecular analysis of acclimation to cold. *Plant Growth Regulation* 29, 47–76.

Pla, M., Vilardell, J., Guiltinan, M.J., Marcotte, W.R., Niogret, M.F., Quatrano, R.S. and Pages, M. (1993) The cis-regulatory element CCACGTGG is involved in ABA and water-stress responses of the maize gene *rab28*. *Plant Molecular Biology* 21, 259–266.

Seki, M., Carninci, P., Nishiyama, Y., Hayashizaki, Y. and Shinozaki, K. (1998) High-efficiency cloning of *Arabidopsis* full-length cDNA by biotinylated CAP trapper. *Plant Journal* 15, 707–720.

Seki, M., Narusaka, M., Abe, H., Kasuga, M., Yamaguchi-Shinozaki, K., Carninci, P., Hayashizaki, Y. and Shinozaki, K. (2001) Monitoring the expression pattern of 1300

Arabidopsis genes under drought and cold stresses by using a full-length cDNA microarray. *Plant Cell* 13, 61–72.

Seki, M., Narusaka, M., Ishida, J., Nanjo, T., Fujita, M., Oono, Y., Kamiya, A., Nakajima, M., Enju, A., Sakurai, T., Satou, M., Akiyama, K., Taji, T., Yamaguchi-Shinozaki, K., Carninci, P., Kawai, J., Hayashizaki, Y. and Shinozaki, K. (2002a) Monitoring the expression profiles of 7000 *Arabidopsis* genes under drought, cold and high-salinity stresses using a full-length cDNA microarray. *Plant Journal* 31, 279–292.

Seki, M., Narusaka, M., Kamiya, A., Ishida, J., Satou, M., Sakurai, T., Nakajima, M., Enju, A., Akiyama, K., Oono, Y., Muramatsu, M., Hayashizaki, Y., Kawai, J., Carninci, P., Itoh, M., Ishii, Y., Arakawa, T., Shibata, K., Shinagawa, A. and Shinozaki, K. (2002b) Functional annotation of a full-length *Arabidopsis* cDNA collection. *Science* 296, 141–145.

Shinozaki, K. and Yamaguchi-Shinozaki, K. (2000) Molecular responses to dehydration and low temperature: differences and cross-talk between two stress signaling pathways. *Current Opinion in Plant Biology* 3, 217–223.

Steponkus, P.L. (1984) Role of the plasma membrane in freezing injury and cold acclimation. *Annual Review of Plant Physiology* 35, 543–584.

Steponkus, P.L., Uemura, M., Balsarno, R.A., Arvinte, T. and Lynch, D.V. (1988) Transformation of the cryobehavior of rye protoplasts by modification of the plasma membrane lipid composition. *Proceedings of the National Academy of Sciences USA* 85, 9026–9030.

Steponkus, P.L., Uemura, M. and Webb, M.S. (1993) Membrane destabilization during freeze-induced dehydration. In: Close, T.J. and Bray, E.A. (eds) *Plant Responses to Cellular Dehydration during Environmental Stress.* American Society of Plant Physiologists, Rockville, Maryland, pp. 37–47.

Terada, M., Kuroda, T., Matsuyama, S. and Tokuda, H. (2001) Lipoprotein sorting signals evaluated as the LolA-dependent release of lipoproteins from the cytoplasmic membrane of *Escherichia coli. Journal of Biological Chemistry* 276, 47690–47694.

Thomashow, M.F. (1999) Plant cold acclimation: freezing tolerance genes and regulatory mechanisms. *Annual Review of Plant Physiology and Plant Molecular Biology* 50, 571–599.

Uemura, M. and Steponkus, P.L. (1989) Effect of cold acclimation on the incidence of two forms of freezing injury in protoplasts isolated from rye leaves. *Plant Physiology* 91, 1131–1137.

Uemura, M. and Steponkus, P.L. (1994) A contrast of the plasma membrane lipid composition of oat and rye leaves in relation to freezing tolerance. *Plant Physiology* 104, 479–496.

Uemura, M. and Steponkus, P.L. (2003) Modification of the intracellular sugar content alters the incidence of freeze-induced membrane lesions of protoplasts isolated from *Arabidopsis thaliana* leaves. *Plant, Cell and Environment* 26, 1083–1096.

Uemura, M. and Yoshida, S. (1984) Involvement of plasma membrane alterations in cold acclimation of winter rye seedlings (*Secale cereale* L. cv. Puma). *Plant Physiology* 75, 818–826.

Uemura, M., Joseph, R.A. and Steponkus, P.L. (1995) Cold acclimation of *Arabidopsis thaliana*: effect on plasma membrane lipid composition and freeze-induced lesions. *Plant Physiology* 109, 15–30.

Wanner, L.A. and Junttila, O. (1999) Cold-induced freezing tolerance in *Arabidopsis. Plant Physiology* 120, 391–399.

Yamaguchi-Shinozaki, K. and Shinozaki, K. (1994) A novel *cis*-acting element in an *Arabidopsis* gene is involved in responsiveness to drought, low temperature, or high-salt stress. *Plant Cell* 6, 251–264.

Yoshida, S. and Uemura, M. (1984) Protein and lipid compositions of isolated plasma membranes from orchard grass (*Dactylis glomerata* L.) and changes during cold acclimation. *Plant Physiology* 75, 31–37.

Yuasa, K. and Maeshima, M. (2000) Purification, properties, and molecular cloning of a novel Ca^{2+}-binding protein in radish vacuoles. *Plant Physiology* 124, 1069–1078.

Index